Elizabeth Blackburn and the
Story of Telomeres

Elizabeth Blackburn and the Story of Telomeres

Deciphering the Ends of DNA

Catherine Brady

The MIT Press
Cambridge, Massachusetts
London, England

For information about special quantity discounts, please send e-mail to <special_sales@mitpress.mit.edu>.

This book was set in Sabon and Meta by SNP Best-set Typesetter Ltd., Hong Kong.

Printed and bound in the United States of America.

Library of Congress Cataloging-in-Publication Data

Brady, Catherine, 1955–
  Elizabeth Blackburn and the story of telomeres : deciphering the ends of DNA / Catherine Brady.
    p. cm.
  Includes bibliographical references and index.
  ISBN 978-0-262-02622-2 (hardcover : alk. paper)
  1. Blackburn, Elizabeth H. 2. Biochemists—United States—Biography. 3. Telomere—History. I. Title.
  QP511.8.B53B73   2007
  572.092—dc22
  [B]
                                                              2007000080

10 9 8 7 6 5 4 3 2 1

For Sarah, Ben, and David

Contents

Acknowledgments

I am deeply indebted to Elizabeth Blackburn for spending hours talking to me about her life and work, always as if nothing else on her calendar was pressing. Her extraordinary generosity also made itself felt in her willingness to speak candidly about her life for the sake of illuminating the particular pressures women scientists continue to face in their careers. So many of the scientists interviewed for this book share Blackburn's passion for their work, making it a pleasure and a privilege to speak with them. I would like to thank the following people for granting interviews: Spyros Artavanis-Tsakonas, Mike Cherry, Titia de Lange, Joe Gall, Dave Gilley, Carol Greider, Carol Gross, Diane K. Lavett, Vicki Lundblad, Elizabeth Marincola, Katherine Marsden, Tet Matsuguchi, Mike McEachern, John Prescott, John Sedat, Janis Shampay, Dorothy Shippen, Dana Smith, and Martha Truett. Dana Smith and Maura Devlin Clancy, both at the Blackburn lab, were resourceful and considerate in providing me with material help as I conducted research for this book. Maura Murphy Clancy provided astute assistance with research. Many thanks to Steven Kahn for suggestions on the manuscript and to David Kahn for his illuminating comments on the science. Barbara Murphy and Meagan Stacey at the MIT Press offered invaluable editorial help as I completed this manuscript. Thanks also go to Caron Knauer, my literary agent, for her steadfast support for this project.

Elizabeth Blackburn and the Story of Telomeres

1 A Certain Sense of Self

Like many other seventeen-year-old girls in 1965, Elizabeth Blackburn listened to the records of the Beatles and Peter, Paul, and Mary and wore the miniskirts that were just coming into fashion, but she felt so shy in the presence of boys that she could not look them in the eye. Rigorously schooled by her mother in polite manners that sidestepped confrontation, Elizabeth was a model student who seemed readily guided by her teachers. But her delight in books exceeded the bounds of obedient studiousness—in particular, she was thrilled by her recent discovery of a biology text complete with detailed illustrations of amino acids, strung together in long chains and then folded up into complex three-dimensional shapes to form enzymes and other proteins. For Liz, these elegant structures had a teasing beauty, promising tantalizing clues to the processes of life and yet also enfolding that mystery. Even the names of the amino acids—phenylalanine, leucine—struck her as poetic. Though she confessed her fascination to no one, she traced drawings of amino acids on large, thin sheets of white paper and then tacked them up on her bedroom wall.

From the start she carefully protected the passion that would shape her life as a scientist, her fierce determination often masked by a polite, acquiescent demeanor. The nice girl who remained silent when confronted or thwarted purchased the freedom of a secret, willful, essential self. Blackburn's first clear memory dates to when she was about three years old. Playing in the yard behind her family's house, she had found a bull ant and was handling it gently, talking to it as it crisscrossed her palm and the back of her hand. When her mother came on the scene, she brushed the insect from Liz's hand and vehemently warned her never to touch these insects, whose bite could result in a painful welt. Surprised by her mother's concern, Liz obeyed. But she remained stubbornly and silently certain that the ant could not hurt her.

Elizabeth Helen Blackburn was born in Hobart, Tasmania, on November 26, 1948, to Marcia Constance (née Jack) and Harold Stewart Blackburn, two general practitioners. Blackburn noted that family lore claims her father's family is descended from Alfred the Great, a ninth-century English ruler. Legend has it that during wartime, when Alfred took refuge in a swineherd's hut, the swineherd's wife asked him to watch over cakes she had placed in the oven, but preoccupied by affairs of state, he forgot his task. In relating her family's history, Blackburn ruefully admitted that "he was famous for burning the cakes, a trait for which I probably inherited the gene."[1] More certain evidence suggests that she inherited from both sides of her family a strong inclination for science. Her father's family had originally come from the north of England in 1882; her great-grandfather, the Reverend Thomas Blackburn, an Anglican minister, moved from Hawaii to Australia to pursue a passionate avocation. Like many Victorians, he was inspired by the publication in 1859 of Charles Darwin's *Origin of the Species*, which had spurred widespread interest in collecting and studying specimens of insects and fossils. Having completed and cataloged an extensive insect collection in Hawaii, Blackburn continued to collect *Coleoptera* beetles in Australia until he had amassed a total of three thousand specimens. In the later years of his life, he corresponded with the British Museum, anxious to sell this collection and certain it could provide his family with a substantial inheritance. Although the family still possesses his letters, there is no record of whether the museum obtained and preserved the collection. Elizabeth's father, Harold Blackburn, born in 1919 and raised in Adelaide, shared his grandfather's scientific bent and earned a medical degree from the University of Adelaide.

Elizabeth's maternal great-grandfather, Robert Logan Jack, was trained as a geologist in Edinburgh. With his geologist son, Robert Lockhart Jack, he traveled through China, surveying for minerals, until they were forced to flee the Boxer Rebellion in 1904, escaping through Yunnan Province. Robert Logan Jack later wrote books about his travels in China and northern Australia, where he continued working as a geologist. Robert Lockhart Jack married Fanny Marr, and they had two children, Bill and Marcia, born in Adelaide in 1918. The family soon moved to Melbourne, where Marcia spent most of her childhood. Like her future husband, she grew up in a middle-class family and attended private school, earning her medical degree from the University of Melbourne during World War II.

Marcia and Harold met during the war, when he served as a captain in the Australian Army, and they married after the war ended. Reticent about their personal lives, neither of them ever told their children the details of their courtship. While they raised a family, Marcia practiced medicine intermittently, working part-time in a large family practice. The first of their seven children, Katherine, was born in 1946 in Adelaide. After the family moved to Snug, Tasmania, three more children were born—Elizabeth (known as Liz) in 1948, Barbara in 1949, and John in 1950. In 1954, when the family spent a year in London, Andrew was born, followed by Margaret in 1957 and Caroline in 1960.

Until Liz was four years old, the family lived in Snug, a small bush town south of Hobart, a coastal city of some 130,000 to 150,000 in southeastern Tasmania, where her father had found work as a government medical officer. Snug was situated on the sheltered North West Bay, with a main street that ran from the highway to the sea. The town's few gridded streets, lined with bungalows, simply ended at farmland or wilderness. Out of necessity, the children were taught to be fearful of snakes, especially the venomous tiger snakes, but other warnings did not deter Liz from squatting on the beach to pick up jellyfish that had washed up on the sand, though some were poisonous. Fascinated by small living creatures, she loved to hold them and even sang to them.

Living nearby in Snug was Liz's godmother, a dental nurse named Cluny Portnell, who went by the name of Bill. A childless single woman, Bill lived near the beach with her sister, and Liz was sent to stay with

Bill in her small house whenever her mother was particularly busy. Despite the pleasure of spending time with Bill and her cats, Liz felt homesick on these visits. After the family moved away, Bill sent Liz a book for every birthday—well-written books by contemporary authors that had been chosen with great care and deliberation, far more advanced than the books Liz was already reading, and a clear and welcome signal of Bill's strong expectation that her goddaughter should excel. Blackburn retained strong ties with Bill into adulthood.

The family moved when Harold Blackburn decided to join a medical practice in a town that offered more educational opportunities for the children. From the time Liz was four until she turned sixteen, the family lived in Launceston, a town of about seventy thousand people in northeastern Tasmania, a move up from the sleepy hinterland of Snug. Situated at the convergence of the North Esk and South Esk rivers, Launceston was set among steep hills, with a mountain range in the distance, dominated by Mount Barrow.

In Launceston the family first lived in a typical one-story Australian suburban bungalow, with a peaked, gabled roof and a veranda that wrapped around two sides of the house. Her mother often took Liz to afternoon tea at the home of their two elderly neighbors, where even at a young age she was expected to sit quietly and politely on their nice chairs. With five children filling up her own crowded house, frequently spilling out to the front lawn to play fantasy games with the children of the neighborhood, Liz delighted in the tidiness of her neighbors' home.

Liz's father was a busy physician, but that did not fully account for his absence from family life. Every morning, her mother sat down to a full breakfast with the children, but her father would appear only briefly before leaving for his office. In addition to his long hours of work, he led a social life separate from that of his wife and children, often spending weekends fishing with his mates. Drinking was the worrisome common factor in these different friendships as well as a strong motive for Harold's frequent absences. Liz recalled "a few occasions of happiness with my father, few and far between, so I clung to them. When I was about kindergarten age, we had a fire going in the front living room, and I was sitting on his lap eating an apple, and I said, what shall I do with the core, and he said, throw it into the fire, and I threw it into the fire." Liz often resented her father's absence, but even more she longed for his love: "Once I remember seeing my father, who had just got home, picking up my little brother, who was about two, and tossing him playfully in the air. My father was silhouetted against the light coming in from the front door at the end of the long hallway that ran down the center of the house, and I remember wishing that he would pick me up and throw me in the air too. But I didn't ask him."

Every night Liz's mother made a point of coming to each of the children's bedrooms to kiss them good night. Marcia provided her children with art, elocution, and piano lessons and was determined to send them to expensive private schools. Although she rigorously instilled politeness in her children, Marcia conveyed tolerance—even fondness— for her second daughter's strong-willed nature and often told amused stories about Liz's stubbornness. Once, when Liz discovered a doll that had been put away at the top of a cupboard, slated to be a present for her or her sister at a later date, she persisted in begging for the doll, yelling at her mother and badgering her until she capitulated.

Liz felt closest to her sister Katherine, who was two years older, and for a time was jealous of her next youngest sister, Barbara, who had been quite sick as an infant and thus commanded much of her mother's time. Liz revered Katherine, and she was one of the few people whose disapproval Liz would tolerate when corrected for not doing her share of the family chores. When Liz was about six, her parents made a nine-month sojourn to London, where her father did postgraduate work. Marcia and Harold left the younger children behind in Tasmania, but Liz and Katherine accompanied their parents to England and stayed briefly with relatives in the English countryside, an experience that drew them even closer.

With the exception of this trip, Liz was educated at Broadland House, a public school for girls. Like most of the nonstate schools, Broadland House Church of England Girls Grammar School operated under the auspices of a church. The school encompassed grades from kindergarten through secondary school. Separate schooling for girls and boys reflected the distinctly divergent expectations for their future roles. During the rough equivalent of middle school, the curriculum at Broadland House split into two tracks, one emphasizing secretarial training, including typing and shorthand, and the other stressing general academics. Although the school did not offer courses in classical languages or culture, it strove to impart a well-rounded grounding in the humanities, strongly focused on British literature and history. Modern history was taught under the heading of social sciences, reflecting the somewhat narrow, colonial range of the school's curriculum and a relative innocence about the world for which it was educating its girls.

At five, Liz was capable of noting, almost with objectivity, how polite she was outwardly, even when she sometimes felt deeply angry. Well-mannered behavior often belied her strong will. On her first day of kindergarten, the children were permitted to draw with pastels, and she drew a big locomotive entirely in black pastel. The teacher, Mrs. Mundy, suggested she shouldn't use so much black, and Liz felt outraged to be told how to draw *her* locomotive. Timidity and training kept her from protesting, but "although I didn't say anything, I didn't like getting what I took to be reproof for what I thought I was doing perfectly well."

Her sister Katherine's memories also suggest Liz had a stubbornly certain sense of self, subtly dissonant with her shy, meek demeanor. Angry at a "mean" first-grade teacher, Liz did not rebel openly but enlisted Katherine to exact a surreptitious justice, mistreating the textbooks this teacher had given her and dangling them out the window on a string. In third grade, Liz got in trouble with a teacher for "smudging her ink" and adamantly insisted she had not done it. "She was so upset to be falsely accused that our mother had to pay a visit to the teacher," Katherine reported.[2] That Marcia would do so says a great deal about her acceptance of and support for her daughter, which helped to preserve Liz's sense of self.

When Liz was in fourth grade, the family moved to another house in Launceston. Built in the early nineteenth century, Elphin House had steeply pitched roofs and a large garden full of wonders: a goldfish pond with a central birdbath of stone, a walk-in aviary for budgerigars and canaries, a henhouse, a vegetable plot, and a row of fruit trees. Liz would sit on the window seat of her brother's bedroom, eating small fresh apples from their own trees and reading Jane Austen novels. In this roomy, magical house, she had ample opportunity to indulge her passionate love for animals. She befriended neighbor's pets as well as her own, and the family owned dogs, a number of cats—including Mugwumpian, whom Liz adored—and some budgerigars that her little brother Andrew once let out of their cage. Liz also kept guinea pigs as her own pets for a number of years. She lavished care on them, raiding the vegetable plot in the back garden for leafy greens to feed them despite the risk of being caught and scolded by her father. When her guinea pigs had babies, she showed them off to school friends, holding the tiny, hairless creatures in her hand, and was mystified by her friends' squeamishness and disgust. By the time she turned twelve, her father decided that she was spending time on her pets that she should have spent on chores and took her guinea pigs away.

When the family lived in Elphin House, they employed a nanny-cum-housekeeper, Flora Douglas, called Douggie by everyone in the family. Constant and reliable, if somewhat severe, Douggie was sought out by a vaguely lonely Liz, who was sometimes treated to a visit to Douggie's large room to look at her books and play with her things. It was not an unhappy childhood, and yet often Liz yearned for a happiness she never quite felt.

Marcia enrolled Liz in piano lessons after she turned seven, and she was not too busy to make the long drive to northwest Tasmania so Liz could participate in a piano competition. Eventually, Liz also learned to play the violin, and her mother enrolled her in Saturday morning art lessons, which she loved, and proudly took her daughter to see her own artwork in an exhibition of prizewinners from a local children's competition. Liz's older sister Katherine recognized Liz's talent for drawing and music, yet she never felt jealous, because "our mother emphasized that she loved everyone equally."[3] If Marcia seemed to view Liz as exceptionally gifted, she conveyed to all her children the strong sense that the family was "somehow a cut above the common herd"—an attitude all the more striking in Australian society, where the concept of "mateship" meant that a man was admired particularly when he did not act superior to anyone else. According to Katherine, Marcia intended to instill in her children that they were "*intellectually* a cut above."[4] Marcia expected each of her children to pursue a profession and taught them to believe that reason and reasonable expectations should govern their

values and actions, never insisting on a single right way but subtly imply-
ing it. For Marcia, intellect and gentility, not money and social position,
were markers of class, and she taught her children to distance themselves
from anything that smacked of vulgarity. When Liz once asked her about
the family's income, Marcia responded euphemistically (they were "com-
fortably off"), and she shunned any display of wealth.

Later, when Liz was in her teens, her mother arranged for her to take
Saturday morning elocution classes from Alison Beattie, the daughter of
close family friends. Alison's father, who had emigrated from Scotland,
was a gentle, approachable man, openly affectionate with his family,
unlike Liz's father. All the Blackburn children were welcomed into the
Beatties' family circle. On school days, when girls commonly went home
at lunch, Liz sometimes went to their house instead of her own. Mr.
Beattie joined the family for a cozy lunch by the fire, often reading books
aloud to the children in his strong Scottish accent.

With her older sister Katherine, Alison had studied at the Royal
Academy of Dramatic Art in London. On Saturday mornings Liz and a
school friend received lessons from Alison in the family's back room,
with the gas fire roaring, studying elocution, enunciation, and pronun-
ciation (in a British accent), with Alison drawing on her stage training
to teach the girls how to exploit breath and posture in their speech, enter-
taining them by letting them practice saying lines from Gilbert and
Sullivan as clearly and rapidly as they could. These lessons provide evidence that Liz's mother, once again, had anticipated what would be
important for her shy daughter.

Liz liked school and, in particular, the relative social freedom of a girls'
school, in which intense friendships and rivalries were formed. In early
middle school, she became enemies with another girl and recalled them
roundly insulting each other in the cloakroom. From the beginning, she
excelled academically and felt driven to succeed. She would get up early
on the day of a test and walk around the garden or the living room,
repeating to herself over and over the terms and concepts that would
appear on the test.

In a school that took girls' sports seriously, Liz played on the high
school's lowest-ranked tennis team, which meant she joined relatively
uncompetitive games with teammates who didn't care if she wasn't a

particularly good athlete. She felt physically and emotionally free: "We would change into our sports clothes, which included voluminous black bloomers, a blue blouse, and a short black inverted pleated tunic, worn with a tasseled girdle, the color of the particular school house we belonged to, tied round the waist. We wore warm cotton wind-cheaters in the winter, and I loved the feel of a new one when its lining was still all fluffy. I remember walking to the sports fields on quite cool days in windy weather and loving the whole feeling of being free in our bodies and in enjoyable company." The girls had only limited opportunities to mix socially with boys, primarily at dances held jointly with Broadland House's "brother" school, Launceston Church Grammar School. Liz went to the dances and even had crushes on a few boys, but always from a distance: "I did not think of myself as very pretty. I was quite shy in general about this sort of thing and not venturesome in pursuing any relationships until I went to college."

Broadland House provided a rich education, and at home Liz indulged her taste for gorgeously illustrated books about science, including one by Jacob Bronowski. Liz was fascinated by these books, particularly by the beautifully colored illustrations that made science alluring. She avidly read an adoring biography of Marie Curie, written by her daughter, Eve Curie. Reading about the life of a woman scientist opened a window of possibility for which Liz hungered: "The yearning for things helped push me into science. I was naturally curious and I loved animals, and I was educated quite well. I believed then and still do that I loved science because it also became a world in which I could escape, in the way that for some people religion is an escape into a world where things are fair and you know where you stand. I remember being able to articulate clearly in my mind at quite a young age—not that I ever discussed it with anyone—that through science I could escape into a world where things were secure and fair."

Liz sustained this interest despite the fact that so many of her secondary-school science classes drew on dry textbooks, offering systematic zoology and botany classifications. Not until ninth or tenth grade did she have an opportunity to do laboratory science in chemistry and biology. Because Broadland House did not yet have a laboratory in Liz's earlier high school years, students attended chemistry lab classes at a neighboring school, Methodist Ladies College, a few blocks away. The girls would take the short walk blissfully free and temporarily unsupervised, and Liz became the ringleader of a group of girls who decided one day to see if they could mix chemicals together to make an explosion. The more abstract and private curiosity she felt about science suddenly became cool and rebellious, and choosing to take this risk granted her a sense of power. After Broadland House acquired its own chemistry lab, Liz continued to indulge a taste for doing slightly dangerous things. From a mixture of iodine and ammonia, she and her friends created "touch powder," which produced small but startling explosions—applied to the inner rim of a desk, it would produce a bang when the lid of the desk was closed. The girls put some touch powder on the steps of one of the school buildings in the hope that the headmistress would activate a loud bang by walking on it. They liked the headmistress perfectly well, but they liked producing a reaction even more and were gleeful when she set off an explosion.

That her teachers identified her as an exemplary student not only granted Liz leeway for these escapades but also nurtured her ambition to excel. "I enjoyed writing quite learned essays," Blackburn recalled. "I would use our extensive library at home and our *Encyclopaedia Britannica*, wonderful big books that I adored, treasure troves of information, and I would write on a card table covered with books open for reference as I wrote." When a high school English teacher singled out one student's essay for praise before the whole class, announcing that the piece was "typical of the exceptional person," Liz at first did not realize the teacher was talking about her, though her classmates had already guessed and were looking at her: "I felt gratified but also slightly embarrassed and worried that my schoolmates might resent my success." In a photograph taken for a school publication, Liz is shown sitting before a chemical balance, dressed in her school uniform and weighing out a chemical reagent with an expression of great care. Her passion for science was fostered by a personable young chemistry teacher: "Closer to us in age than most of our teachers, she was a contrast to the fuddy-duddies and seemed to enjoy us, calling us 'a motley crew.' She made chemistry fun, and I came to see science as an interesting and serious thing to do."

Before Liz finished secondary school, her parents separated. The tension in the marriage, aggravated by her father's drinking, had become increasingly apparent to the children, who were not shocked when their mother broke the news one day at lunch. Soon after this announcement, Marcia and the seven children moved to Melbourne to live in the home she had inherited from her late parents. Until the upstairs tenants left, the family crowded into the one-bedroom downstairs flat, a painful change from Elphin House, which for Liz had held an aura of magic. Marcia supported her children on her own until the divorce proceedings were complete and she resumed a more regular medical practice, but the resourcefulness and energy her children had always relied on began to fail her. Struggling to sustain her family by herself, she suffered from bouts of depression severe enough to require hospitalization. Because of her mother's illness, Katherine, who had been studying in Adelaide, arranged to continue her premedical studies at the university in Melbourne. Liz became withdrawn: "I rarely had friends over to the house. I did not want anyone outside the family to be part of this complex life, for which they might judge or, worse, pity me. I did not know anyone who had a comparable experience."

The family could no longer afford private school tuition, so the children attended the coed University High School. During this last year of high school, when her mother struggled with depression, Liz's teenage world was "self-circumscribed" by her choice to immerse herself in schoolwork: "School and studies were an escape from the anxiety and uncertainty that were always in the background, owing to my mother's illness." Liz concentrated so fiercely on academics that her sister Katherine felt she couldn't "really raise issues in the family with her." When Katherine chided Liz for her failure to take on a larger share of the housework, "Liz got furiously angry," and though she was instantly remorseful, she didn't give in. Katherine noted, "I was much keener to please than Liz was. She wasn't scared off by what other people thought about her."[5]

At this crucial time in her life, Blackburn received warm encouragement from teachers, but she did not discuss with her friends her interest in biochemistry, an intellectual fascination grounded in her aesthetic attraction to those illustrations of amino acids and in "a feeling that because biochemistry was about molecules and the workings of cells, it offered a doorway into the mystery of how life worked. I felt you had to go to that level of detail to understand life." She was grateful to Mr. Stuttard, a math teacher who encouraged her interest in science. Once, when he wrote a difficult algebraic problem on the board and asked if any of the students could solve it, Liz volunteered and solved the problem elegantly. He asked her how she'd done it, and she answered, to her surprise, that she didn't know: "I had intuited the process but could not articulate how I had gotten from A to B. At that moment, I was aware that I was able to intuit things in a certain way. I didn't consciously set about problem solving but responded to the problem with a creative processing I couldn't fully access logically." When Mr. Stuttard asked her what subject she would pursue in college, she promptly answered, "Biochemistry." He spoke about her to a biochemistry professor at Melbourne University, where she would be accepted for the following year.

Complete concentration enabled Liz to do exceptionally well in the end-of-year final statewide matriculation exams; in three of the subjects, she earned the highest scores in the state of Victoria, an achievement known as "winning the exhibitions." At the time the results were posted, a local reporter arrived at the house to interview Liz and take her picture for the paper. Her mother waited for Liz to arrive home from her summer job at a bakery and sent the reporter out to announce the news, watching from the front window, giddy with excitement. Liz took the news calmly but worried about whether to pose for the photograph with her recently acquired guitar, "all the better to try to look normal and conventional and not too much of a nerd." Her sister Katherine recalled that the actual photo featured dressmaking, not a guitar, but confirmed the impression that the photographer emphasized her sister's feminine hobbies, as if her intellectual achievement had to be "normalized."

Without talking to anyone about her choice, Liz decided to major in biochemistry in college. At an orientation meeting, as Liz discussed with a college adviser whether she should take a difficult course, she mentioned quite diffidently that she had won the exhibition in the subject. Again, she did not seek advice on which courses to take; she had become used to making decisions on her own, especially since the onset of her mother's illness, and did not worry about whether science was a feminine pursuit. She owed this beneficial insularity at least in part to her mother, who never assumed any such limitation on professional aspirations, but it was also her mother, perhaps out of a desire to protect her single-minded daughter, who once challenged Liz's ambition by remarking, "You know, there are other things besides going to university." That her mother would contemplate, even hypothetically, any alternative made Liz fiercely indignant. But she never openly declared her certainty to her mother or anyone else. Early in her first year at college, while she and some friends visited acquaintances in Hobart, their host, a schoolteacher, asked about their academic interests. When Liz answered that she was studying biochemistry, the man said, "What's a nice girl like you doing in science?" Liz made a polite, noncommittal reply.

2 Shedding Encumbrances

Like her great-grandfather, whose passion for the defining science of his era led him to cross hemispheres in search of *Coleoptera* specimens for his collection, Elizabeth Blackburn would be drawn to the defining science of the latter half of the twentieth century: molecular biology. She arrived at the University of Melbourne, the leading university in the Australian system, when modern molecular biology was still in its infancy. From the start, she took most of her classes in science. Her first-year curriculum consisted solely of math, chemistry, physics, and genetics, and her course work soon focused intensively on biochemistry.

Students at the University of Melbourne could join residential colleges modeled after those of Cambridge and Oxford universities, and Blackburn had been admitted to Janet Clarke Hall for women. This residential college offered tutoring to supplement lectures, a library, and meals, and students were required to wear academic gowns over a dress at dinner. Unlike the vast majority of Janet Clarke Hall students, who had come straight from private girls' schools, Blackburn had spent her last year of high school at a state school. This unusual distinction provoked speculative gossip, and Blackburn was relieved to overhear another student pronounce her "quite nice." Her family's financial difficulties meant that she attended college on a scholarship, which accentuated her distance from her well-to-do classmates, especially when, in her first semester, she discovered that residence fees were due at the beginning of the term, though her stipend would be paid only at the end. Never considering that she could call on her father and unwilling to burden her mother with an additional worry, Blackburn prevailed on an elderly relative to advance the several hundred dollars she required. Her preoccupation with the sciences and the need to keep in close contact with her family reinforced a sense of isolation from her peers; she felt a duty to support her younger sisters Margaret and Caroline, still living at home and young enough to be vulnerable to the uncertainties of their mother's health. In her first years as a student, Liz took a tram to the other side of the city to visit her family every few weekends.

Social life at Janet Clarke Hall became a welcome refuge from the worry Blackburn felt for her family. She made lifelong friends at the residential college, but her social relationships eventually centered on the Biochemistry Department, where in her fourth (honors) year, she spent much of her time and socialized with graduate students and a number of the women faculty. Women professors in the science departments—one in microbiology and another in botany—commanded respect, Blackburn noted, "though they were also seen as old maids, outside the norm for women."[1] Occasionally, a junior researcher would make a casual remark that implied Blackburn, too, was regarded as a bit odd: "I was aware of this—not terribly bothered by it or uncomfortable, but I certainly noticed it." She may not have been entirely impervious—she took care to mention that she had four boyfriends while at the university—and as a young woman, she conformed to the norms of femininity, though in a perfunctory way. In Melbourne, where people dressed formally and women typically wore skirts, Blackburn managed to dress stylishly on her limited funds by shopping at fashion-outlet stores. She would shed her undergraduate garb of skirt and sweater as soon as she left for her doctoral work in England. By not challenging norms, she insulated herself against conflict—a defensive measure that preserved a cerebral inner life she regarded as "genderless."

When Blackburn eventually chose to pursue a master's degree at the university, she discovered that her thesis adviser, Frank Hird, had only ever had one other female graduate student, but Blackburn recalled Hird, along with many other professors, as supportive and encouraging: "I felt I was perceived as a pleasant young woman and a promising student in the department." Hird often met with students for lunch or dinner at the university, and once he asked each person at the table, "What do you want?" Blackburn gave a simple answer: "I want to understand how living things work." This was the first time someone had asked her this question, and she had been surprised into responding with succinct clarity: "In answering that question, I had identified only as a genderless person to what I took as a genderless question."

During Blackburn's undergraduate years, exploring the pathways by which proteins did their work in the cell was one of the hot topics in biology, and gene sequencing was just beginning to attract intense interest. A biologist might approach the study of the complex pathways and systems of proteins in the cell in three primary ways, each a fairly distinct discipline at the time. A biochemist emphasized breaking down chemical reactions into their component stages with the aim of eventually synthesizing the data, often relying on the quantitative analysis of purified substances in the test tube. In advance of discoveries confirming that DNA contained the genetic material, biochemists focused on proteins, not genes, so that classical genetics constituted a separate type of inquiry—one that depended heavily on acute visual observations of organisms. A geneticist studied the physiological change wrought by random mutations in living organisms and worked backward from the phenotype to make inferences about underlying mechanisms, analogous to working backward from effect to cause. By the 1950s, the interest in genes gave rise to molecular biology, and molecular biologists studied how genetic information was decoded in living cells, employing a more holistic approach than biochemical analysis.

Blackburn was drawn to biochemistry because understanding came in the form of deep knowledge of the smallest possible subunit of a process. At each stage understanding was thorough and precise, if still partial. Her classical training stressed painstaking quantitative analysis, in which sharply delimited questions could be answered in a rigorous, unequivocal way. Living with an uncertain family situation, Blackburn was temperamentally inclined to exactitude, and biochemistry appealed to her childhood notion of science as a discipline in which the answers were "fair"—in which logic and empirical reasoning would always be rewarded.

Blackburn's first real research experience came during her fourth year in biochemistry, when her course work incorporated a thesis project. Under the supervision of her coadvisers, Theo Dopheide and Barrie Davidson, she worked on characterizing and purifying an enzyme from bacteria. This research would demonstrate that chorismate mutase-prephenate dehydratase carries out two of the sequential chemical steps (reactions) in building up an amino acid rather than just one, as is typical for enzymes; a previous study had identified just one other enzyme with this unusual property. Quantitative biochemical analysis of this purified enzyme provided a means of understanding how it carried out any of the steps for biosynthesis of the amino acid phenylalanine. Blackburn assayed for the enzyme's activity in order to purify it from extracts of bacterial cells and then set out to determine its mass by analyzing its behavior under a purification process known as electrophoresis. After the protein solution was carefully layered on top of a gel in a cylindrical tube, both ends of the tube were dipped in a salt solution. A positive electrode was placed in the salt solution at one end of the cylinder, and a negative electrode at the other. Moving through the gel under the influence of this electric field, the protein, depending on its size and whether it was richer in more positively or negatively charged amino acids, would run through the gel at a particular rate, which could be compared with that of other known proteins that had been included as standards. By comparing these rates, Blackburn could deduce the molecular weight of the enzyme, providing a clue to how it worked in a living cell.

Her painstaking process produced a seemingly useless set of comparative measurements; the protein she was using as a standard ran on the gel at a rate that made no sense. One day, as she sat at the lab bench, Davidson stopped by to commiserate with her on the difficulties. She appreciated his kindness, but at a certain level she also dismissed it; sympathy didn't solve the problem at hand. Shortly afterward, she searched the department's library and found a review on the properties of this protein. She discovered that two molecules of this protein often stuck together; this would double the apparent molecular weight she had assumed for her standard. Blackburn dashed back to the lab to determine whether the molecular weight fell into a meaningful pattern in light of this property. It did: "At the moment I figured out the molecular weight of my protein, when it had all fallen into place, I knew I was hooked on research, convinced that such work would shed light on things. On the one hand, it was not an earthshaking problem. On the other hand, I had a feeling of real accomplishment and satisfaction that my persistence had solved the puzzle. This gave me confidence I could succeed at something that was perceived as difficult." This work resulted in her first scientific paper, coauthored with Davidson and Dopheide.[2]

While completing her undergraduate honors degree in biochemistry, Blackburn decided to continue her graduate work in biochemistry at Cambridge University. Not only had she confirmed her vocation, she had acquired confidence in her talent. Graduate students from Australia commonly went overseas for their doctoral work, and both of her undergraduate advisers had encouraged her to pursue a doctorate at their respective postdoctoral institutions. She decided on the bucolic setting of Cambridge, Davidson's alma mater, not entirely frivolously, after Dopheide, who had studied at Rockefeller University in Manhattan, described packing up his apartment to return to Australia and finding the tops of all his books black with soot. Cambridge had the additional attraction that her father's sister, Suzanne Oram, lived there with her family.

Because Cambridge required that a PhD candidate from outside England complete a year of research as a prerequisite for admission, Blackburn enrolled in a master's degree program in the Biochemistry Department at the University of Melbourne. Her thesis adviser, Hird, had become interested in the metabolic and anabolic pathways by which amino acids are built up from simpler chemicals in the body and then broken down to produce energy or be changed into other amino acids when required. For her research project, Blackburn measured the rates at which a particular amino acid, glutamine, was changed into another, glutamate, in the liver. This reaction, which releases ammonia, has implications for how nitrogen is used as a nutrient in the body. She recalled that "amino acids were starting to be a theme. I had loved their names and their structures as a high schooler. For my honors project I had worked with an enzyme intimately involved with how an amino acid was synthesized—phenylalanine, by coincidence one of the favorite names I used to roll off my tongue as a teenager." Although her master's degree research struck her as akin to a series of exercises, the work taught Blackburn the value and discipline of planning experiments, performing them reproducibly, and determining the validity of any measurement, always asking, Was it a realistic measurement, or were the conditions of the experiment such that quantitative measurement was not really possible?

Characteristically, Blackburn may have provided a modest estimate of her work with Hird. Many years later, when her mother wrote to Hird to report on Blackburn's accomplishments, Hird responded in an effusive letter: "How could I forget Elizabeth, who had about the best integrating mind I have ever seen? To be honest with you, I used to wonder what Elizabeth would do when she was on her own. Quite often one sees brilliant minds like hers fail to achieve because there is no strong ego to drive them along. On the other hand, I always thought if somebody supplied that for Elizabeth or if she woke up one morning with a desire to do something, she could be a great and novel contributor."[3] Hird's letter suggests he did not perceive Blackburn as driven or ambitious, implicitly a prerequisite for success.

In 1970, while she was still completing her master's degree, Hird introduced Blackburn to a visitor, Fred Sanger. Hird had met Sanger while completing research at Cambridge shortly after World War II. Sanger, a Nobel Prize–winning biochemist, led a lab at the Laboratory of Molecular Biology (LMB) at Cambridge University. The LMB, a center of intense exploration of the genetic code, was funded by the Medical Research Council (MRC), a government agency. Hird led Sanger to Blackburn, working at a bench in the lab, and explained that she wanted to complete a PhD at Cambridge. Blackburn remembered being grateful that she had dressed nicely that day, in a demure pink skirt and a pink-and-white-checked blouse. She was pleased to meet Sanger, but not particularly awed; like most Australians, she adopted an informal, casual attitude toward the famous. (Much later, observing reverential visitors to Sanger's lab in Cambridge, she would wonder at her offhandedness.) Of the work being done in Sanger's lab, she knew only that it was world renowned and she did not yet even comprehend the vast difference between Sanger's research in molecular biology and the biochemistry she had done so far.

On the strength of Hird's recommendation, Sanger informally accepted Blackburn as a PhD student in his lab. Blackburn then applied to the program in biology and to a residential college at Cambridge, a requirement even for students accepted to graduate programs. In 1971, Blackburn was accepted to Cambridge University's PhD program in biology and to Girton College, a women-only residence. But the fabled, much-sought-after Girton College primarily housed undergraduate students and was several miles from the town center, at the opposite end of Cambridge from the LMB, and the college also imposed a strict evening curfew on its student residents. Blackburn, expecting to work late at night in the lab, rebelled at the absurdity of climbing over the college walls to sneak back to her room in the evening. She chose instead to live at Darwin College, a coed, graduate-student-only college, and after she wrote to Girton College to decline its offer, she received a huffy letter of reply, informing her, much to her amusement, that she had forfeited the right to attend a particular sherry party held by the college.

The molecular genetics revolution had begun in 1953, when Francis Crick and James Watson solved the structure of DNA by model building based on the X-ray crystallographic information obtained experimentally by Rosalind Franklin, Maurice Wilkins, and their associates. Efforts to determine the genetic code surged forward in the wake of this discovery, spearheaded by scientists such as Crick and Sydney Brenner, both still at the MRC when Blackburn arrived there. Crick and Watson's model of the double helix accurately predicted its chemical structure but left open the question of how information was transferred from the gene to proteins in the cell. By the mid-1960s, biologists had fully mapped out the relation between the codon of three DNA bases and the amino acid it specified, with the operating assumption that one gene, made up of many codons, coded for one protein, made up of many amino acids. In the early 1970s, molecular biologists were just filling in the picture of how enzymes replicated DNA.

At this time, researchers at the MRC pursued methods for RNA and DNA sequencing as the most informative way to identify pathways in the cell. Rather than measure quantities of molecules, they were intent on abstracting from their observations models for cellular mechanisms, an approach that drew on the methods of physics, which had informed Watson and Crick's model. The first protein had been sequenced in 1951 by Sanger, and the protein researchers at the MRC also investigated primary structures, in this case the sequence of amino acids, as a factor that determined the shape in which a protein folded. These researchers had far less interest in the kinds of questions quantitative biochemical analysis might answer, such as how rapidly an enzyme reaction took place. In fact, mutual disdain often characterized the attitude of molecular biologists interested in sequencing and traditional biochemists, as Harrison Echols reports in his history of molecular biology, *Operators and Promoters*:

> The information people, mostly physicists and geneticists, formulated hypotheses or "models" with a number of assumptions about how it all worked, and they hoped that they could figure out the code without doing too much biochemistry, which many of them regarded as mindless drudgery. The protein synthesizers largely followed the biochemical fashion of the time, attempting to figure out a pathway with the fewest assumptions. They viewed the information crowd as naive and overbearing. . . . In fact, the fusion of these two approaches is the basis of contemporary molecular biology: precise models based on genetic and physical insights, combined with a rigorous biochemical analysis.[4]

Slated to begin work in Sanger's lab in October 1971, Blackburn traveled to England by ship. Because the Suez Canal had been closed due to the conflict in the Middle East, the passenger liner followed a route around the Cape of Good Hope, stopping in Cape Town, the Canary Islands, and Lisbon before arriving in Southampton after four weeks. Aboard ship, Blackburn began reading scientific papers by Sanger and his colleagues for the first time. Her traditional training in biochemistry had so far focused on metabolism, the reactions carried out by protein enzymes in cells, and not on the other large molecules in the cell, DNA and RNA. Though Sanger was originally a biochemist, no one in his lab thought of using quantitative analysis as a primary means of investigating the workings of the cell. Blackburn began to register that the elegant, sexy gestalt of molecular biology was far removed from her own training: "The culture of biochemists differed from that of molecular biologists, who valued most the elegant solution and the clever deduction rather than the painstaking quantitations of proteins and measurements of rates of biological reactions catalyzed by enzymes in the test tube that characterized biochemistry."

When the ship arrived in Southampton, Blackburn was met by the Australian boyfriend of her English cousin Felicity. He took her up to Cambridge, where she stayed for a few weeks with the family of her aunt, Suzanne Oram, who lived just up the road from the LMB. Blackburn kept in close touch with Aunt Zanne and her family throughout her years in Cambridge.

The LMB, opened in 1962, was a blocklike building with large windows and an exterior fire-escape staircase that wittily imitated the form of a double helix. Those associated with the biology labs at the LMB always referred to them simply as "the MRC." A position as head of a lab left gifted scientists free to focus on research, granting them a rare dispensation from teaching and other academic responsibilities, and the MRC boasted a legendary roster. When Blackburn worked at the MRC, Crick was a frequent presence, at the time focused intently on how DNA and proteins became wrapped around one another in the chromosome. Brenner concentrated on the genetics of the nematode as a model eukaryotic system. Like a number of others at the MRC, Max Perutz, who had been the first to identify the three-dimensional protein structure of hemoglobin, investigated protein structure. Aaron Klug, in a lab down the hall from Sanger's, worked on virus structure; he had worked on RNA genome viruses with Franklin after her DNA X-ray crystallographic work had made it possible for Watson and Crick to deduce the structure of DNA.

An equally illustrious member of the august company at the MRC, Sanger had worked with A. C. Chibnall researching the protein chemistry of insulin. Sanger developed new methods for amino-acid sequencing and used them to deduce the complete sequence of insulin (the first amino-acid sequence of a protein ever determined), work for which he received the Nobel Prize in Chemistry in 1958. On the staff of the MRC since 1951, he had become interested in nucleic acids and the problem of deriving a method to sequence DNA, which was fundamentally the same as for proteins.[5] Efficient DNA sequencing would exponentially increase the pace of molecular biology, and Sanger's lab competed intently to derive the best method first.

At the time that Blackburn joined the MRC, the generic structure of DNA had been identified. The genetic code has an astonishing simplicity that led biologists to hope that its elegance might hold true for other aspects of cell biology. Its building blocks consist of only four bases—adenine, cytosine, guanine, and thymine, known familiarly as A, C, G, and T—each of which pairs with only one other base to form the "rungs" of the ladderlike chain of DNA. Segments of this information are replicated in messenger RNA (mRNA), which copies just one strand of DNA and then delivers this information to the ribosome, a template for the creation of proteins from a string of amino acids. To create amino acids, the bases are "read" by the ribosome in order in groups of three; a complete gene might be made up of hundreds of thousands of these building blocks, in most cases coding for a specific enzyme. Since the bases are read in groups of three (codons) and four possible bases might occupy any position in this order, sixty-four amino acids are theoretically possible, but in nature, somewhat arbitrarily, only twenty amino acids exist—all that are required for richly varied life. Just as an alphabet of only twenty-six letters can allow for a multitude of words and sentences, this language, requiring only four letters for its alphabet and composed into words of only three letters, some of which are "synonyms" that create the same amino acid and three of which are stop codons that function like punctuation to signal the end of an amino acid chain, can generate potentially millions of "sentences" of varied order and length.

No one had yet hit on a means to determine directly the sequence of DNA. Watson and Crick's modeling of the structure of DNA had demonstrated that identifying the nature of DNA structure could provide insights into its function, the replication of hereditary material. Operating on this assumption that form would illuminate function, the molecular biologists in Sanger's lab were attempting to identify DNA sequences, which represented another crucial aspect of the DNA structure, in order to piece together the larger puzzle of the function of genes. Although ultimately the methods Sanger devised for sequencing DNA differed from those used for protein or RNA sequencing, at that time he and his group of researchers at the MRC explored a variety of methods as they sought the most workable solution. This was "discovery" science; rather than testing a formal hypothesis, the researchers adopted a provisional "let's look and see" approach.

So far, biologists had focused on sequencing RNA, a near-exact copy of DNA, with the single difference that in an RNA molecule, the base uracil (U) is substituted for thymine (T). The conceptual methods that Sanger had developed to sequence protein were being applied to sequencing RNA molecules, and these had already become the standard in this tiny but growing field. Sanger used digestive enzymes, or nucleases, to break up a chain of molecules into small fragments called oligonucleotides ("a few" nucleotides). Each nucleotide is made up of a base and its sugar-phosphate backbone, which links the bases in a linear chain. (For amino acids, a different chemical backbone links the chain, which then folds into three-dimensional structures to form proteins.) To sequence a small fragment, researchers tried various methods for cutting it into still smaller segments, using RNA enzymes that "snipped" the chain at known bases. Messenger RNA constituted a more experimentally approachable model than DNA because its segments, and the bases to be read, were shorter than for a complete strand of DNA.

As her doctoral adviser, Sanger assigned Blackburn the project of applying these sequencing methods to small (by current standards) stretches of RNA, made by copying the DNA fragments of the genome of a small bacteriophage, phiX174. PhiX174 was first isolated in 1935 from the sewers of Paris by scientists interested in determining whether these bacteriophages could be used as antibacterial agents—an idea that rapidly fell into disfavor once the spectacular success of antibiotics became evident. Bacteria and bacteriophages, minute viruses that insert their genome into the host bacteria, have proved to be the ideal model systems for understanding the central mechanisms of gene replication and expression. For the researchers in Sanger's lab, the relative simplicity of the DNA in the organism phiX174, one of the smallest naturally occurring DNAs known, provided a set of training wheels for the first forays into sequencing DNA in increasingly complex organisms.

By copying DNA into RNA and then determining the sequences of these fragments of RNA, Blackburn would indirectly map the DNA sequence. Bit by bit, reconstituting such fragments would enable researchers to determine the full sequence of DNA. To achieve results with this conceptually simple process, a researcher had to create experimental conditions for painstakingly cutting apart an RNA molecule in order to identify the sequence of its nucleotide bases. A comparatively short RNA molecule might typically be seventy bases long; often, these molecules consist of thousands of bases and can even be as long as hundreds of thousands of bases. To dismantle this chain and determine its sequence, Blackburn employed nucleases. For an unknown fragment of RNA, she might employ a particular nuclease that cut to the right of every G in the RNA molecule, breaking it down into a number of smaller fragments. This would tell her how frequently G recurred in the chain and show where some were closer than others along this chain. She might identify fragments such as ___ G, _____ G, _ G, and ___ G, but she could not yet fill in the blanks or deduce the order in which these fragments occurred. Next, each of the smaller pieces might be separated ("fractionated"). Like the proteins that separated under electophoresis at different rates depending on their charge and size, oligonucleotides also separated out from each other under an electric field. (Though in this case, the solution, or gel, was not placed in a test tube but affixed to a type of blotting paper that wicked up oligonucleotides at different rates, depending on their molecular properties.)

The fractionation principle that Blackburn relied on remains the basis for direct DNA sequencing. In order to detect the different lengths of fragments, she employed biochemical tricks, including the use of radioactive isotopes to label specific sequences or bases, so that when an X-ray film was placed over the gel, radiation from the DNA appeared as darkened areas on the developed film. On this autoradiogram, relative lengths of fragments were inferred from the positions of bands of varying darkness. Oligonucleotides that ran at a similar rate were further separated using chromatography (two-dimensional, or 2D, separation). Once each oligonucleotide had been purified by this means, a different nuclease, such as one that cut after every C or U, might be employed in order to segment these fragments again and fill in more of the blanks.

To piece together a unique solution to the original sequence, Blackburn would have to try still more tricks, such as using additional enzymes known to chew away the RNA fragment from one end or employing special chemicals that cleaved specific bases of the oligonucleotides. This painstaking, often tedious process might be compared to filling in a cross-word puzzle in an unfamiliar foreign language on the basis of partial clues that make it possible to guess only one letter at a time. Where a native speaker might readily guess from the context that a three-letter English word beginning in *t* probably ended in "*h-e*," in this case, only a laborious process of elimination could determine what followed the *t*.

Sanger's laboratory was staffed by six postdoctoral fellows, two profes-sional technicians who held long-term positions at the lab, and two doctoral candidates, Howard Chadwell and Blackburn. Among the postdoctoral fellows were John Sedat, Blackburn's future husband, and Edward Ziff, fresh from earning a PhD in the Chemistry Department at Princeton University, who would become a lifelong friend. Sedat, who had come to the MRC in 1970, had previously worked with the bacte-riophage phiX174 as a graduate student at Caltech in the laboratory of Bob Sinsheimer, known as the grandfather of modern molecular biology. Sedat had persuaded Sanger that phiX174 would be "a good DNA" on which to develop DNA sequencing methods.[6]

As one of the most elite research institutions in the world, the MRC could count ambitious, gifted scientists among its doctoral and post-doctoral researchers. Paradoxically, this elitism, since it so clearly derived from merit and presumed an intense absorption in the work itself, resulted in an egalitarian camaraderie and collaborative exchanges among the researchers. From Blackburn's perspective, the Sanger lab was amazingly free of hierarchies:

I have a vivid early memory of speaking with Mike Fuller, who was in charge of the lab supply store room. I introduced myself as one of Dr. Sanger's students, and he corrected me—"Oh, we don't say Dr. Sanger; we all call him Fred"—in a way that clearly conveyed I was expected to follow this custom. Later, I was bemused to discover that in the United States, things could be *more* hierarchi-cal, despite the stereotypes of Britain and the United States that would seem to imply the reverse. Shortly after I arrived at my postdoctoral laboratory at Yale, Joe Gall, my postdoctoral adviser, told me it would be fine for me to call him Joe, since I was no longer a graduate student. I didn't mention to him that it hadn't even occurred to me, since my days at the MRC, that I would call him anything else.

Sanger so disliked anything that smacked of unnecessary fuss that once, investigating a new centrifuge, he complained that it had "too many knobs"; there were two. On her arrival at Cambridge, Blackburn had been concerned about whether her undergraduate work had fully prepared her for a molecular biology lab, but when she asked Sanger for a list of additional readings, he remarked that this would not be necessary and instead showed her how to do the lab-bench manipulation for dissolving solutions of dried oligonucleotides, fragmented from much longer RNA molecules, and then transferring them into tiny reaction test tubes devised for the miniscule reaction volumes needed. He was careful even to instruct her how to seal the ends of the fragile glass tubes to keep the solutions from drying out. Blackburn plunged directly into bench work, and she recalled Sanger's intense focus on the immediate demands of their work as characteristic: "Fred liked to be in the lab and not have people interrupt with what he thought of as boasting. Mark Ptashne, a visiting professor from Harvard and the proud owner of a recently acquired Guarneri violin, brought the violin to Fred's own little lab to play it for him, so Fred could appreciate the sound. I looked into Fred's room from the hallway and saw Fred, seeming rather unappreciative of the interruption and the loud sound of the violin, trying to reach past his visitor to continue the task of loading his DNA-sequencing gel."

Sanger's advising style granted wide latitude to the biologists who worked in his lab. Equipped with a ready, dry sense of humor, Sanger was low-key and mild tempered, unless he felt someone was foolish with lab equipment or careless of procedure. Blackburn appreciated the independence that his approach cultivated: "Fred's advising style was hands-on in the sense that he was always in the lab and he gave me a very specific research project to work on without any delay. But he was hands-off in the sense that only when I asked him did he offer specific suggestions about experiments. I got much of my information about how to do experiments from others in the lab—Bart Barrell, Fred's long-term technician, who worked at the lab bench across from me, and other post-doctoral fellows, mostly those in Fred's lab." Sedat, a frequent source of helpful advice for Blackburn, also didn't hesitate to challenge her: "At one stage, John made a suggestion, and I remarked that it might entail a lot of work. John immediately answered, 'Graduate school is not a rest home.' I was quite favorably struck by this—it gave me a wake-up call, and I found it quite invigorating. I never felt I was left to drift, and I liked the atmosphere in the lab a great deal."

Like the other researchers in the lab, Blackburn worked long, hard hours; according to Spyros Artavanis-Tsakonas, a fellow doctoral candidate at the MRC at the time Blackburn worked there, graduate students working in the lab at midnight often found Brenner also at the lab, smoking and chatting.[7] Blackburn relished the routines of laboratory work despite stretches of inactivity, as when she had to wait for electrophoresis of her oligonucleotides to be completed. After she had spotted the short oligonucleotides on to blotting paper and dunked them under the organic solvent, they were allowed to run through the paper under the influence of the electric field. Blackburn had to wait in a room full of tanks of organic solvent under which an electric field of a few thousand volts had been set up: "Many a summer evening I would be waiting in that third-floor room, leaning on the windowsill, looking out over the fields near Cambridge, wondering why I was not enjoying things in the outside world on such a beautiful evening. There was always also this slight underlying tinge of alertness during my many hours there, because the room was alarmed in case a fire broke out in the high-voltage tanks. If the alarm was set off, you had to get out fast before a big metal fireproof door automatically slid closed and trapped you in that room."

Freedom, conditioned by "an underlying tinge of alertness," marked Blackburn's years at Cambridge. She had received an Australian federal government scholarship, which funded study in another Commonwealth country but would support her for only two of the three years of graduate work required to earn a PhD. Although she later applied for and received a Commonwealth scholarship that supported her third year of studies, she planned to eke out her scholarship allowance over three years. She calculated that she could survive if she did not spend money on anything but bare necessities. Characteristically, Blackburn embraced this means of shedding encumbrances, because the simplicity of living on so little bolstered freedom of mind:

This was quite different from my life before, which, while not wealthy, had allowed me to buy things like books, clothes, or records that took my fancy. So this simplifying factor allowed me to think only of spending my time and

energies in the lab, since to do much else would have cost money I was not planning to spend. It also freed me from having to do something on which I would
feel judged—my taste in whatever I chose to purchase would reflect on me. I was
simply a poor graduate student, and I could wear the same pair of corduroy jeans
(with periodic washings) throughout the rather chilly first summer I was in
Cambridge, with no fear that any comment would be elicited about the fact that
I was not dressing up nicely.

At Cambridge, men far outnumbered women students, which made
women like Blackburn suddenly sought after. At first she socialized with
shipboard acquaintances, none of whom worked in related scientific
fields, but her life quickly centered on the MRC lab, a paradise for her.
The hardworking researchers showed little inclination to seek entertain-
ment outside their work, and the sleepy college town offered relatively
few temptations. Most of the researchers were foreigners, lacking a
natural social circle in England, so they formed their own circle, a "lab
rat" culture in which they found their family substitutes and friends.
They congregated for morning tea, lunch, and afternoon tea at the cafeteria on the top floor of the four-story LMB building; on sunny days,
they sat outside on a large deck opening off the cafeteria. At any table
a member of any lab might happen to sit down with postdocs, the few
graduate students, or the more senior or junior scientists, and where
anyone sat differed from one day to the next.

Crick often took part in cafeteria conversations, and he placed great
value on them, as noted by Soraya de Chadarevian in her history of
molecular biology, *Designs for Life*: "Acoustic tiles were fitted on the
[cafeteria] ceiling, after Crick complained that it was too noisy to follow
a conversation and that it was the wrong place to economise, given the
importance of scientific interaction over lunch."[8] The free-ranging discussions about scientific ideas were both egalitarian and elitist, as Sid
Altman's recollections of his time at the MRC demonstrate:

Everybody went to tea, according to the English custom, mid-morning and mid-
afternoon. These "gods" of molecular biology were there, sitting with everyone
at tea. They encouraged everybody to participate freely in discussion and they
treated everybody equally. . . . You had to develop yourself in two ways. On the
one hand, you had to train yourself not to say anything that was superfluous or
stupid. The "gods" made it very clear if they felt that you were frivolous or you
were thinking in a less than coherent way . . . if you were able to engage in the
discussions, take advice, and approach the gods only when you had really something to say, it was fantastic.[9]

Blackburn's colleague Artavanis-Tsakonas also recalled a mixture of awe and freedom in his interactions at the MRC:

We were at the forefront of a revolution. To go to a place like the MRC as a graduate student meant you were framed by people who were very prominent, very smart, yet approachable. You were not afraid to ask their opinion; they were very smart and critical but not intimidating. Being in that atmosphere and being given great independence to pursue your own goals and follow a finding where it led—not being confined in a very rigid way as often PhD research might be—was fantastic schooling. There were people who broke their necks—who never made it in science after all. But they were not very many. In the class that Liz and I were in, almost every one of our contemporaries made a good career in science.[10]

Blackburn recalled that in the cafeteria she often "soaked up the conversations in silence." She was one of the few women working in the labs; all the postdocs in Sanger's lab were men, and only a few women postdocs and career scientists worked at the LMB. During the early 1970s, very few women pursued doctorates in science; in 1973, U.S. women earned only 17 percent of doctoral degrees in the life sciences and constituted only 10 percent of faculty in the field.[11] The first woman scientist joined the MRC staff in 1954.[12] By the early 1970s, only 12 percent of the senior staff and workers at the MRC were women, and by 1989 this number had risen to only 22 percent.[13] Early in his long tenure at the MRC, Max Perutz had bemoaned the "lack of initiative and originality of most women research workers (there are some notable exceptions!)"[14] In his memoir *The Double Helix*, Watson recalled attending a lecture by Franklin, and "wondering how she would look if she took off her glasses and did something novel with her hair."[15] These attitudes had not changed dramatically by the 1970s. When molecular biologist Joan Steitz arrived at the MRC with her husband, Tom Steitz, to do postdoctoral research in the late 1960s, she found lab space had been provided only for Tom, and Crick blithely suggested she work instead on a theoretical project in the library. Fortunately, she scrambled to find lab space, where she ultimately did important work on messenger RNA.[16] In all the scientific conversations in which Blackburn took part during her tenure at the MRC, no one ever discussed the failure to grant Franklin credit for her role in Watson and Crick's discovery.

Blackburn consistently portrayed herself as a respectful and quiet junior colleague. Having become close to Sedat in her first year at

Cambridge, in late 1972 Blackburn moved in with him, and they shared a rented row house on the edge of Cambridge with Ed Ziff. Sedat and Ziff would spend long hours discussing their scientific experiments, and again Blackburn mostly listened. Though she registered their inattention when she offered her own comments, she didn't mind particularly at the time: "People were generally friendly, and I think they liked having a young woman around who was involved in the sciences and didn't make demands—I listened and learned a lot, and didn't feel I had to contribute to discussion. All this is not to say I was shy and retiring—I was involved and enjoyed our social gatherings a great deal. At the MRC, I did not sense the hostility I had felt in Australia as a woman who was intelligent and accomplished."

Blackburn did not have to act the part of a nice girl; Blackburn *was* a nice girl. But she may have been less unassuming than her own recollections imply. If she was polite and deferential, the observations of Artavanis-Tsakonas suggest that she also earned respect for her intellectual candor and acumen: "Like Fred Sanger, Liz was understated—very smart, very critical when you asked her to be critical. But I never remember Liz saying anything negative about anybody. It's a quality not many people have."[17] In some respects Blackburn cultivated "gender blinders"; her social life and confidence as a woman could be set aside in the lab, where she would pass for one of the guys: "I could be free not to feel the need to act feminine. But at the same time I felt, without calculating it, that my male colleagues at work would feel less threatened by me than by another man, with whom they'd feel more competitive, if I were friendly and undemanding, which became a protective coloration. Not acting *too* feminine, on the other hand, was a protection against something I did not want—I had a sense they would have their most substantive conversations about science with me if I was seen as one of them."

If she adapted protective coloration, Blackburn's relative indifference to gender also betrayed a character trait rooted in her essential, "genderless" self. Her mind was attuned almost exclusively to science, to the extent that she did not seek the companionship of women but formed her close relationships primarily within the Sanger lab, a decidedly male social world. She chose to pair off with Sedat, who had grown up in the remote highlands of Guatemala, where his U.S. parents worked as missionaries. Like Blackburn, he was disinterested in pop culture and consumed by his scientific work. Working at the MRC with Ziff and Francis Galibert, he would publish the first DNA sequence of any kind. Unlike Blackburn, Sedat was "rich"—his U.S. fellowship dollars translated to comparative wealth in Great Britain. If they worked hard, writing several papers a year, they also felt free to take in London art museums and music performances and to travel for four or five weeks at a time. "Which wouldn't happen today," Sedat noted. "It's become too hard-nosed a culture."[18]

Like many of her peers, Blackburn believed that deriving a clear view of the sequence of DNA would lead to an epiphany about its function. This had already proved true for the transfer RNAs, which mediate how the sequence of mRNA is translated into the protein sequence. Efforts to deduce its structure and sequence had yielded information about how the transfer RNA itself was folded up and also about how it acted as an adapter. She was a little disappointed when, after she had sequenced the first little fragments of a few tens of continuous bases of DNA, no epiphany occurred. She even showed the sequence of her little fragment to her gifted cousin John, eleven years old at the time, to see if his unusually talented mathematical mind could discern anything of interest. Nothing emerged. But the idea of mining DNA for mathematical patterns was simply ahead of its time. Only when many more and much longer sequences were identified did it become possible to make sense of the patterns of bases, often over stretches of bases that were much longer than those Blackburn was analyzing. She was applying her theory to a too-short sequence.

As a doctoral candidate, Blackburn felt relatively free of the burden of competition; for the postdoctoral candidates, the stakes were higher. Most of the Americans worried about obtaining a position at a good research university, preferably in the United States, after they completed postdoctoral work, but to do so, they had to produce scientific papers beyond the merely competent. Blackburn was still at liberty to lose herself in her work. For her, a feeling of awe for the mystery of life, symbolically embodied in the "secret code" of the genes, was wedded to curiosity and a methodical effort to dismantle that mystery. She trusted her passion for her work and the excellent training she had received at the MRC to open doors for her in the future. And because Sanger had assigned Blackburn an approach to sequencing that no one else in the lab was trying, she felt shielded from the intense rivalry among the post-docs, who aligned in different clusters to compete to derive the best method for sequencing DNA directly. Yet ambition was a double-edged sword, not merely an encumbrance but also a powerful engine.

In Sanger's lab, Blackburn was exposed to two significant influences on her scientific career. First, Sanger imparted a curiosity-driven, pragmatic approach to research: "just get in and find the sequence." No elaborate justification was needed in advance of this exploratory process. According to Blackburn, "I have found that such elaborations are important to do post hoc, in building up a rationale and a story in order to communicate the science in written papers or talks, but the drawback is that one can just sit down and endlessly find reasons why some experiment in biology might not work. So the courage to just wade in and try, if you thought it was a good idea, came from Fred's general philosophy." Ultimately, Sanger would develop a process for sequencing DNA that derived from his own practices rather than the methods tried by others in the lab. This process would become the universally used DNA-sequencing method and win him a second Nobel Prize, shared with Walter Gilbert. But methods developed jointly by Sedat and Ziff and by others associated with the MRC lab became key to Blackburn's later success in sequencing telomeric DNA when she did her postdoctoral research.

Second, Blackburn's career was influenced by the sense that henceforth she would carry the mantle of the prestigious LMB:

The institute was widely referred to as "*the* MRC," despite the fact that there were other MRC institutes in England. To the members of the biological sciences world, this must have been an irritating tenet to encounter, although many also subscribed to it, if the number of people who made a pilgrimage to the lab (which we regarded as its natural due) is any indication. But this feeling that the MRC was at the epicenter of it all gave me an unspoken confidence, and that consciousness continued even after I left the MRC. Throughout my life, I felt an identity as an "alumna" of the MRC lab far more strongly than I ever did for my undergraduate university. Part of it is because I was so much happier at the MRC.

Blackburn's term of "protective coloration" might be applied to her reluctance to attribute her early successes to her own excellence. By the time she left Cambridge, Blackburn had written two scientific papers, one alone and one coauthored, both immediately accepted for publication by the *Journal of Molecular Biology*. Both papers constituted solid science, Blackburn felt, but they were not "exciting enough." "I was modest in my expectations," she noted. Soon after she left Cambridge, she was formally invited to give a research talk at the Cold Spring Harbor Laboratory on Long Island, because "people were interested in what someone who had come from Fred's lab had to say."

Even as she began to establish a reputation among her peers, Blackburn continued to turn to her mother for confirmation and essential emotional support. During the years she spent in England and after she moved to the United States, Blackburn received regular letters from her mother every two weeks, blue airmail envelopes that kept coming even after her mother grew old and ill and had to dictate short letters to Blackburn's sister Barbara. Her mother would respond in detail to Blackburn's last letter, offer copious news about family and friends, share details of her domestic life, and even report on the weather—reassuring reminders of ordinary life for a daughter who had dispensed with so many distractions. Throughout her life, Blackburn would write or phone her mother to tell her of her accomplishments as a scientist; for so long, her mother had been "the only one whom I wanted to be proud of my achievements."

3 One of Gall's Gals

As she completed her degree at the MRC, Blackburn knew she wanted to continue sequencing DNA during her postdoctoral training. Postdoctoral training typically lasted from two to three years in the 1970s, providing not only further education but also a litmus test of one's worth as a scientist. While a doctoral candidate pursued projects determined by a laboratory head, a postdoctoral fellow was expected to carve out an area of specialization and expertise that would provide the basis for a future career as well as qualify him or her for a research position in an academic institution. Blackburn recalled the anxieties of this transition period: "When you come to a new lab in a different research area, you are going from having been practically a world expert to a novice. You have to prove yourself all over again. That is the first anxiety. The second worry is whether you'll get results that mean something and are novel and interesting enough that they are perceived as producing a real advance in the research area. Because there could be blind alleys or insurmountable technical difficulties."[1]

After consulting with colleagues at the MRC and researching current scientific journals, Blackburn arranged to do postdoctoral work with Herbert Boyer and Howard Goodman at the University of California at San Francisco (UCSF). At a scientific conference held at a former monastery in Belgium, she had met and interviewed informally with Boyer while they walked in the garden. Boyer and Goodman wanted to sequence the small DNA genome of Simian Virus 40, a virus with a DNA genome similar in size to bacteriophage phiX174. A number of scientists had chosen to study Simian Virus 40, which caused tumors in rodents, because they hoped it would generate insights into the nature of human cancer, but for Blackburn, "the virus was DNA waiting to be sequenced, and sequencing any interesting region of DNA was what intrigued me then."

Having identified a research area and found a lab, Blackburn wrote a proposal for a fellowship to fund her research. Postdoctoral fellowships, which for non-U.S. citizens were sponsored primarily by private foundations, conferred not only essential support but also prestige, so that while a number of postdoctoral fellows received funding directly from a lab, many laboratory heads made the acceptance of postdoctoral fellows contingent on the applicant's ability to obtain a fellowship grant. No such condition had been set for Blackburn's acceptance to Boyer and Goodman's lab, but they assumed she would apply for a fellowship and, with her credentials, would be successful. She was not surprised when she received funding from the Anna Fuller Foundation.

But Blackburn's carefully laid plans would change. During her last year at the MRC in Cambridge, John Sedat had left for an eight-month appointment at the Biochemistry Department of the Hadassah School of Medicine in Jerusalem. Even before he left for Jerusalem, he'd been accepted for a subsequent, salaried senior postdoctoral research position, with space provided in the Yale University lab of William Summers. Summers worked on a bacteriophage, as Sedat had done in Sanger's lab, but in Jerusalem John had worked with *Drosophila* fruit fly DNA methylation, and he had decided to shift to a new research area once he got to Yale. Having pioneered sequencing, he considered the problem essentially solved; instead he would continue working with *Drosophila* to explore the structure of chromosomes in living cells.

Blackburn had visited Sedat for six weeks while he was in Jerusalem, and during the visit, they became engaged. The relationship took precedence over her previous plans, and she decided to seek a postdoctoral position at Yale instead of UCSF. When Sedat recommended Blackburn apply for a position in the lab of cell biologist Joe Gall, citing his excellent reputation as a scientist and a mentor, Blackburn wrote to Gall seeking a postdoctoral position in his lab. In retrospect, Blackburn could not recall having any second thoughts or fears she would jeopardize her career; she'd felt confident "something would work out."

Blackburn did not hear back from Gall before she and Sedat departed on a trip. She and Sedat had traveled together in Ethiopia in 1973, and in November 1974, freed from academic responsibilities for a few months, they journeyed to Nepal, at the time a favorite destination for hippies as well as serious mountaineers. Unable to afford an organized trekking agency, they found a Sherpa-run cooperative in Katmandu that provided guides for trekking (a necessity on the unmarked mountain trails). Blackburn and Sedat, a guide, and two porters, all laden with food, tents, and sleeping bags, trekked to the Everest Base Camp. From Nepal, Sedat went directly to Yale, while Blackburn continued to Australia to visit her family for several weeks and to await a visa that would enable her to join him in the United States.

To Gall's astonishment, Blackburn wrote to him announcing her arrival at Yale in February 1975, and when she got there, she simply stopped by his lab. Since she had not explained in her correspondence that she was coming to New Haven regardless of Gall's decision, Gall believed that she had audaciously assumed she had been accepted, sight unseen, into his lab. She had refrained from explaining her personal reasons for coming to New Haven: "At some unspoken level I felt that my personal life was, and should be, irrelevant to professional matters. I was also reluctant to inject the rather lame-sounding reason that I was coming—namely, that I had followed John to Yale." Blackburn had to correct the oversight and assure Gall she had not presumed she'd have a space in his lab.

Why, when postdocs usually joined a lab only after a long courtship and careful planning, did Gall agree to take on Blackburn? If Blackburn had committed a social gaffe, she was far from naive when it came to the science. She had done her homework before she first wrote to Gall and knew of his interest in chromosomes. In his early work, Gall had focused on giant lampbrush chromosomes in the egg cells of amphibians and studied the DNA of genes coding for ribosomal RNA in amphibians, insects, and protozoa. With his student Mary-Lou Pardue, he had developed the technique of in situ hybridization, which became widely used to localize DNA and RNA sequences on chromosomes visible under the microscope.[2] He had recently begun investigating highly repeated DNA sequences found on parts of the centromere. The centromere does not function as a gene, coding for a protein, but has a structure targeted to proteins in the nucleus during cell division. By this means, a chromosome is aligned along the mitotic spindle, ensuring that once a cell divides, each daughter cell will receive a copy of each chromosome. Gall had been trying to piece together the sequence of the centromere, using a variant of Sanger's method for sequencing RNA, and Blackburn had turned up at his door with state-of-the-art credentials from the Sanger lab.

When Blackburn next met with Gall to discuss the possibility of joining his lab, Gall informed her he had received a letter of recommendation for her from Sanger. With characteristic euphemism, Blackburn recalled that Gall spoke of this letter as "helpful." In addition, Gall had quickly determined that her fellowship could be transferred to another institution, providing she obtained approval, and she wrote a short new research proposal that fit the context of Gall's lab. After their inauspicious beginning, for a long time during her stay in New Haven, Blackburn felt a little shy around Gall, afraid she might make another inadvertent mistake: "This was no fault of his and rather serves as a statement of my lack of confidence, as it quickly became evident to me that he was the soul of kindness and supportiveness and courtesy and indeed set an example as a scientist and a lab head which to this day I think of as the one to try to emulate."

Not only did Gall have excellent credentials as a scientist—in 1983, the American Society for Cell Biology would give him its E. B. Wilson Award for his research contributions to cell biology—but also he fostered a collegial spirit in his lab, an important boon for someone like Blackburn, at best ambivalent about head-to-head competition. Gall believed this made for better science: "The competition among several groups working on the same problem is obviously good when science moves ahead faster. It is bad when it leads to overlapping effort, stealing data, withholding data, and the like. I've always told people in my lab they would get much further ahead by cooperating and collaborating than by competing."[3] Gall had an air of assuming that those he mentored were good scientists, and he assured them of this whenever their confidence flagged. Martha Truett, a lab technician in Gall's lab at the time Blackburn worked there, was encouraged by Gall to participate in presentations of scientific papers at weekly lab meetings, though lab technicians elsewhere were not routinely included in these meetings. She especially appreciated Gall's willingness to share credit on a scientific paper, treating her just as he would have treated a grad student or a postdoc. Truett viewed Gall as exceptionally generous in crediting his students when it came to publishing their findings: "Because a lab head provided grant money to support research and often the initial idea as well, his or her name would typically appear as author on any papers coming out of the lab, sometimes as first author, a position that signaled primary responsibility for the findings. I authored a paper with Joe, reporting research I had done under his tutelage, and I was touched when he told me I would be the first author. This was very characteristic of Joe. Ginger Zakian was sole author on a paper that came out of her graduate research under Joe."[4] With the support of Gall and her many women colleagues in the lab, Truett decided to pursue a graduate degree in the sciences and went on to a successful career in the biotechnology industry.

Gall's fair treatment of all his lab members provided an exceptional opportunity for women to flourish. Truett noted that over several decades, an amazing number of successful women scientists have "gravitated to Joe's lab. I never had the sense that he felt this was his mission. It just happened because his open, collegial way of working with others attracted women."[5] A number of women who had worked in his lab and gone on to brilliant careers became known as "Gall's gals," an appreciative nod to Gall's role as a mentor. The jacket copy for Gall's book *A Pictorial History: Views of a Cells* notes that Gall served as president of the American Society for Cell Biology, which he helped to found, and also lists three of his former students, all women, who held this prestigious post as well: Pardue, Susan Gerbi, and Blackburn.[6]

That these women could be dubbed Gall's gals reflected not only the influence of their mentor but also their own status as exceptions to the rule. In the mid-1970s, when Blackburn was at Yale, women were still a rarity among research scientists. From 1961 to 1974, Yale granted a total of seventy-one doctorate degrees in molecular biology and molecular chemistry; only eighteen of these degrees were granted to women.[7] During the same period, fifteen women at Yale received doctorates in biochemistry and biophysics, out of a total of forty degrees granted.[8] By 1979, women receiving doctorates in the biological and physical sciences at Yale still made up only 10 percent of the total, though by the late 1990s, their numbers would rise to 38 percent.[9] Nationally, the representation of women among recipients of doctorates in the life sciences grew from 17 percent in 1973 to 40 percent in 1995.[10] Despite this increase of women earning doctorates, their representation among tenure-track faculty in the sciences and engineering did not increase proportionately; from 1979 to 1995, men held a steady 14 percent advantage.[11]

Gall's lab at Yale differed dramatically from the virtually all-male environment at the MRC. Among Blackburn's colleagues were Virginia (Ginger) Zakian, just completing her doctorate, and grad student Kathy Karrer. It made a difference for Blackburn to be working in a lab with other women. She recalled Zakian as "forthright about women's issues, open, and clearly very good at science." And Karrer took Blackburn under her wing, teaching this newly arrived Australian basic lessons on domestic life in the United States. Karrer explained to Blackburn that "when you need things for the house, you go to Sears," and Blackburn was conscious that "women do this kind of thing more than men." If she registered that feminism was in the air—*Our Bodies, Ourselves*, a feminist manual on women's health and sexuality, had only recently been published—Blackburn didn't see it as particularly relevant to her professional life: "In the lab, these things weren't supposed to matter." At Yale Blackburn came to know Joan Steitz, who along with her husband, Tom, was now an assistant professor in Yale's Molecular Biochemistry and Biophysics Department, but they never discussed the discrimination Steitz had encountered at the MRC and during her subsequent job search. (Before they took posts at Yale, Steitz's husband, hoping they would both find posts at the University of California at Berkeley, was warned that a woman would be unlikely to obtain an academic position there.)[12] "We talked about science," Blackburn said.

For Blackburn, preserving this cherished freedom to focus on the work meant resolutely shutting one's eyes to evidence that gender could be a handicap, just as she denied other obstacles. In contrast to Blackburn's perception that "in the lab, these things weren't supposed to matter," another female colleague at Yale in the 1970s portrayed the university as a tough place to be a woman. Diane K. Lavett (then Juricek) had taught briefly before coming to Yale for postdoctoral work, and she emphasized that discrimination was the norm:

My teaching position was temporary, and I was on the search committee for my replacement. The two final candidates were a man and a woman, and everyone agreed the woman was the better candidate. Yet the committee voted for a man, even though I said, "Let me explain about affirmative action," and pointed out that if the candidates were merely equal, under the law the woman should have been hired. When I was a doctoral candidate, I was advised by a guy on my thesis committee to use an initial, not my first name, when I tried to publish my dissertation. He told me the double whammy of being from the South and being female would ensure my papers did not get published. This was helpful—even accurate—advice. At that time in academia, there was no room for women to maneuver, and I'm not sure there's any now. It wasn't necessarily that these guys were intentionally sexist. They just didn't know better.

As the mother of two children, Lavett could not stay late at night doing research, and she noted that "Liz had a rightfully deserved reputation of being exceedingly hardworking. I didn't, because I didn't stay until ten or eleven at night. If you were a married woman with kids, you were dead." Lavett spoke of the atmosphere at Yale at this time as "contaminated": "When we discovered that we would not have enough room at a conference to accommodate the audience, a primary investigator, a basically nice guy, said we should set up a TV in another room and have all the women watch the lectures on TV. In another instance, I had researched and written a paper on my own but discovered that a primary investigator had added the names of several other researchers as coauthors. When I told him that I felt I had been mugged, he answered, 'Well, at least you haven't been raped.' "[13]

Although sexism was not directly discussed in the lab, Blackburn told a story that revealed Gall's concern with its impact on his students' careers and betrayed her own inability to admit this: "One day Joe came into the lab waving a piece of paper and announcing that Mary-Lou Pardue had just received tenure at MIT. I couldn't quite make sense of his excitement, and I said, 'Well, of course she should get it—she's very good.' The tenor of Joe's reply suggested to me he well understood that for a woman, to be very good might not be enough, and so Mary-Lou's triumph was especially sweet. It was an important message to convey to the lab." Gall had in fact worked hard to help his former student succeed against the odds, and in his opinion, some of the same difficulties hold true today: "Mary-Lou Pardue was one of the first female faculty—one of few—at MIT in the seventies. It is unclear whether women were being discriminated against overtly, but clearly they were and still are being discriminated against in terms of promotion. Things have changed, but it is still not an even playing field."[14]

The timing of Blackburn's arrival in Gall's laboratory could not have been more perfect. She had anticipated she could use her exciting new abilities to sequence DNA, and her passion for sequencing was so great ("sequences were cool—every one a treasure") that determining *what* she might sequence came almost as an afterthought. But Blackburn had become curious about DNA at the chromosome end regions, specialized structures that capped both ends of a chromosome. And Gall had just discovered how to purify the linear minichromosomes in *Tetrahymena*, a single-celled protozoan, away from the rest of its cellular DNA. He had also determined that rDNA genes, which encode ribosomal RNA, were housed on these very short, linear minichromosomes. Each cell had tens of thousands of these minichromosomes, thus enabling the organism to produce enough ribosomes to provide protein synthesis for so large a cell.

Typical chromosomes are millions of bases long, and their end-region DNA constitutes too small a fraction of the whole to be purified for analysis by the methods then available. In contrast, *Tetrahymena* produces plentiful minichromosomes with two end regions for every twenty thousand base pairs, a much larger proportion of the genetic material, thereby making it possible, for the first time, to separate out the end regions for further study. In the 1970s, Ray Wu and collaborators as well as Ken and Noreen Murray had sequenced a few bases at each end of the linear DNA in the lambda bacteriophage, far simpler DNA structures than the chromosomes of more complex organisms. So the genetic material of *Tetrahymena* offered a chance to explore a true eukaryotic organism, with potentially far-reaching implications. Though eukaryotic end regions had been defined cytologically (under the microscope), no one at the time was exploring their molecular nature. Although Blackburn first considered studying the short minichromosomes of another type of ciliated protozoan, Gall pointed out that *Tetrahymena* would be easier to culture in the quantities needed for direct molecular analysis. *Tetrahymena* offered one too many advantages for Blackburn to pass up: "I was fresh from a world where we'd found all these new methods for sequencing DNA, and here were these juicy molecules, their end regions ripe for the picking."

In the 1930s, the natural ends of chromosomes had attracted the attention of Hermann Muller and Barbara McClintock, both Nobel Prize–winning geneticists. Working in advance of the 1944 discovery that DNA was the genetic material, Muller and McClintock had both found that the DNA at the chromosome end regions was specialized to prevent chromosomes from fusing end to end, which could cause the destruction of the cell. Muller, working with the *Drosophila* fruit fly, had shown that genetic mutation could be caused by X-rays, and he studied the broken chromosomes that resulted from irradiation, noting they could "reattach" to other broken chromosomes. McClintock had developed powerful cytological methods for observing the whole chromosomes of maize under a light microscope, focusing on how mutations in the genes were later reflected in the altered phenotypes of the maize plants. McClintock also discovered an interesting phenomenon that she had not set out to investigate. Blackburn had read an early report, printed in 1931, in which McClintock noted that X-ray irradiation frequently produced translocations, deficiencies, and ring chromosomes. Like Muller, McClintock deduced that these alterations in the chromosomes resulted from the fusion of broken ends with each other. One statement leaped out for Blackburn: "No case was found of the attachment of a piece of one chromosome to the end of another [intact chromosome]."[15] While Muller did not coin the term *telomere* until 1938, McClintock had already presciently recognized that the natural end of an intact chromosome had distinct functional properties that preserved the integrity of chromosomes. Her work on the fate of broken chromosomes underscored this mysterious property. Studying germ cells and endosperm in maize, she observed that even if broken chromosomes fused with each other, they were repeatedly broken each time such cells divided—a process she termed the breakage-fusion-bridge cycle.[16]

By the 1970s, when gene replication was more fully understood, speculation about the function of these end regions, or telomeres, grew intense as scientists puzzled over the problem posed by replication of the ends of linear DNA molecules. During cell reproduction, the complex array of protein enzymes that replicate DNA are unable to copy the DNA strand all the way out to the end, which in theory means that DNA will grow shorter and shorter each time a cell reproduces. Clearly, something in the cell compensates for this predicted loss. By this time biologists had determined how the problem of DNA replication was neatly sidestepped in the lambda phage. Prior to replication, its DNA molecules formed a ring, which presented no problem for DNA replication enzymes. Influenced by the current predilection for universal, minimalist solutions that would bear out the elegant simplicity of DNA itself, molecular biologists surmised the process might well work the same way in other organisms.

Blackburn now had a heady opportunity to explore the problem of DNA replication, though with no certainty that sequencing the genetic material would provide any answers. Looking through a microscope at *Tetrahymena thermophila* (*T. thermophila*) in Gall's lab, Blackburn "fell in love at first sight." Like its better-known relative *Paramecium*, *T. thermophila* is a ciliated protozoan that lives in pond water; some species of *Tetrahymena* can grow to four-fifths of a millimeter in size, relatively large for a protozoan. The laboratory species of *T. thermophila*, only a tenth of that size, is still visible to the naked eye and looks like a speck of dust in water. Poetically named "swirlers," these single-celled organisms swim in pond water in a corkscrew path by rhythmically beating rows of hairlike cilia that cover the plump, pear-shaped cell. With their chubby cells and lively, graceful motions, *T. thermophila*, informally referred to as *Tetrahymena*, are visually appealing to humans. Blackburn commented that "they tend to inspire affection from their laboratory handlers, who have been known to refer to them as 'cute,' and in my lab at Berkeley, it wasn't unheard of to sing Ravel's *Bolero* to them to encourage them to mate."

Blackburn raved about the fascinating traits of her organism: "*Tetrahymena* has seven sexes. And one cell can be up to four of them at a time. After mating, it has to go through a number of cellular divisions before it reaches sexual maturity and is able to mate again, and it even has a recognizable adolescent period. The last, somehow to me endearing, accomplishment I will mention is that when it is starved, it reabsorbs its cilia, becomes long and lithe in body shape, puts out its cilia in new rows in a racing configuration, and grows a long caudal (back-end) flagellum so it can cover great distances swimming fast to look for food. Need I say more?" Biologists often cherish the idiosyncratic qualities of a favorite experimental organism, and ultimately idiosyncrasy (of the organism *and* the observer) can prove valuable to scientific endeavor. As Blackburn noted, "Biology sometimes reveals its general principles through what appears to be the arcane and even bizarre. But in evolution most things have stayed essentially the same mechanistically at the molecular level, and what changes is the extent and setting in which the various molecular processes are played out. So what looks like an exception often turns out to be a manifestation of a molecular process that is actually fundamentally well conserved. This is the case with the ciliated protozoa."

Tetrahymena's distinctive properties were perfectly suited to Blackburn's purposes. In Gall's lab, Karrer had shown that the entire rDNA minichromosome was a palindrome—its bases were arranged in an order that reversed itself at midpoint, as in "madam I'm adam." Karrer had also found that these minichromosomes could be cut with a restriction endonuclease that separated a large central fragment from its two small end-region fragments. Not only did Blackburn have a source of purified DNA to analyze, she could also conjecture—rightly, but for the wrong reason—that the two end regions on a palindromic molecule would be alike (although in inverted orientation). Using an electron microscope, Gall had observed that the minichromosomes sometimes formed circles, providing an important clue that these end regions might resemble those of the lambda phage, sequenced in the early 1970s. Consequently, Blackburn could start off on her quest by applying end-labeling techniques similar to those Wu and the Murrays had used for the lambda phage.

Just as Gall's shared excitement had launched her on her project, the technical expertise of other molecular biologists at Yale often helped Blackburn to carry it out. Before she could begin sequencing, Blackburn had to learn from her peers in the lab, from Karrer in particular, how to cultivate *Tetrahymena,* purify the rDNA minichromosomes, and examine them by electron microscopy, a process still new to her. Thanks to new methods, including those developed by Sedat and others at the MRC, Blackburn could now sequence DNA directly, though it was not yet possible to clone a gene to make greater amounts of DNA. For access to the technology necessary for analyzing nucleic acids, she relied on Sid Altman, a Yale faculty member who had been at the MRC a few years before Blackburn; Altman had set up the sequencing equipment Blackburn would use at Yale and could provide occasional pointers. Blackburn again drew on her MRC connections to obtain help from Joan Steitz, and she was also aided by a younger faculty member, Alan Weiner.

Because sequencing was such a new field, Blackburn had to jury-rig some of the equipment she would need for her experiments. Truett, intrigued by this "new and different" equipment, still recalled much of it years later. Blackburn installed heavy glass tanks, two to three feet high, to hold organic solvents for separating oligonucleotides. She fitted these tanks with grooves to hold glass rods, over which she draped the blotting paper that wicked up oligonucleotides. "Scientists had to be creative in terms of the equipment," Truett said. "Only later did it become more of a production line, with specialized companies producing scientific lab equipment. For the early DNA-sequencing efforts, the shop at the lab had to make the glass plates to which you attached the gels."[17]

Blackburn was also being exposed to a new way of working. In a cell biology lab, visual observation, as opposed to biochemical analysis, played a crucial role, even when, as in Gall's lab, researchers were exploiting a variety of approaches. The diverse organisms studied in the lab were chosen based on their suitability for observation under a microscope: *Drosophila, Tetrahymena, Paramecium, Physarum* (slime mold), and *Xenopus* (African clawed frogs). Each organism lent itself to certain experimental questions, as Truett explained: "*Xenopus* had nice, fat eggs (oocytes) that at certain stages in development have big chromosomes that you can spread on a slide for microscopic viewing, and you can actually see the RNA being transcribed off the DNA in the lampbrush chromosome—the whole structure looks like a Christmas tree. That was one of the first things Joe showed me, and I thought, wow, what a fantastic structure! And you're seeing what it really looks like, at once molecular and structural and unique to that organism."[18] Gall's enthusiasm and curiosity extended to many species, so long as they could be viewed under a microscope. Blackburn recalled that once, after an outbreak of head lice at a local school, a lab member collected a louse from her daughter's hair and delivered it to Gall, confident he would happily identify the species by microscopy. Gall had a particularly aesthetic appreciation for his work; his *Pictorial History: Views of a Cell* is a compilation of slides of stained cells viewed under the microscope—colonies of algae as beautifully tinted and intricately patterned as Moroccan tiles, the spirals of chromosomes in an onion root cell reminiscent of a mobile by Joan Miró.

As Truett's interest in her equipment suggests, Blackburn had a distinct niche in Gall's lab—one in which she didn't compete directly with other lab members, even those studying the same organism. Like Karrer, Truett also worked on *Tetrahymena*, studying the replication of the rDNA in the micronucleus of the organism, using the electron microscope to see if she could get a visual image that might reveal at which points along the DNA strand replication began. Every chromosome has a number of these sites, called replication origins, and Zakian was also using electron microscopy to observe replicating cells in *Drosophila* and infer where replication started along the chromosome. These replication origins, where DNA strands are forced apart by enzymes, appear as "replication bubbles," growing larger as cell division progresses. Though the chromosome is so long and tangled in the cell that researchers couldn't visually determine exactly where these special sites occurred along the DNA strand, Zakian discovered that they often clustered. Blackburn's first year in the Gall lab was Zakian's last, but their careers would intersect again when Zakian's interest in DNA replication led her to conduct original research on telomeres. Another colleague, John Preer, on sabbatical from Indiana University, had come to the lab with his wife, Louise ("Bertie") Preer, to learn molecular techniques for studying ciliates; in an arrangement that reflected the times, Bertie worked with John in his Indiana lab, but only her husband held a position as a professor.

At Yale, as at many other universities, the social life of the researchers revolved around the lab. In the small town of New Haven, Yale University was a privileged enclave bordered on one side by a poor neighborhood, and the laboratory building was literally yet another ivory tower within the ivory tower. Kline Biology Tower, a large, modern, multistoried building with rounded corners and tall, narrow windows, was surrounded by lawns, but the weather in New Haven did not often invite sitting outdoors. In addition to the long hours spent in the lab, most postdoctoral researchers did not share the concerns or social networks that other graduate students did, and they were not allied, as faculty were, by shared responsibilities to their community. Efforts had been made to organize postdocs in order to relieve their isolation, but Blackburn attended only one meeting of this group. While she "appreciated the sentiment," Blackburn did not feel that she had the time or that she knew "how to benefit from extending my contacts beyond the existing troupe of colleagues and associates, which included John's colleagues in his department."

Blackburn quickly settled in to the lab-rat routine at Yale. As she had done at the MRC, she dressed casually, with an eye to comfort rather than style. Lavett recalled her "slouching down the hall, scuffing her feet, a cup of tea nearly always in her hand, with her long hair pulled back at the nape of her neck to keep it from falling into things."[19] While at the MRC, Blackburn had become accustomed to a workplace culture in which scientists worked in the lab around the clock, and she was mildly scornful on discovering that at Yale, people went home in the evening; the standard hours were a comparatively reasonable schedule from late morning until nine or ten at night. Every weekday, the members of Gall's and Peter Rae's labs would eat lunch together at a big round table or a couple of adjacent tables in the large cafeteria on the top floor of the biology building. Gall's lab and equipment rooms, which occupied about half a floor, were situated next to Rae's lab, where he conducted research on *Drosophila* that related to Gall's interests, and they held joint weekly lab meetings. Members of this social group literally and scientifically breathed the same air.

Over lunch one day, Blackburn reported to her companions that recent work in Sanger's lab had yielded a surprising finding: a short segment of phiX174 DNA consisted of a sequence containing the genetic information in three overlapping phases of the genetic triplet code, all read, but out of "phase" with each other, from the same stretch of nucleotides. To her astonishment, her tablemates refused to believe that this was possible. She realized then that the MRC and sequencers there were not held by all biologists to be godlike; no one in the strictly molecular biology research world would have questioned a result coming from Sanger's lab. Yet Blackburn came to feel at home in both worlds: "I felt as though I belonged to a small group of sequencing aficionados, and I felt very legitimized by my connections with Fred's lab at the MRC and very accepted into this group. Later, when I attended my first Cell Biology Society annual meeting, I felt I had found yet another community of scientists who appreciated the kind of training I had received in Joe's lab and, by extension, the kind of scientist I was."

To have a foot in both worlds would have enormous impact on Blackburn's work, since, as Truett remarked, so many scientific questions crossed the boundary lines of what had once been distinct disciplines but were now complementary approaches:

Depending on whether you're a developmental, cell, or molecular biologist, you make a particular kind of observation about a phenomenon and think about how it works in an organism, whether the phenomenon might extend to other organisms or be unique. Some things are already understood, and then all of a sudden you come up against a question you can ask and you don't know the answer. Then you begin to think about how to go about answering this question experimentally. Maybe it's complicated and has to be broken down into smaller questions. Good scientists try not to be limited by their own specialty but to let their minds wander and hit on whatever might be relevant or useful in other disciplines. The fact that you are a cell, developmental, or molecular biologist might influence how you frame a question, but how you answer it might require a blend of approaches.[20]

Blackburn's exposure to three very different research environments as an undergraduate, a graduate student, and a postdoc granted her great flexibility and range in how she approached experimental problems.

Blackburn, who rarely even drove a car, did not require much beyond the world of the lab. She and Sedat had married soon after she arrived at Yale, and their domestic arrangements were minimalist. After first renting a small attic apartment near campus, they rented a furnished beach cottage for nine months of every year on Long Island Sound, just outside the city of New Haven, which they forfeited during the summer season. Fortunately, many faculty left Yale during the summer. Still without a stick of furniture, Blackburn and Sedat took their pick of apartments available for rent and spent the summer working long hours in the air-conditioned labs, a welcome escape from the muggy heat of a Connecticut summer. They stayed at Yale for roughly two and a half years (from February 1975 to October 1977), and yet when some friends suggested they consider buying a house, "we looked, but could not imagine taking the time to work on it and to live so far away from our work." Although she and Sedat managed a second trip to Nepal, "work was the dominant theme of our lives," Blackburn reported, so much so that she felt "jolted out of my cocoon" when this complete devotion was challenged: "I remember my doctor advising me that medically it would be best to get pregnant before I was thirty, but all I felt was indignation that he was trying to tell me what to do with my life." Blackburn had become good at shrugging off gender rules that would impinge on her freedom. But neither she nor the doctor challenged the assumption that the choice he posed was either/or. The grueling work ethic of research science—and its absorbing satisfactions—did not seem to accommodate family life.

Satisfactions proved plentiful for Blackburn while she was at Yale. Her work on sequencing rDNA quickly produced exciting results. She succeeded in selectively labeling the end regions of the rDNA with the DNA repair reactions she carried out, using chemical scissors that cut the DNA chain every time an A or G occurred, thus destroying these bases, and leaving intact any Cs and Ts that were grouped together. Along with the enzyme that replicated the genetic material (a polymerase), Blackburn added a radioactive label for specific bases, enabling her to distinguish C from T. By June, after 2D separation (electrophoresis plus further separation) of the oligonucleotides, she obtained her first autoradiogram. A string of four Cs in a row (a "C4 run") was apparent. This told Blackburn that at a minimum, the DNA must contain the sequence A (or G) CCCCA (or G).

Dramatically, her experiments demonstrated that this DNA did not resemble that of the lambda phage. First, the end-region fragments were heterogeneous in length. For any one type of lambda phage, in every viral particle the DNA molecule is exactly the same—a perfect carbon copy—as in another viral particle. The operating assumption in molecular biology at the time was that all chromosomes were exactly copied. But the replication of the *Tetrahymena* rDNA was not exact; though all the central fragments were exact copies, the end regions varied from one molecule to another, with some shorter and some longer. (Blackburn would need to conduct further experiments to prove conclusively that the number of repeats in the end regions differed from one another.)

Second, a much longer stretch of DNA was available for radioactive labeling in the end regions of *Tetrahymena* than in those of lambda phages. Chromosomes in the lambda phage look much like a ladder with one leg slightly longer at its base, and the opposite leg slightly longer at the top. So, unlike the double strand of paired bases that make up most of a chromosome, these ends are single-stranded extensions, twelve-bases long and bearing only one of each base pair. The sequence of bases at one end complements the sequence at the other, so that the bases on these "sticky," single-sided extensions will bond to each other to form a ring prior to DNA replication. (For example, a sequence ACAGGT at one end will correspond to the sequence ACCTGT at the other end, inverted as the chromosome curls to form a ring, as shown in figure 3.1.) When Blackburn measured how much radioactive label was incorporated into the very end regions of the rDNA molecules, she discovered far too much had been incorporated to be accounted for by a simple, single-stranded end structure like that of the lambda phage, raising the possibility that *Tetrahymena* end regions were double stranded. If there had been just one C4 run in each end region, only a tiny amount would have been labeled. On the autoradiogram, the abundant C4 run stood out like a beacon; the same sequence must be recurring many, many times in each end region, a startling anomaly.

Delighted just to observe something so "strange," Blackburn did not even wonder what model might account for this. She had been thoroughly indoctrinated in Sanger's approach to sequencing—"get in and find the sequence"—which gave priority to developing a descriptive

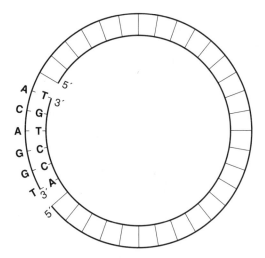

Figure 3.1
End regions of the lambda phage. During replication, complementary bases at the single-stranded 3′ end of either strand act as "sticky ends" that bond to form a ring chromosome. *Illustrator*: Alan Stonebraker.

account before a theoretical one. She just wanted to discover what this strange structure actually was. Her results might mean only that these DNA end regions were different in conformation rather than in length, so first she needed conclusive evidence that what she inferred was in fact true. To rule out any artifact, she would need to look for this same repeated sequence in living cells because "the cell won't exaggerate."

Blackburn set to work to validate that her labeling accurately reflected the rDNA sequence and also to determine the number of C4 runs per molecule. Instead of using enzymes in test tubes to label the end regions after the rDNA molecules were purified from the cells, she would now radioactively label the molecules in living cells. Since most radioactivity got taken up by the rest of the cell and very little got into the rDNA, an excess of the radioactive label had to be present to make the end-region rDNA visible on an autoradiogram. With some trepidation, she asked Gall if she could order a larger than normal amount of radioactive inorganic phosphate salt to be added to the cell growth medium for labeling. Gall agreed that this was the only way to get results. In October 1975, working in the lab's "hot room," set aside for doing work with radioactive isotopes, she added a highly radioactive solution to her cell culture, beginning with two millicuries of radioactivity, and several days later, when that was not effective, she repeated the experiment with five millicuries. With respect to the unusual amount of radioactive material she used, Blackburn was insouciant, as if this were on a par with her high school forays into making explosive powder: "People in the Gall lab thought I was being a little extreme to be handling so much radioactivity in one experiment. They would look in at me through the window set in the door of the hot room as I worked; either they wanted to make sure I didn't contaminate the room with radioactivity, or they were just curious about this departure from lab norms."

By October 22, 1975, Blackburn had purified the radioactively labeled rDNA and could triumphantly record in her lab notebook her plans for this precious sample. Piecing together the sequence, as always, required careful puzzle solving. One by one, a cacophony of nucleases was brought to bear on this terminal region of the rDNA in order to sequence it. She developed her X-ray film in a darkroom, on semitransparent blue sheets the same size as film used for chest X-rays. In the early stages of purification, the gel separation produced bands; subsequent separations produced a gridlike pattern of spots on the blue background, fluctuating in size and shading from gray to dark black, depending on the concentration of the oligonucleotide. Blackburn recalled that "the smell of photographic developing fluid still conjures up anticipation for me. Working under the red light, you might not be able to see the film very clearly, but you looked anyway for the first tangible evidence of your experiment."

Carrying the developed film to the office she shared with graduate students Brian Kay and Martha Wild, Blackburn laid it out on the shiny white surface of a lab table that had been pressed into service as a desk. Analysis of the 2D separation was not truly observational, as studying material under a microscope would be. An autoradiogram provided both a spatial and quantitative "translation" of what had occurred in the cell. To read it accurately required visual acuity, the ability to determine that a pattern signified particular common properties rather than technical contamination or an artifact. The length of exposure was determined by a researcher's *assumption* about how much radioactivity would be incorporated—if Blackburn expected a lot, she exposed the film for a shorter time—and this estimate in turn influenced the appearance of the autoradiogram. Working from the patterns of silver grain that appeared on the film, Blackburn had to imagine the minuscule world of the actual cell, down to its atomic level, in three-dimensional space. Abstract conceptual thinking supplied an essential foundation, but grappling with the problem ultimately required imaginatively reconstructing the data in visual and spatial terms, and at her desk she drew and discarded many diagrams and models to account for the incorporation of radioactivity in these end regions.

Often on Saturday mornings, when Gall came into the lab to do his own experiments, free from the weekday distractions of his academic duties as a Yale professor, Blackburn would tell him about her progress. Gall's memory of these conversations is tinged with amused exasperation:

Liz jumps right into the middle of an argument or the middle of what she's going to tell you, and you may need to be reminded about the context or what she is actually aiming at. I can remember any number of conversations where she would bring me a gel and start telling me the interpretation of the data or its implications, and I didn't even know what she had done to get the gel. At times, because she was thinking of several different things at the same time, she was way ahead of me. I get that when I talk with her even now. If you ask, What are you getting at? she can come down from the heights and give you the full story. She doesn't think the way everybody else does. A lot of people who think this way then have trouble going into a laboratory and doing a simple experiment, but she doesn't at all. It's not that she's off in the clouds thinking about such abstract things she can't do an appropriate experiment. Instead, her thinking focuses her incredibly on the exactly right thing to do. That's a really rare combination. She's also got an amazing ability to extract the right answer out of the data, where somebody else given the same data might not see what was there.[21]

In her notebook on April 8, 1976, Blackburn confirmed that she'd discovered an unprecedented repeat sequence. She identified the motif CCCCAA, denoted as (CCCCAA)n, with *n* standing for the number of repeats. The C4 runs recurred roughly *fifty* times per end region, making these end regions much, much longer than those of the lambda phage. By August 17, Blackburn routinely referred in her notebook to the (CCCCAA)n sequence, and she also verified that these end regions were double stranded, unlike lambda's sticky ends. In other words, both of the base pairs recurred in the sequence:

```
C — G
C — G
C — G
C — G
A — T
A — T
```

The last few hundred bases at each end of the minichromosome consisted of this simple sequence repeated over and over to the very tip. Blackburn hypothesized that the end regions varied from one molecule to the next because they differed in the number of times the C4 run was repeated. The data confirmed her hypothesis, and she also identified the degree of variation—the C4 run was repeated as few as twenty times in shorter molecules and as many as seventy in the longest ones. In further experiments, Blackburn and fellow postdoctoral researcher Meng-Chao Yao found the first evidence that this same repeated sequence also occurred in the other (much longer) chromosomal DNA of *Tetrahymena*.

Blackburn's original experiment also turned up an intriguing structural anomaly in these end regions. In the chromosome, DNA is bound up with an array of proteins, known collectively as chromatin, which regulates the access of other proteins that activate transcription (the translation of DNA into RNA) and protects DNA from enzymes that degrade it. At intervals within this sheath are nucleosomes, in which DNA is tightly wound around a core of related proteins called histones, so that the chromosome "bunches" at these regions. To cut her DNA into fragments for sequencing, Blackburn had used micrococcal nuclease, which cuts between nucleosomes. But this enzyme had left intact end regions from 145- to 200-bases long, evidence that these repeat-sequence regions were bound not within a nucleosome but instead with a specialized complex of proteins.

The telomeric DNA had another unanticipated property, and its biological significance is still not understood. Concentrated toward the tip of the end region, small breaks occurred along the strand with the CCCCAA sequence. In other words, at intervals on this leg of the DNA ladder, sections between one rung (a pair of bases) and the next were simply *missing*. (The ladder still held because its other leg was intact and because of the chemical attraction between bases on both strands.) While puzzling, these breaks had been a boon; if they had not existed, the sequence would have been inaccessible to methods then available for purifying and sequencing oligonucleotides.

No one had dreamed DNA molecules would have this kind of repeat sequence at their ends or that they would vary at all, and no one had ever scientifically documented it. What biological mechanisms could possibly account for these properties? In this uncharted territory, Blackburn had no template or lab protocol book to work from as she painstakingly tried to piece together a model that synthesized her observations. Since biologists always try precedent first, she attempted to relate her results back to the model of the lambda phage. She continued playing with pencil-and-paper diagrams, but the rDNA molecules would not fit the existing model, so she had to attempt to create a new one that could take all the available information into account.

More than just one operating assumption was at stake in the difference between these double-stranded end regions and those of the lambda phage: the repeat sequences were an anomaly, and their varying lengths were not consistent with a polymerase copying existing DNA, presumed to be the only means of replication. Blackburn did not theorize extravagantly. At this stage of investigation, her method was far more cautious, and she was still focused primarily on gathering descriptive data: "I was in the business of discovery—of finding sequences. Hypothesizing may be a useful discipline, but most of the exciting, really new biological discoveries are not hypothesis-driven. By the time you make a hypothesis, the interesting thing has already been discovered. What is it like? Let's describe it in every molecular way we can and then infer what the structure is. Eventually, you do have to account for the phenomena you see." Not all scientists operate this way—Crick comes to mind as someone who accurately theorized in advance of the data—and many of those who do still don't share Blackburn's willingness to regard something as an open question until exhaustive evidence has been uncovered. Blackburn occupies this particular mental space rather gleefully; she recalled being "excited," not frustrated or anxious, when she failed to find a conservative or any other explanation for the data she had accumulated on these end regions of *Tetrahymena*. She may have hit on something big or she may have failed to recognize all the factors in play. Either way, she was interested.

Blackburn's efforts provided a picture that didn't immediately suggest a solution to the problem of DNA end replication but made it possible to frame new questions in highly specific terms. Why would there be hundreds of base pairs made of the same repeating sequence, and how might this prove to be key to preserving the integrity of chromosomes? Why would the end regions vary from one chromosome to the next? She could be certain only that she'd begun to map the distinctive features of this new terrain.

Blackburn knew she'd produced valid and accurate results, based on sound observations, and she also knew that the implications of this work were decidedly not business as usual. She prepared a paper, working with Gall, her coauthor, who added editorializing material to a draft she showed him. In what would become habitual practice, she saved early drafts of the paper; tangents and corollary questions that had to be screened out to sharpen the focus of the paper documented the process of thinking and could provide valuable triggers for new experiments at a later date. Blackburn and Gall submitted the paper to the *Journal of Molecular Biology*, a highly regarded periodical that published the few DNA-sequencing results produced at the time. This was the first scientific paper ever on the molecular structure of the end regions of chromosomes.[22] Had the far-reaching biological implications of Blackburn's discovery been clear, she and Gall would have submitted the paper to the more widely read journal *Nature*. They didn't yet realize the significance of these findings, as Gall explained:

The work Liz was doing in my lab was technically advanced, because she was using techniques learned in Sanger's lab, and virtually no one else in the world was doing that kind of work. We were interested in sequencing partly because it was something you could do and we knew there were unusual features about these molecules. At the time, we didn't say, Eureka, we found what the ends of chromosomes are like. She had made a discovery whose significance we didn't yet appreciate completely. I knew Liz was extremely good, but I didn't know she was a superstar until she started doing her own independent work.[23]

When it became time to seek a job, Blackburn was forced to turn her attention from the thrill of her work at the bench to the pragmatic and fraught process of establishing her career. Feeling ill-equipped to negotiate this task, she operated under a double handicap: as a woman, she had a lower chance than a man of being hired, and she was coordinating her efforts with Sedat's, when finding two jobs in the same region would be harder than finding one. She quickly accumulated a stack of rejection letters from universities that had declined even to interview her. Her capacity for protective coloration—a willed refusal to acknowledge that such peripheral issues as gender and career politics could impinge on her work—did not shield her from the vicissitudes of the job market. Her confidence in her work evaporated under the stress of the search:

I remember waking up in the night in a cold sweat of anxiety, listening to the waves on the beach outside our rented house, wondering how on earth we would become employed. As seems usual in my case, my insecurity made me overdo everything. Thus I wrote to apply for just about any assistant professor job advertised, regardless of where in the United States it was. Then at some point, Joe talked to me in his office and recommended I withdraw my application to a small southern university. He said, "You and John are first-class scientists, and you could do better than that." I was greatly encouraged that he used the term *first-class scientists.* Joe was not given to hyperbole.

Blackburn planned to do a routine talk on her work, presented to the other members of Gall's and Rae's research groups, a trial run for the talk she would give for job interviews. Despite the fact that she had plenty of solid, even novel results, she felt "desperate to convince a skeptical world about the validity of my scientific conclusions":

I poured every bit of my data into the practice talk so that it was unintelligibly dense. From the muted reaction of the lab group audience, I sensed the talk went badly. I went back to my shared office and succumbed to despairing tears. After tactfully waiting a bit, Joe came by and said something encouraging—I forget what. Then Diane Juricek [later Lavett], a visiting postdoc in Peter Rae's lab, volunteered to help me work my talk into shape, taking out the excessive quantities of data description and making it into a coherent whole. This she did; I remember rehearsing the revised talk in the big old-fashioned lecture hall in the neighboring building for an audience of one: Diane. This enormous kindness to me, simply as a colleague whom she didn't even know terribly well, made a huge impression on me. I realized what an academic scientific community could be about.

Lavett offered her own analysis of these rehearsals: "Liz's main problem was she lacked confidence. She was terrified of talking. But she only needed to be told, 'This point is fine,' or 'Here you need to punch up your point.' If she needed someone to listen and to help her feel better, who wouldn't help a person who was so kind and generous?" Eventually the revised talk would help Blackburn to get a job. And Lavett would note with satisfaction the course of Blackburn's career in the following years: "I don't see how Liz did it. She was up there battling with the big boys, though it seemed from the outside that she didn't even have to battle."[24]

Sedat and Blackburn interviewed at universities on the East Coast and in California. After they visited San Francisco, they decided they wanted to live there. Sedat accepted a position as an assistant professor in the Biochemistry Department at UCSF, beginning in December 1977. Blackburn had applied for a position as an assistant professor at the University of California at Berkeley, across the bay from San Francisco, but the application was still pending. While she waited to hear from Berkeley, Blackburn was offered a research position at UCSF, provided she wrote a successful grant application to the National Institutes of Health (NIH) to fund her salary and research. At the same time Sedat began his position, she would begin work at UCSF as an assistant research biochemist, a nontenure track "soft-money" position with no teaching responsibilities. For now, Blackburn could continue her research, even if her new position was tenuous and temporary: "I so liked the idea of being in San Francisco, and John had a job, so I just accepted and thought, somewhat implicitly, that something would work out." Blackburn and Sedat planned a third trip to Nepal before starting their new positions; trekking in Nepal, Blackburn said, "was our rebellion against the pressure."

When Blackburn and Sedat began packing for their move, she knew that she still faced a prolonged job search, made more difficult by the limitation that in order to stay with Sedat, she could look for teaching positions only in the San Francisco area. Her paper on her research on telomeres had been accepted, though not yet published, by the *Journal of Molecular Biology*. Although she was fired up by the enormously exciting results and implications of her work on telomeres, she felt uncertain about her ability to persist at the tension-filled and often political process of job seeking: "When we were in the process of packing up to leave Yale, I thought I might be pregnant. I remember a feeling of relief that I could escape from the pressure and anxiety. I would just be able to have a baby and not have to worry about all this."

4 Revelations

While waiting to hear whether she would be hired by the University of California at Berkeley, Blackburn began work at UCSF, where two departments provided support for her research. Herb Boyer, director of the genetics graduate program in the Department of Microbiology and Immunology, paid Blackburn's salary until she obtained grant funding. The Department of Biochemistry provided lab space in the Health Sciences East Tower research building, next door to the lab of molecular biologist Bruce Alberts, with whom Sedat and Blackburn had trekked in Nepal before their arrival in San Francisco. In his lab, Alberts studied chromatin as well as the DNA replication mechanism used by the T4 phage. He had devised a method for identifying the proteins that bound tightly to DNA, an expertise of immediate relevance for Blackburn, who was about to undertake a study of the proteins that bound to the telomeres, the telomeric chromatin. Proximity made it easy to exchange ideas on their related work, and Blackburn also joined in Alberts's and Boyer's lab group meetings. Both she and Sedat quickly made friends in their new setting; Assistant Professor Lou Reichardt, who lived down the street from the Sedats in their newly purchased house, sometimes commuted to work with them and shared their enthusiasm for the Himalayan region, where he had been among the first Americans to scale K2, the second-highest mountain in the world. Sedat and Blackburn were not unusual among their peers in fleeing an intensely high-pressure environment for remote wilderness that offered different but equally extreme tests of endurance.

Early in 1978, Blackburn interviewed for a teaching post at Berkeley. She gave a seminar wearing her "best clothes," a red cotton long-sleeved top with a kangaroo-like pouch pocket ("which I thought of as my Australia connection") and a skirt. Later, when she interviewed individually with faculty members, Harrison (Hatch) Echols was receptive when she suggested a good experiment for him, and Ellen Daniell, an assistant professor who had been hired three years earlier, told Blackburn she would support her candidacy. Blackburn left the campus feeling hopeful but uncertain; characteristically, she decided not to think about her chances unless and until she received a job offer.

When Blackburn was offered the job at Berkeley in March 1978, Boyer came along with several other colleagues to celebrate over lunch at a restaurant near the Parnassus Avenue campus. Boyer, who had graciously provided important practical support to Blackburn, was then a figure of controversy at UCSF. With Stan Cohen of Stanford and his lab group member Annie Chang, Boyer had recently figured out how to engineer recombinant DNA by splicing DNA from a frog into *E. coli*, a bacterium that would function as the lab rat of recombinant DNA. While maintaining an active research lab at UCSF, in 1976 Boyer had joined with entrepreneur Rob Swanson to form the first biotech company, Genentech. Boyer, seeking start-up capital, had invited fellow faculty members, including Sedat, to invest five hundred dollars each. Though this investment would turn out to be wildly profitable, Sedat chose not to participate because of reservations about the speed at which bioengineering was developing into a commercial undertaking.

Boyer's new enterprise departed from the norms of biological scientists at the time, although connections between chemists at universities and industry had been commonplace for years. In the idealistic atmosphere of the 1970s, some of his colleagues felt that Boyer's venture, with its taint of profit seeking, threatened the integrity of biological research. Boyer eventually retired from UCSF to focus on Genentech, and when he grew wealthy from his corporation's success, he donated money to UCSF. Although Blackburn fell into the idealist camp, she felt loyal to Boyer and was intrigued by his pioneering effort: "Much later, when I returned as a professor to UCSF, I was amused to note that Herb's generous donations to the university had the effect of completely silencing any criticism of him by his formerly outspoken colleagues."[1]

Blackburn started her own lab at Berkeley in July 1978—a heady experience. When first offered the job, Blackburn had consulted a former colleague at Yale for tips on negotiating the offer, and this colleague encouraged her to ask for the terms in writing, to the surprise of Art Knight, the courtly department chair at Berkeley. Blackburn was more concerned to bargain for adequate resources for her research than to negotiate a higher salary: "I had no skills at negotiating, but I knew one thing: what it would take to get my lab up and running." The department carpenter built her a small enclosed area to house a tissue-culture hood for manipulating tissue cultures of *Tetrahymena* under sterile conditions, and she happily ordered equipment and availed herself of the excellent resources available for setting up a new lab, including the carpentry shop and an electrical and electronics workshop where simple lab equipment could be built. Only a few days after she started her new position, she climbed onto the lab benches in her white lab coat to clean the grimy walls of her new space, "happy as could be" and overwhelmingly grateful that she had been entrusted to run her own lab.

Blackburn was then one of only three untenured assistant professors in the Department of Molecular Biology at Berkeley. Steve Beckendorf worked on fruit fly genetics, and Daniell on adenovirus. Their clearly differentiated areas of specialization mitigated against any competition, and the three of them served on each other's graduate student committees and read drafts of each other's papers. Daniell, the only other woman in the department, mentored Blackburn and quickly became a friend and ally. A vivacious person with interests in the arts and literature as well as science, Ellen often cooked dinners for friends and led the carol singing at holiday parties in her Berkeley home, and as an enthusiastic backpacker, she organized camping and driving trips in the Sierras and northern California with Blackburn and her husband. Echols also proved to be a supportive colleague.

For the first year, assistant professors did not have any teaching obligations, a common agreement for such positions, but Blackburn willingly agreed when she was asked to conduct a graduate seminar, seeing this as a means to get students interested in the research taking place in her lab. She quickly found recruits, though students told her that during the talk she had given for her job interview, her Australian accent had left them wondering about her frequent references to "eye" DNA (rDNA). Two graduate students, Liz Howard and Peter Challoner, joined her lab to work on long-term projects, and she also relied on bright and competent undergraduates, hungry for research experience. She was doing experiments with her own hands in the lab for as much time as possible, but undergraduates could be assigned to do specific tasks, and several of them, including Aliza Katzen and Gordon Cann, demonstrated such initiative that they were named as coauthors on papers during Blackburn's first years at Berkeley.

The involvement of undergraduates suggests how much Blackburn's small lab was truly a cottage industry, in desperate need of extra pairs of hands, especially given the ambitious range of the work. When Mike Cherry, a graduate student who arrived not long after Howard and Challoner and stayed until 1985, proved adept at computers, Blackburn immediately put him to work on a newly purchased Sun Microsystems computer, adapting programs designed for microbiology to the specific needs of the lab. But Cherry noted that it took a while before anyone else got the hang of the new technology: "Once I found Liz and Peter Challoner sitting together. Peter was reading off a DNA sequence he'd determined in triplets, and Liz was looking up that particular triplet on a table, doing it all by hand and writing down the amino acid it coded for. They were sitting right next to the computer, which had been in the lab for a month. They could have typed in the sequence and gotten the computer program to do these translations for them."[2]

Teaching began in earnest in Blackburn's second year on the faculty, when she was expected to teach the first quarter of the main lecture course on prokaryotic molecular biology for undergraduate majors. She would fill in only for a year, while the faculty member who usually taught the course was on sabbatical, so she had to prepare material she would not teach again and that fell outside her area of expertise in biochemistry, though she had gleaned some knowledge of the molecular biology of bacteria and bacteriophages during her years at the MRC. Conscious of her inexperience, she asked whether she could team teach the course but was told students would fare better with just one lecturer.

Guided by topic headings for the course, she read fat texts and monographs on the subject. Commuting early in the morning to arrive on time for her 9:00 a.m. class, she often recited her lectures, the precepts of her childhood elocution lessons coming back as she fought her tendency to mutter too fast. In her anxiety to do the job well, she covered the chalkboard with screeds of written information as she lectured and presented far too much material, fascinated by the detailed in-depth information she had discovered in her reading and anxious to prove that she knew her stuff. Having been warned that students would try to take advantage of a new professor, she treated students warily, which further undercut her effectiveness.

At the end of the term, she made the mistake of reading the student evaluations as soon as they arrived. Although the evaluations were anonymous, she recognized the handwriting of most of the students. Predictably, students who did well rated her more favorably. But many students criticized the density of the material and the rapid pace of the course. Not long after, on a lunch picnic with Daniell and Helen Wittmer, administrative assistant to the department chair, Blackburn vented her frustration and disappointment, even though she recognized some justice in the comments. She had yet to figure out how to translate the kind of thinking about science that was natural to her into the organized form of an introductory lecture course. Sipping red wine from a plastic glass, Blackburn, who had obediently accepted the difficult assignment, raged about the students who had made negative comments: "If on some dark night I ever came across any of them, I would kill them." This was an unusual outburst for the mild-tempered Blackburn. She was unaccustomed to failure, especially when she made conscientious efforts, and the critical student evaluations may have shaken her bedrock, insulating belief that her work would prove itself. Her companions responded sympathetically, and Blackburn resolved never to read evaluations until many months had passed. She could not help wondering how much difference good mentoring might have made during this first term of teaching.

In contrast to her trials by fire in the undergraduate lecture room, Blackburn's research continued to be exciting and challenging as she pursued the implications of research begun in the Gall lab. She was still attempting to identify the complex of proteins that associated with the *Tetrahymena* telomeres. But with the technology available at the time, she could not purify telomeric DNA in amounts large enough to enable her to isolate and identify ("get out") these associated proteins, though her experiments confirmed their specialized function. Working with lab technician San-San Chiou, Blackburn inferred from their different packaging that the proteins associated with the telomeric DNA were nonhistone, distinct from the proteins packaged with DNA in the nucleosome, though they functioned in an analogous manner to protect the telomere from digestion by nucleases in the cell.[3] In the late 1970s, researchers had found that the double-stranded linear genome of certain viruses, particularly adenoviruses, was associated with a terminally bound protein, which was attached by a covalent bond to the DNA end. Blackburn's observations suggested the telomere proteins did not have the same tight molecular bond, formed by shared electrons. Using sensitive labeling methods, her postdoc, Marsha Budarf, found no covalently attached protein at the ends of the rDNA chromosomes.[4] Not until 1984 would Dan Gottschling and Tom Cech, working on the somatic chromosomes of another ciliate, *Oxytricha*, identify the first telomere structural protein, which is tightly, but not covalently, bound to the telomere.[5] Gottschling subsequently corroborated this finding in collaboration with Zakian.[6]

A more productive line of inquiry soon became Blackburn's central focus. To demonstrate that the repeat sequences of telomeres might be a conserved structure and not just an idiosyncrasy of *Tetrahymena*, Blackburn sought evidence of telomeric DNA in other organisms. She sequenced the telomeres of the chromosomes in a related ciliate called *Glaucoma*, purifying the DNA molecules from the living cell, as she had done at Yale. Blackburn could now adapt for telomeres new DNA-sequencing methods that required fewer steps. She would leave a dried sequencing gel overnight for exposure to X-ray film and the next morning drive across the Bay Bridge, with eagerness kicking in as she exited the freeway near the giant Spenger's seafood restaurant and approached the campus. As soon as she arrived at her lab in the Stanley Hall building, she developed the autoradiogram. She felt lucky in having an "amazing job": "I would drive to work, anticipating results, almost daily it seemed, with such excitement. And with each new result would come new revelations or a new question."

If the details at the very end of every *Glaucoma* DNA molecule resembled those of *Tetrahymena*—CCCCAA/GGGGTT repeats, ending in the sequence GGGG at the end—then sequencing would generate a pattern on the X-ray film. If the end of one molecule differed randomly from another—one molecule ending in GGG, another GGGT, another TTGG, and so on—the fragments of DNA Blackburn had targeted radioactively would obscure each other on the gel, providing an unreadable mess on the X-ray film. She vividly recalled the morning she arrived in the lab to discover that the repeat sequence read loud and clear on the gel. The precise molecular end structures of *Glaucoma* macronuclear chromosome telomeres were just like those of *Tetrahymena*, proving what she had long suspected: *Tetrahymena* rDNA was not just a freak exception.

By 1982, Blackburn's continued research on *Tetrahymena* led her to deduce another unusual property of its telomeric DNA. As is characteristic of ciliates, *Tetrahymena*, which can reproduce itself through single-cell reproduction and also by mating, has two nuclei. The somatic nucleus, which harbors the rDNA minichromosomes, has a plastic genome and can rearrange its genetic material to form new chromosomes. In the germ-line nucleus, the cell stores a "safe copy" of its genetic material, passed on from generation to generation. In a process unique to ciliates, the somatic nucleus functions like a factory, making genetic material that produces all the proteins needed by the cell. As DNA is reshuffled in the somatic nucleus, new telomeres are added to fragments of DNA so they can function as complete chromosomes.

In a pretty spectacular leap from the available evidence, Blackburn hypothesized that the new telomeres were not copied from existing DNA but somehow created de novo ("from scratch") during cell division, and by 1982, she and her lab members had found some evidence for this, using molecular analyses that compared the "before" and "after" versions of sample regions of the genome where such a change had occurred.

As Blackburn put it, "The organisms were screaming out a message," but no existing model of what biologist Gunther Stent called "DNA transactions" could account for the structural features she was discovering in such exquisitely precise molecular detail. Was this property peculiar to ciliates? What process could account for it?

In her lab, Blackburn talked with her three graduate students about these ideas, partly to train them and partly out of a hunger to bounce ideas off someone. Who better than her students, who knew the work intimately? Blackburn had already instituted informal discussions of science—"like the kind we had at the MRC"—and now the members of her lab engaged in intense debates that Challoner dubbed "theory of the month" conversations. The lab was small, Blackburn was not much older than her students, and unlike many primary investigators, she worked at the bench right alongside her lab members. Her disregard for hierarchy shaped an easy intellectual give-and-take in the lab, as Cherry suggested: "Liz was quite young, and she really understood what would help motivate us. She was very open with us. At one point, there were only three graduate students in the lab, Liz Howard, Peter Challoner, and me. Liz left her office unlocked, and we were free to look at her desk. The 'top-of-the-pile' rule meant that we could look at the top of the pile of papers to see what she was working on, to learn things. We'd see letters asking, Have you ever seen this particular phenomenon? Or, do you have this mutant? Most communication in science is through published papers, but a lot of information passes through personal communications too."[7]

Blackburn's excitement about her results also spilled over into her social life. Most Friday evenings, she left the lab "early" and battled rush-hour traffic across the Bay Bridge to San Francisco to join Sedat and a small group of fellow biologists who were regulars at a pub near UCSF, where the Iranian immigrant owner informed them that he was honored to have such learned conversations going on in his establishment. Blackburn welcomed the chance to discuss her work at this informal roundtable: "I was always eager for such conversations, hoping some new insight might come from them. Those whom I regaled with my new results would listen with interest, but did not have much to offer except an ear—they just didn't fit anything."

During her first five years at Berkeley, Blackburn worked ferociously hard, with the ultimate goal of gaining tenure. Not only was she learning how to teach but also she had to learn how to supervise a lab technician, students, and postdocs in the lab, manage the lab operation, engage with her colleagues at Berkeley, and establish her credentials as a researcher, which required active participation in scientific conferences. More than ever before, molecular biology demanded participation in overlapping communities. At a genetics conference, Blackburn had sat with Cech and Karrer, "molecular people" who, like her, were junior faculty and still initiates in the genetics world, grateful for the willingness of ciliate geneticists to "explain their jargon to us." Cech, Karrer, and Blackburn had much in common: all were interested in ciliates and connected via the Gall lab, since Cech had studied with Gall's former student Mary-Lou Pardue. Reflecting on her frantic schedule, Blackburn recalled, "It was exhilarating in a macho kind of way. Whenever anyone asked did I ever think of quitting, I would answer, 'Only about once a week.' But I was so deep into the whole life and so convinced that it was the kind of life that was worthwhile that I did not reflect seriously that there could be any other way of going about a career."

Blackburn recognized that total devotion to research was "culturally important" to convey—not that she had to fake it. When research groups in several campus departments had applied for a joint grant from the National Science Foundation (NSF) to fund the purchase of an expensive cell sorter, Blackburn wrote a portion of the grant, describing how this would facilitate her research. Part of the review process included a site visit from an assessment team of NSF officials and scientists, and in anticipation, Blackburn had carefully prepared a scientific talk and a flip chart. The morning of the team's scheduled visit to her lab, Blackburn decided to proceed with her work while she waited. The visitors arrived to find her working in lab coat and rubber gloves, manipulating a radioactive reaction in the fume hood, and they had to wait while she finished the manipulation and shed her gloves. Confident she was demonstrating devotion to her work and had made a good impression, she didn't register any breach of protocol and shrugged off her visitors' quizzical looks. As Blackburn wryly noted, "Clearly my political radar was pretty cloudy. I don't think the grant was funded, but I never wanted to find out why—some reticence about hearing what might be potential criticism kept me from actively inquiring."

An unexpected event derailed Blackburn's assumption that she had only to do good work—with feverish devotion, of course—to earn tenure. One day, a couple of years after she had come to Berkeley, Daniell phoned to tell Blackburn that she had been denied tenure. Blackburn was outraged:

In my view Ellen more than met the criteria for tenure. She was a truly excellent scientist who did fine research and was a brilliant teacher, and she was an interactive, actively contributing member of the department. I was so angry and shaken that I used the "f" word—a rarity for me. This was the first time that it really dawned on me that Ellen was different from the rest of the faculty (all male except me). Ellen was a lively conversationalist who often talked about her interests outside science, and if she felt strongly about something, she was willing to express her emotions forthrightly. Once, when a male colleague had refused a request to help teach a lecture course, Ellen confronted him and asked why he was being such a bad colleague—the way a woman would confront someone, asking why he had acted as he did rather than veiling the personal aspect of the conflict. I strongly sensed that it was this difference in style that had been her undoing.

Blackburn's suspicion had some basis in fact; from 1984 to 1985, women accounted for only 7.7 percent of new appointments in "ladder rank" faculty in the life sciences throughout the entire University of California system.[8] In a published account of her experiences at Berkeley, Daniell recalled that when she applied for the job, she was warned, "That department will never hire a woman."[9] At her initial interview, "several faculty members had been obsessed or uncomfortable with my being a woman"; Daniell later learned that a woman postdoc interested in applying for a subsequent opening had been told "we already have a woman."[10] In its report, the committee that denied tenure to Daniell cited her "lack of standing as a productive research scientist" and merely "adequate" performance as a teacher.[11] Daniell, who characterized the report as having "a negative slant," found it unfair on several counts: it ignored positive student evaluations, failed to mention laudatory reviews of her recent research, and omitted or misrepresented facts, including faulting her for failure to produce findings she had in fact published.[12] The university rejected Daniell's formal appeal of the committee's decision, and she quit research and academia, "totally disenchanted."[13]

Still a few years away from tenure herself, Blackburn felt helpless, scared to protest to her colleagues at Berkeley, and worried about her own chances of getting tenure: "This experience probably reinforced my already set tendency to act more and more like a man—to camouflage myself. Ellen stood out as the living proof that 'feminine' personality traits did you no good in academia. The only road I knew to try was the one I was on: to devote myself to science—which was what I wanted to do anyway—and to try to fit in, which I decided, unconsciously, meant keeping my head down and trying not to make waves." Blackburn's choice reflected the common strategy of many of her female peers. In a 2001 account of bias in the sciences, women scientists reported that they perceived their field as a "bastion of sexism," and "for most women, survival means they must deny that the biases exist." In the same report, education researcher Anna MacLachlan affirmed that "denial still plays a very large role in the lives of successful female scientists, and accommodation plays a part too. . . . This is not just the culture of polite, Anglo-Saxon male scientists we're talking about, but a historically built set of dictates about how you approach thinking about things and how you act and accommodate others to accomplish your goals."[14]

In 1980, at a Gordon Research Conference on Nucleic Acids, Blackburn presented her research on telomeric DNA structure. These conferences, held each summer on the vacated premises of New England prep schools and colleges, were attended by about 150 scientists, focusing on specific aspects of biology and chemistry, and cutting-edge results were often unveiled at these meetings, which most molecular biologists attended. Walking across the lawn between presentation sessions, Blackburn talked with Jack Szostak, a yeast molecular geneticist at what was then the Sidney Farber Cancer Institute at the Harvard Medical School. Szostak, who like Blackburn had only recently set up shop in his own lab, studied genetic recombination in yeast. Genetic diversity arises not only from mutation but also from recombination, in which segments of DNA can be exchanged between two chromosomes, essentially rearranging existing genetic information. Under the microscope, McClintock had observed this phenomenon in maize, and molecular biologists had studied it extensively in bacteriophages since the 1950s. During meiosis (cell reproduction in a gamete-producing cell, such as a fertilized egg), segments along the chromosomes of the parent cells are exchanged, thus yielding unique new genetic material in the offspring. Scientists were experimentally manipulating yeast cells before or during meiosis in order to understand the molecular mechanism for this recombination. At the time, researchers studying the molecular genetics of eukaryotes favored yeast as an experimental model because, as Blackburn put it, they are "small, cheap, obedient . . . and they don't cry."

Szostak wondered if *Tetrahymena* telomeres, which Blackburn could isolate as purified DNA from the rDNA molecules, might function as telomeres if genetically inserted into yeast cells and thus provide a means for isolating yeast telomeric DNA. Szostak was interested in telomeres as a possible product of recombination, while Blackburn had begun to doubt this model held true. The chance to find out intrigued both researchers, and Blackburn believed each of them brought to the collaboration expertise that the other needed: "Jack knew yeast genetics and I knew as much as, if not more than, anyone in the world about the molecular nature of telomeres."

For this novel experiment, Blackburn and Szostak relied on the methods developed by Boyer, Chang, and Cohen for engineering plasmids (small circular strands of DNA that can replicate independent of chromosomes) as a means of transferring extraneous DNA into a cell. Recombinant fragments of DNA could be used for experimentation, including inserting bacterial DNA into an organism's DNA. These tiny rings with a bare minimum of DNA made it possible to "harvest" genes in a cell "factory"; the human insulin-encoding gene had been successfully grafted on to bacterial plasmids normally harbored in a bacterial cell to produce insulin in these cells. So far, researchers had engineered only ring-shaped plasmids. Though scientists could manipulate yeast DNA by using enzymes to chop it into fragments, linear fragments without telomeres are destroyed by natural enzymes in the cell, and fragments without replication origins, at which DNA copying begins, do not get replicated and are ultimately lost. If a fragment of DNA was first circularized, forming a ring-shaped plasmid, when reintroduced into a yeast cell it would reproduce during cell division, provided it contained the necessary replication origins and its ring shape bypassed the need for telomeres. If they could devise a means to add telomeres to plasmids, Blackburn and Szostak might create a linear minichromosome that would replicate. They would conduct the experiment in two stages: in the first, they would attempt to create a linear minichromosome; if they succeeded, in the second stage they would attempt to engineer this minichromosome so that it could harvest yeast telomeres.

The chances of successfully creating the first linear plasmid seemed low. It was a long shot to imagine that *Tetrahymena* telomeres, introduced into yeast DNA to create minichromosomes, might provide the bait by which scientists could finally isolate and identify yeast telomeres. Scientists had already demonstrated that a DNA segment that worked as a replication origin in a different species would not function as a replication origin in a yeast cell. Similarly, a centromere worked only when reinserted in its own species, driving home the likelihood that a telomere from one species would not function in another, and hence the experiment would fail. Blackburn considered this risky experiment well worth the gamble:

One would be unlikely to try to persuade even the average graduate student to perform the experiment. One had to give graduate students a rationale for an experiment, and how could one do so if similar experiments had failed to work? Every now and then in science, you come up with a cockamamy idea for the fun of it. There's a tension between being very rational and being prepared to do things simply because they might work—they're far out, but if the experiment works it provides rich information. This experiment would imply that the process for end replication of DNA was highly conserved, since yeast and *Tetrahymena* belong to two different kingdoms, and all biologists love what's conserved, hoping to capture fundamental truths.

Blackburn and Szostak divided up the work of the experiment. Using restriction enzymes, Blackburn prepared the right fragments of the rDNA, containing telomeric CCCCAA repeat regions. Szostak prepared a DNA plasmid, using the same restriction enzyme to convert it from a ring DNA into a linear DNA. The linearized plasmid DNA, with its ends cut by restriction enzymes, would be unstable and lost during yeast cell reproduction, unable to survive unless it was recombined with another broken end or insinuated itself into a chromosome. At Szostak's lab, in the test tube each end of the plasmid could be joined to Blackburn's telomeric restriction fragments, thanks to the "stickiness" of the restriction enzyme-cut ends. Once the plasmid's bases linked with those of the telomere fragments, Szostak could stitch the two fragments together using a special enzyme called a ligase. This created a linear minichromosome: the originally circular plasmid plus two ends, the *Tetrahymena* telomeric fragments. The minichromosome was then introduced into *Saccharomyces cerevisiae* yeast cells to see if it could maintain itself autonomously. To Blackburn and Szostak's delight, the experiment worked. The *Tetrahymena* ends behaved like telomeres in yeast, stabilizing the minichromosome, which was able to replicate intact, without breakage or fusion.[15] For Blackburn, this confirmed that the properties she had observed in *Tetrahymena* were a fundamentally conserved principle.

Blackburn set out to sequence these end structures, to see if anything had happened to them during their residence as foreigners for many generations in yeast cells. Depurination offered some partial but critical sequence information about these ends. Blackburn found that as in *Tetrahymena*, these telomeres had single-stranded discontinuities and a pronounced run of C bases, but now they occurred in runs of three rather than four (CCC rather than CCCC). If the sequences had been copied by a DNA polymerase, they should have remained exactly the same. This astounding evidence immediately suggested that the *Tetrahymena* telomeres had been altered by some substance in the yeast cell.

Once Blackburn and Szostak knew yeast telomeres must resemble those of *Tetrahymena*, they could continue to use *Tetrahymena* telomeres to "fish out" from the mix of genetic material in the cell those fragments of yeast DNA that functioned as telomeres. Using a restriction enzyme, Szostak cut off the telomeric fragment at one end of the artificially created minichromosome and then cut up all the chromosomes in the yeast cell. Of the many thousands of fragments, most made up the internal parts of the long chromosomes of yeast, but at random a small fraction, elusive as a needle in a haystack, would be yeast telomeric fragments, analogous to those of *Tetrahymena*, which could attach to the broken end of the truncated minichromosome. These rare constructs in the whole mix would have an advantage over the great majority, because they would survive and replicate, while the other reconstituted fragments would be reabsorbed and disappear. Szostak successfully fished out such rare fragments, thanks to the bait he and Blackburn had devised. Graduate student Janis Shampay, who had previous research experience with yeast, had just joined Blackburn's lab, and her work confirmed that these plasmids had acquired a yeast telomeric fragment. Now these telomeres could be sequenced, and Blackburn could begin to characterize the true molecular nature of yeast telomeres as well as decipher how the *Tetrahymena* telomeres had been altered in the yeast cells.

When Blackburn was working with Szostak, an older colleague at Berkeley warned her that she must be vigilant to ensure she received full credit in any collaboration with a male scientist. Although her colleague Daniell's experience had heightened Blackburn's awareness of gender politics in academia, she fought such anxieties when it came to how she conducted her research. Wariness, even if realistic, could constitute a handicap in a field dependent on the open sharing of ideas. Blackburn referred to such caution as "paranoia," something she resisted even though her collaboration with Szostak had introduced her to a far more competitive field: "By entering the crowded field of yeast molecular genetics, away from the small and underpopulated *Tetrahymena* field, I would be entering a new arena of competitiveness. Being there first and getting credit for it was enormously important, but I knew I could get the work done."

By persisting in her notion of her work as "genderless," Blackburn preserved a sense of freedom and autonomy in her research. Her fierce and tireless pursuit of a big result trounced any reservations about collaboration. She and Szostak would publish two jointly authored papers on their research (in 1982 and 1984). Blackburn also undertook a series of collaborations with a *Tetrahymena* geneticist, Ed Orias, a professor of biological sciences at the University of California at Santa Barbara. She felt at ease in her collaboration with Orias: "Ed was immensely kind and generous with all his scientific knowledge and resources. As a starting assistant professor at Berkeley, I asked if I could come to Santa Barbara to learn about *Tetrahymena* genetics. He and his wife, Judy, took me into their home, and he treated me to long and invaluable tutorials on the genetics of *Tetrahymena* as the three of us ate breakfast in their kitchen. It was to be a most important set of collaborations with Ed for several years, in which I never felt threatened, even though Ed was a master geneticist—as with the collaboration with Jack, we had complementary areas of expertise, and one needed the other." Though the collaboration with Orias initially focused on sequencing an unrelated tract of DNA, Blackburn immediately put to use in her telomere research everything she learned from him about the life cycle of *Tetrahymena*.

Blackburn's life and work had settled into a routine by this time. Student evaluations of her courses had improved, and students had begun to seek out her lab. She now supervised three graduate students and two postdoctoral fellows as well as a couple of undergraduates and a technician. Consistently successful in obtaining NIH grants to fund her research, she was publishing solid scientific papers at a steady clip. In the early 1980s, Blackburn and Sedat pooled resources with four friends to purchase a thirty-six-foot sailboat. On weekends, they sailed on the San Francisco Bay, which Blackburn viewed as an essential way to recuperate from her exhausting workweek: "One Saturday afternoon on the boat was the equivalent of a week's vacation away." She recalled sailing with her friends one afternoon, talking casually, "when it dawned on me that our preoccupations outside of work were becoming more and more yuppielike. Even though our jobs were all that we consciously took seriously, I could vaguely feel that there was more to life, and this discontent with terminal yuppiedom was an expression of that unease and longing for more to life."

The tenure review process might have overthrown Blackburn's settled life as well as undermined her hard-won sense of her growing abilities as a scientist. Though she had made substantial research progress, she felt apprehensive as a matter of course, "because I could never assume that I had done enough." And the specter of Daniell's denial of tenure hovered in the back of her mind. Following a by-now established pattern, once Blackburn submitted her tenure application, which required copious documentation, she "let the matter lie low," hoping that she had been accepted as "one of the boys." During the months when the university tenure committees decided her fate, her fears did not erupt into full-blown paranoia until one day in May 1983, two months before the July 1 deadline for a decision on tenure and promotion to associate professor. She collared her friend Helen Wittmer, who worked for the department chair, and demanded to know her fate. To Blackburn's intense relief, the department chair soon stopped by her office, bringing a bottle of champagne to celebrate her promotion.

At the same time Blackburn applied for tenure, Sedat had endured his own agonizing wait. When both received tenure, the months of strain undercut any celebratory mood. But in time Blackburn would relish the freedom she'd earned: "In our field, tenure doesn't mean you relax. It means you give full expression to a fanatical personality. Tenure made me bold. At first I just went on with the work, but a sense of relief started to penetrate, and I started thinking about entering a whole new era of research."

Blackburn was fully absorbed by the question of what process could account for the alteration of the *Tetrahymena* telomeres that had survived several generations in yeast cells. She continued the work begun in collaboration with Szostak and enlisted Shampay to sequence the *Tetrahymena* and yeast telomeres at either end of the linear plasmid. Blackburn had deduced, though not yet proved conclusively, that at their very tip, telomeres ended in short regions of single-stranded DNA (an overhang). Both Shampay and Blackburn were familiar with a study by Szostak, Richard Walmsley, and Tom Petes proposing that a "left-handed" DNA strand was formed at the *Tetrahymena* telomeric ends after their maintenance in yeast (a proposal not borne out) and suggesting recombination as an explanation for this data.[16] Walmsley, Petes, and other collaborators had also found evidence in yeast telomeres of a sequence rich in G and T bases.[17]

If Shampay and Blackburn were not working entirely in the dark, the technical difficulties of isolating the yeast telomere slowed their progress. Unlike a typical restriction fragment, telomeric fragments have peculiar properties that make it especially tricky to get them into a form that will allow them to be sequenced. Blackburn drew on all her knowledge of telomeric end structure to help Shampay insert the DNA molecules into recombinant bacterial plasmids so that she could clone the molecule to obtain the many copies needed to perform sequence analysis. Methods for sequencing by then had improved by leaps and bounds; Shampay would sequence the end regions using new methods for direct DNA sequencing developed by Sanger, Alan Maxam, and Walter Gilbert. Seeking to identify how a protein bound with a specific site on the DNA strand in *E. coli*, Maxam and Gilbert had discovered a means to experimentally cleave the DNA strand only at certain bases. Capitalizing on this, Sanger, using specific fragments of DNA cut by restriction enzymes, devised a method for controlled replication that would provide adequate amounts of the desired DNA to be sequenced using electrophoresis and autoradiography. (Later this method would be further refined.)[18]

Shampay conducted extensive analyses proving that the linear plasmid, which they now regarded as a true minichromosome, had acquired yeast telomeric ends. To do this, Shampay sequenced both the *Tetrahymena* and yeast telomeres at either end of the minichromosome. Her results provided a startlingly clear picture of the pattern that Blackburn had already noted. Reading from the interior of the minichromosome toward its end, the tract of its characteristic *Tetrahymena* telomeric CCCCAA repeats suddenly stopped. Instead, a completely new kind of repeat sequence took over, eerily familiar on the one hand—Cs and As strung together as they were in *Tetrahymena* telomeric DNA—and on the other hand, strikingly different in their sequence arrangement. Along the tract of about two hundred bases that Shampay sequenced first, the sequence CA was found often and was frequently mixed with the sequence CCCA or sometimes CCA. She eventually reduced the repeating sequence to the shorthand formula $(C_{1-3}A)n$, which indicated a variation of from one to three Cs in a given sequence, with *n* standing for the number of times the sequence was repeated. Most exciting of all, this $(C_{1-3}A)n$ sequence proved to be exactly the same type of sequence that made up the yeast telomeres capping the other end of the minichromosome. Shampay, Szostak, and Blackburn published these findings in *Nature* in 1984.[19]

A sequence of this type had never been identified before in yeast, and thus Blackburn could conjecture about the universality of telomeres on normal chromosomes in more widely diverse organisms than ciliates. By the end of the 1970s, scientists in the new field of telomeric research had identified repeats of C_4A_4 (CCCCAAAA) in another ciliate. In 1981, Alan Weiner and his student Herschell Emery had found that in slime mold, a species that like *Tetrahymena* has palindromic rDNA minichromosomes, telomeres had yet another irregular repeat sequence of $C_{1-8}T$.

Now Blackburn and Shampay had found a similarly irregular sequence at the ends of typical chromosomes in yeast, an organism regarded by many biologists as more conventional than ciliates or slime molds.

The *Tetrahymena* telomeres in yeast had been extended by some unknown process that produced the sequence $(C_{1-3}A)n$—a crucial finding in line with Blackburn's earlier hypothesis that telomeric DNA was created, not just copied, in *Tetrahymena*'s somatic nucleus. No known enzyme that acted on DNA could account for these observations. If some substance in the cell were adding a yeast telomeric sequence to the end of the *Tetrahymena* telomere, this would imply a completely new process at work—a risky hypothesis given the scientific community's justifiable reluctance to accept anomalous findings that can't be explained in terms of any known precedent.

In the field of yeast molecular genetics, debate flourished over the interpretation of Blackburn, Szostak, and Shampay's findings. Molecular biologists generated a variety of hypothetical models based on some of the known processes for replicating and shuffling DNA molecules. Some linear viruses accomplished end replication by using a neighboring viral DNA strand as a primer—a mechanism that might occur in eukaryotic cells. Another model proposed that telomeres were palindromic, so that following chromosome replication, a single-stranded telomere could fold back, base pairing with itself, thereby accounting for the double-stranded structure. Other researchers inferred that telomere lengthening could be accounted for by recombination, the most popular model at the time.[20] According to telomere researcher Titia de Lange, even Szostak first interpreted the results of his collaboration with Blackburn as evidence for this model:

This was a fabulously productive collaboration between two people with very different perspectives. Jack was interested in making artificial chromosomes, and Liz in these curious DNA sequences, which weren't really appreciated as telomeres yet, just initially in the ciliates. And because ciliates did all this funky stuff with their DNA, telomeres could be seen as an oddity of the ciliate genome. Then Liz and Jack found the same thing in a nonciliate, yeast, and found that yeast added its own telomeric DNA to telomeres that came from ciliates. Jack's first reaction was that this was a recombination event. But once that possible cause was excluded, they had to propose there was a new enzyme that could make this DNA. A shocking proposal. Sort of a far-fetched, speculative conclusion, but they were spot on.[21]

Blackburn has a strong capacity to attach only loosely to conclusions or hypotheses and instead let the data drive her thinking, opportunistically pursuing novel or unexpected findings. In general, good scientists excel at this, but Blackburn's method is distinctive in several respects. Her mind works laterally, in relation to exploring a single experimental question from every angle and to synthesizing a wide range of potentially relevant information. She tends to "roam" the data—as anyone who has ever listened to her discuss her work will attest—and then sharply snap a model in place, shifting gears from a patient, cautious stance to a bold one that synthesizes information in unexpected ways. She noted once that "after you make a big leap, you have to revert to caution again"—veer back toward a careful yet still speculative analysis of the data in order to test a new model. Her former graduate student Carol Greider portrayed Blackburn in a way that suggests her originality is bound up with her skepticism toward established truths: "Many people will go to lectures or read texts or scientific papers and say, 'OK, that must be true,' and file the information away as fact. Liz was never afraid to ask, 'What is the evidence that this is really true?' and 'What is a completely different way of thinking about it?' Just because there was some established dogma in the field didn't mean she would abide by it. If you don't believe something is really established, then you can think about more possibilities."[22] Blackburn's remarks on the tension between the rational and the risky in experimentation apply particularly to her own manner of thinking, in which both these qualities stem from the same impulse to interrogate conventional explanations.

The results of her collaboration with Szostak and Shampay marked a tipping point for Blackburn. She had first suggested the need for a new model in a 1982 article, and early in 1984, in a short review in the journal *Cell*, she proposed this possibility more forcefully.[23] By then, researchers had sequenced telomeric DNA in organisms ranging from slime molds to higher plants, and Blackburn knew that such findings were likely to be general across eukaryotes. Studies of the molecular details of the newly formed telomeres did not suggest any traditional known mechanism at work. Her own work in Gall's lab and later with others in her Berkeley lab had shown that telomeres in ciliate minichromosomes vary in length, and David Prescott and his collaborators had observed telomere lengthening in a hypotrichous ciliate. As part of the reshuffling process that occurs in *Tetrahymena*'s somatic nucleus during a particular part of its life cycle, telomeric DNA tracts are added to freshly cut DNA no matter what its sequence. This dovetailed with the results of McClintock's earlier cytological research, reported in 1939. At a certain time in the life cycle of maize, which corresponded uncannily to the time in the ciliate life cycle when somatic nuclei reshuffle DNA and add new telomeres, the breakage-fusion-bridge cycle could be halted, with the broken ends of chromosomes "healing" and thereafter becoming as stable as any other natural chromosomal end. Blackburn considered McClintock's observations to be "arguably the first formulation of the idea that new telomeres can be formed by an active process."[24] Further evidence that stability could be acquired came from more recent studies on the roundworm *Ascaris*; at a similar phase in the roundworm's life cycle, its chromosomes fragmented to form new chromosomes with stable ends.[25]

Research published in 1984 unexpectedly supplied yet another strong clue. In *Trypanosome*, the sleeping sickness parasite, a gene codes for a special protein that determines how well the parasite can flourish in its host, and this gene happens to lie near a telomere. Piet Borst and his collaborators, observing the gene over a period of time as these tiny parasites proliferated, happened to notice that the nearby telomere was steadily lengthening.[26] Although Borst and his collaborators attributed this lengthening to recombination, a last confirming clue that this might not be so came from the field of cytogenetics. In a conversation and a letter in 1983, McClintock reported to Blackburn an old finding of hers. McClintock had long ago identified a maize mutant that lost the capacity to heal broken chromosome ends early in plant development, suggesting that a gene regulated this behavior and could be mutated so that it did not function.[27] Blackburn concluded that healing was not just a lucky accident but the result of a deliberate process, possibly the action of an enzyme that added repeat sequences to the telomeres.

In the wake of her collaboration with Szostak, Blackburn was determined to look for a new enzyme that could synthesize telomeric DNA on to the ends of DNA molecules. In her version of how she went about this, Szostak simply drops out of the picture. In fact, he was racing to achieve the same thing, working with Vicki Lundblad to see if he could get out the enzyme using genetic screens. Blackburn tends to describe her experimental work as if it occurs in an enclosed space—as if it takes place off in a quiet corner, and her ego is tested solely by whether or not she can solve the problem immediately in front of her. If, like other scientists, she cares very much about getting there first, she tends not to measure her progress against that of the competition. But a lot of good racehorses wear blinders.

To identify this new enzyme, Blackburn relied on a biochemical approach that had not yet been applied to the problem. She wanted to capitalize on the reshuffling process in *Tetrahymena*, which added new telomeres, including thousands of rDNA telomeres, immediately after cells mated. The work of David Nanney, Peter Bruns, Orias, Sally Allen, and their colleagues in the ciliate genetics field had made *Tetrahymena* into a powerful model system; an entire cell population could be synchronized for studying developmental processes following mating. Because ciliate cell reproduction took place in a matter of an hour or so rather than over the many generations it would take to observe yeast, *Tetrahymena* once again proved to offer the perfect opportunity at the right time.

Blackburn and Szostak had observed changes to telomeric DNA in live, intact yeast cells, which made it virtually impossible to single out the reaction they were studying from all the other enzymatic activity in the cell. A biochemical assay would enable Blackburn to screen out some of this "background noise." Blackburn would extract the cellular contents from masses of cells right at the moment in their life cycle when the cells were adding new telomeres and such an enzyme would be expected to have maximal activity. Blackburn already had some clues as to how to proceed, thanks to recent work by Challoner in her own lab. He had incubated DNA molecules in extracts of mating *Tetrahymena* cells, adapting a reaction mix that Cech and his colleagues had used to study rRNA transcription and processing. Challoner had been exploring whether he might create palindromic DNA in the test tube to learn more about how *Tetrahymena* accomplished this, but as often happens, he discovered something he hadn't set out to look for: the end regions of restriction fragments grew longer and their lengths varied.[28]

The autoradiogram Challoner produced can be read vertically and horizontally, like a graph with columns and rows. Read from top to bottom, in a given column each lane indicates the separation of different fragments of DNA according to their length. When the autoradiogram records consecutive separations, one can observe successive changes to a given lane, read as a row from left to right. Researchers infer from this any changes to a given fragment, and by now members of the Blackburn lab were attuned to a telltale signal of telomere lengthening: telomeric bands grew fuzzier over time. The neat dark band first recorded in the lane grew larger, acquiring a halo of "fuzz." This did not provide unambiguous evidence of the addition of telomeric DNA, but it did give Blackburn a good lead on how she might construct her biochemical assay.

To get more precise results would require careful recalibration of the extract, the assay, and the substrate (the substance on which the enzyme acts—in this case, the DNA molecules). Blackburn tinkered with the reaction mix in an effort to provide exactly what the new enzyme might require to do its work, planning this experiment as if she were laying siege. She baited the presumed enzyme with DNA restriction fragments, some with telomeric tracts already on their ends and some without any such tracts. The latter might control for random events unrelated to the activity she sought, or they might also be acted on by the enzyme. Before she introduced the mixture of DNA restriction fragments, grown as plasmids in bacteria, she devised an extract of everything she thought might help the enzyme to flourish in the test-tube reaction: DNA and RNA base-building blocks (NTPs), an enzyme that would pump the energy-providing molecule adenosine triphosphate (ATP) into the soup for energy-hungry chemical reactions, salts, and preservative chemicals to keep destructive oxidative damage from destroying the enzyme once Blackburn broke open the cells to make the extracts. She grew up some big batches of cells and set them all to mating in synchrony. At the critical moment in mating, she extracted the cellular contents.

In early 1984, Blackburn performed the first experiments in which the reactions could be analyzed in the test tube, adding radioactive labels and chemicals to enable the hoped-for products of telomere addition to be fractionated and detected and taking time points. She ran the products out on gels to separate them by length (knowing exactly the lengths of the restriction fragments she had introduced into the cell) and identify their sequences. She asked intently focused questions:

Would these fragments grow longer and would they do so by acquiring more CCCCAA repeat DNA sequences? I used special molecular probes to ask. Many fragments had acquired this repeat DNA sequence. I had a visiting graduate student rerun my previous reaction samples on longer separating gels so they could be probed with all kinds of other control DNA molecular sequence probes to get a better picture of what was going on. The amount of added repeat sequences increased the longer the possible enzyme had been allowed to play with its DNA toys. In a way that was visually highly suggestive of telomeric DNA, these newly extended fragments were heterogeneous in length, just like ciliate and yeast telomeres. Looked and smelled like a new telomere was being made.

On the autoradiogram, Blackburn observed added repeat sequences where a probe of radioactively labeled DNA had been bound ("hybridized"), darkening a particular band, and she also saw that a new band appeared in a later separation. Long familiarity and a developed sensitivity to visual cues enabled her to view this as "unequivocal" evidence of an enzyme adding to the DNA.

Fired with excitement, Blackburn made slides of the autoradiogram and presented her preliminary results at an April 1984 UCLA/Keystone Symposium on Genome Rearrangements in Steamboat Springs, Colorado, including them at the end of a talk on genomic reshuffling.[29] Walking to the next session, she overheard a senior scientist cast doubt on her work: "He was quite sure I was going about it the wrong way. Part of me immediately feared he might be right, although I could think of no compelling reason why. But the other part thought: 'Hah! What does he know!' I was all the more determined to prove him wrong. In biological research the proof is in the pudding. He could criticize my experimental approach as much as he wanted, yet only the facts would prevail, and I was determined to get them."

To prove her hypothesis conclusively, Blackburn had to refine the experiment in order to definitively sort out the activity of this particular enzyme from the innumerable other enzymes in the crude cell extract. In the absence of any precedent—any "cookbook" for concocting the right mix to entice the enzyme to act on a DNA molecule—she had baited the enzyme with a complex mixture of restriction fragments (its substrate) to prepare for any eventuality, but now the assay could be simplified in order to see exactly which ends had telomeric repeats added. Obtaining more conclusive results would be a tall order, however. A postdoctoral fellow in the lab, Jim Forney, had written up a series of experiments to explore this, but he was already engaged in other experiments. Greider, a PhD student in her second year in the molecular biology program at Berkeley, had chosen to pursue her graduate work at Berkeley rather than Caltech as the result of a chance meeting with Blackburn, and now Greider was eager to dive in and see if the *Tetrahymena* extracts could be made to yield anything more.

Convinced that the enzyme they pursued was not an ephemera, Blackburn was intently engaged in the daily progress of the experiments. Greider reported that "Liz very much wanted to know the details of experiments as they were happening—day-to-day bits of data. She'd come looking for the 'dripping gels' while they were still wet from development. Sometimes I'd take the results and go sit in another room so I could formulate my thoughts before I had to present them to her." Keeping up with Blackburn could be taxing, if exhilarating, as Greider recalled: "The breadth of Liz's general knowledge really made an impression on me when I was working in her lab. She'd come back from a conference and tell us about all these fields. I'd research new ideas to score a point with myself if I could dig up something she didn't know. And I recall that happening only once."[30] Playfulness leavened an intently ambitious atmosphere in the Blackburn lab; Cherry recalled Greider tinkering at the computer to generate banners designating him as a computer dork.[31]

Blackburn and Greider's first stabs at a simplified assay proved impractical, requiring a complex process of purification and a long wait before results could be obtained. In addition, the activity of other enzymes in the cell could obscure the action of their target enzyme if conditions were not ideal. In one instance, all the ends of the fragments became labeled, probably as a result of the action of DNA repair enzymes. So they

modified the substrate, using a single type of restriction fragment, purified away from contaminating nucleic acids or other enzymatic activity
that synthesized nucleic acids. This time they observed only tiny signals
of activity, not nearly so pronounced as before, but the assay yielded
another small scrap of evidence that the enzyme went to work on restriction fragments that had a particular overhang on one strand (a 3′ overhang). Greider, recalling these frustrating months, remarked that
Blackburn's "boundless enthusiasm" helped.[32] Still, it would have been
easy to quit: "In the nine months that I was doing this, I kept getting
negative results. How much longer would I have gone on getting negative results before I would have decided, 'Well, it probably doesn't
exist?'"[33]

But Greider, like Blackburn, was persistent; Cherry remembered her
as "a dynamite person" who would "still have so much energy at the
end of the day that she'd ride her bike uphill to get home, the iron-man
athlete type."[34] The breakthrough came when Blackburn decided that
instead of using the laboriously purified plasmid fragments, which they
could recover only in meager amounts, they might try synthetic DNA
oligonucleotides as a primer, which would enable them to obtain a thousandfold higher concentration of DNA. Their research so far supplied a
strong hint as to which DNA sequence was required for synthesis—
GGGGTT. (In other words, the enzyme went to work on the G-rich
strand rather than the complementary C-rich strand.) Blackburn had
learned in Sanger's lab to take a worthwhile gamble and worry later
about the rationale. When she walked into the lab to suggest this
approach, she found Greider "sitting at the bench, valiantly slogging
away. Carol agreed immediately. She had been thinking of this possibility too. Suddenly, we were hopeful. The great thing about Carol was that
she wouldn't list fifty-five reasons you couldn't do something, but
responded yes, let's try it. She doesn't have inertial barriers. Disposition
and temperament as well as intellectual capability are important in
science."

Seeking far more detailed results than she'd gotten previously, Blackburn worked with Greider to construct an assay and time points so that
an autoradiogram would show the enzyme at work—bands would indi-
cate not only the order in which bases were added, thus clearly showing

any repeat sequence, but even the time periods at which the enzyme went to work. As Blackburn put it, the autoradiogram she'd shown at her Keystone Symposium talk "was a view from the rooftop," and by comparison, this autoradiogram would offer "a view through a magnifying lens."

Within weeks they obtained results. On Christmas Day 1984, Greider developed her autoradiogram of the gel. When Blackburn came into the lab the next day, Greider showed it to her. On the blue background of the X-ray film, the five lanes that had the G-rich primer showed a repeating band pattern that extended like a ladder all the way up the gel, with a compact regularity that made Blackburn think of stripes on a tiger's tail. Attuned to thinking about repeat sequences, Blackburn could deduce by eye that the evenly spaced bands were too far apart to represent a random mix of DNA molecules.

The clear pattern on Greider's gel recurred every six bases, and it told Blackburn and Greider that an enzyme had added six-base tandem repeats, with pauses that gave rise to the bands, a regulated pattern of periodicity. The bands were stronger in the samples made from the cells right at the stage when new telomeres were added, and weaker but still visible in the reaction products of the other extracts, made from cells that were just growing. As Blackburn recalled, both she and Greider were ecstatic: "Carol had gone home and danced when she first looked at the film. Just as she had, I knew by looking at it that it had to be right. Thoughts rushed through my mind that this discovery of a new enzymatic activity was as important as the discovery of DNA polymerase. DNA polymerase had garnered a Nobel Prize for its discoverer. I said nothing about this, not wanting to raise unrealistic hopes. But we both knew that this was it."

Haunted by the fear that their findings could have resulted from some artifact, Blackburn and Greider undertook the long process of testing their conclusions. Blackburn noted that the two women took turns suggesting "some depressing possibility that might account for the findings." According to Greider, "In one case we had a long discussion about an experiment I was going to do, and she argued that of course it would come out one way, and I argued very strongly that it would come out the other way. Then, the next morning we came in and found we had

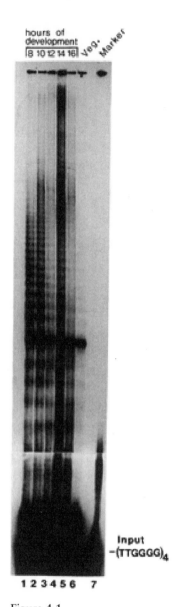

Figure 4.1
The autoradiogram providing conclusive evidence that telomerase adds to the telomere. The Christmas Day 1984 autoradiogram shows six-base repeat patterns added to the GGGGTT DNA fragments in the extract made from cells. Reprinted from Carol W. Greider and Elizabeth H. Blackburn, "Identification of a Specific Telomere Terminal Transferase Activity in *Tetrahymena* Extracts," *Cell* 43 (December 1985): 405–413, copyright 1985, with permission from Elsevier.

both flipped sides. We each had convinced the other."[35] In a comment that reveals the tenacity of both women, Greider spoke of struggling to hold her ground in these exchanges: "If you've ever had to argue with Liz, you know it is hard to argue with her. She is *very* forceful about getting her point across."[36]

Ruling out artifacts is not merely routine follow-up, nor is it a matter of crossing off items on a checklist. Often, it constitutes the greater part of experimental work, and biologists' as yet incomplete understanding of complex cellular mechanisms makes it difficult even to be certain what elements one might need to rule out. This stage of experimentation, which requires a researcher to devote great imaginative effort to undermining her own inferences, can yield important clues to new or more precise questions that an experiment might ask. With respect to the creativity demanded, ruling out artifacts and generating a new model are two sides of the same coin. "Our excitement built as experiment after experiment excluded the possibility of another artifact that could account for our results," Blackburn reported. "There was no expertise in this new activity, although there was expertise in DNA polymerases, and I talked to those experts to see if something like this had ever shown up but had not been published as an artifact. Nothing had."

Blackburn and Greider had to rule out with certainty the possibility that conventional DNA polymerases added these new stretches of telomeric DNA. In the entire crude extract containing long stretches of repeat DNA and nucleic acids, the enzyme might merely copy a sequence that existed elsewhere on the strands of DNA (internal, or endogenous, CCCCAA repeats). To control for this, Blackburn and Greider employed micrococcal nuclease, an enzyme that chopped the restriction fragments into smaller bits. If the enzyme were merely copying preexisting DNA, it would now create much shorter telomeric additions—the bands that showed up on a gel would be fewer and shorter. Blackburn and Greider repeatedly tried this experiment and discovered that long repeat sequences were still being added, just as before; the enzyme was adding telomeric DNA de novo. (They were incredibly lucky to get this result, since micrococcal nuclease, which cuts both RNA and DNA, might have inactivated the enzyme by destroying its RNA subunit.)

In June 1985, Blackburn and Greider completed the experiment that persuaded them they had successfully ruled out any artifact. They decided to add a yeast telomeric sequence to the *Tetrahymena* restriction fragments they were using as a substrate, essentially the converse of Blackburn and Szostak's experiment in which the enzyme in yeast had added to *Tetrahymena* telomeres introduced by a plasmid. In an analogous fashion, this time the yeast telomere was elongated by the *Tetrahymena* enzyme in the regular six-base repeat pattern of *Tetrahymena*. In the process, Blackburn and Greider made an additional discovery. While a *Tetrahymena* telomeric fragment ended in four Gs, a yeast telomeric fragment ended in three Gs. When the enzyme added to *Tetrahymena* fragments, it routinely added TTGGGG repeats. But when the enzyme added to a yeast telomeric fragment, it began by adding a G before producing the subsequent TTGGGG repeats. In other words, the enzyme could somehow recognize and correct for the "deficiency" in its yeast substrate. This provided the first clue that the enzyme might contain an RNA template.

Increasingly confident about their results, by summer 1985 Blackburn and Greider submitted a paper to the high-profile journal *Cell*. The reviews were favorable, suggesting a few more experiments to confirm the validity of the findings before the paper could be published. After Blackburn and Greider carried out these experiments, the paper appeared in the December 1985 issue of *Cell*, almost exactly a year after Greider had looked at her X-ray film on Christmas Day.[37] This paper opened by boldly asserting that they had found the answer to the problem of end replication, proposing that a "novel" enzyme "is involved in the addition of telomeric repeats necessary for the replication of chromosome ends in eukaryotes."[38] Blackburn and Greider also offered a model for how the enzyme did its job: the enzyme "recognizes" a telomeric sequence on the single-stranded 3′ overhang and adds to it, and DNA polymerases subsequently fill in the partner strand, using the extended G-rich strand as a template.[39] (To this day, this remains a default assumption, since scientists have yet to nail down this process conclusively.)

If Blackburn had high hopes when she and Greider first looked at the gel that showed a clear repeat pattern, she declared that she felt noncommittal when their paper did not cause a stir: "There was little response. It didn't even occur to me to care one way or the other. I had presented the results at a Cold Spring Harbor Symposium, and people had seemed convinced by the data but not particularly excited. It seemed as though they just thought it a *Tetrahymena* weirdness. Only I knew that organisms don't invent enzymes out of whole cloth—these processes evolve. If an enzyme exists in one species, it is likely to exist in many more."

Why wasn't telomerase instantly recognized as a dogma-breaking enzyme? In the first place, as Greider and Blackburn noted in an account of their discoveries, "this work had little impact beyond the—then quite small—circle of people interested in telomeres and chromosomes."[40] Second, biologists hesitated to generalize from the quirky *Tetrahymena* to other species. As yet, telomeres had been sequenced in only a few simple eukaryotes, so these findings might have the same novelty value as *Tetrahymena*'s seven sexes and would not generate widespread interest until telomeres were sequenced in human cells. Bacteriophages had other mechanisms for end replication, which offered a further reason to reserve judgment on whether this enzyme might be fundamentally conserved among eukaryotes, and suggestive evidence still supported competing models. As late as 1989, Zakian and Ann F. Pluta proposed recombination as a method for the synthesis of yeast telomeres.[41] Confusion, in this case, stemmed from the difficulty of distinguishing a primary pathway from an alternate one, as recombination turned out to be. Third, Greider and Blackburn were reporting the results of in vitro assays, and this did not prove conclusively that the same process occurred in living cells. Blackburn would not feel she had provided formal proof of the enzyme's existence until a study in her lab, published in 1990, demonstrated its activity in living cells.

Finally, a potentially deceptive blend of boldness and caution characterized the 1985 paper. These qualities are reflected in an assessment of the paper that Greider offered much later, in January 2006: "We were actually going out on a limb rather than being cautious. We were very careful to have all the evidence be very clear. There weren't any holes in the argument."[42] And if Blackburn and Greider made a novel claim in this paper, they withheld another. Their first paper gives no hint that the enzyme might be a reverse transcriptase, containing an RNA template, a truly astonishing possibility, since reverse transcriptases were thought to be an oddity confined to viruses. Instead, Blackburn and Greider cautiously identified the enzyme as a "telomere terminal transferase," so it read as if the enzyme would be a deoxynucleotide terminal transferase enzyme, which, unlike a reverse transcriptase, had precedent as a more common enzyme in eukaryotic cells.

In the series of three coauthored papers (in 1985, 1987, and 1989), Greider reported that a balance was struck between Blackburn's conviction and her own relative caution: "I would say Liz is more interested in big ideas and whether things might be going in an unconventional direction. She's very optimistic scientifically, extremely optimistic. She wanted to make the more optimistic interpretations in our papers, and I was the one saying, 'Let's wait. We have to make sure that we really show this is in fact the case.' She's very systematic, but we were trying to determine if what we thought we were seeing was a fact as opposed to something that was fooling us. That goes to her more creative side, what's most interesting to her."[43] Blackburn remembers herself as more cautious than this: "It's hard to try to remember, twenty years after the fact. But I think I was more nervous then. When you've taken the risk, you have to be very rigorous in showing it's not an artifact. I think I was more nervous then about publishing something that might be an artifact." Despite her high expectations for this discovery, Blackburn could say she "didn't much care" about the immediate reaction to the paper because she could take a long view about the implications of her findings, confident the paper met her own criteria for "sound science" and already in hot pursuit of yet another amazing property of this enzyme.

Probably only in retrospect could the significance of this paper be fully measured. And probably it takes someone other than Blackburn to underscore the fact that she had beaten out the next person in the race. De Lange, whose studies on telomeric proteins have been important for both basic science and cancer research, pointed to the publication of this paper as a watershed moment: "Liz and Carol Greider's discovery in 1984 that an enzyme maintained telomere DNA broke open the field.

Vicki Lundblad, working with Jack Szostak, was approaching the same question in a genetic way. Liz went after the enzyme using biochemistry in *Tetrahymena*, and Jack used genetics in budding yeast, and Liz got there first. Liz knew so much about the biology of her system. She knew the ciliates had to make a bunch of new telomeres after sexual interaction, so she knew exactly where to look. Later Carol came to her lab, and Carol really made it happen."[44]

5   Opportunism

Though Blackburn and Greider had yet to name the enzyme they'd identified, their explorations in this tiny patch of the genetic code solved the mystery of the end-replication problem, which had defied the best efforts of molecular biologists. Blackburn and Greider's next imperative was to learn more about how telomerase functioned in the cell. How did the enzyme know to make perfect six-base repeat sequences for *Tetrahymena*? What mechanisms in an organism (or its environment) influenced or regulated the enzyme's production? Would it function in similar ways in other ciliates and different species, including human beings? To search for and identify telomeric repeat sequences in more complex life-forms, while not intellectually exciting, would confirm whether the activity of telomerase was universal. Throughout the late 1970s and early 1980s, evidence emerged that telomeres existed in other eukaryotes, from simple to complex—in the slime molds *Physarum* and *Dictyostelium*, hypotrichous ciliates such as *Oxytricha*, and the *Trypanosome* parasite. All these organisms had tandem repeat patterns, and other researchers also observed the single-stranded overhang at the 3′ end of the DNA strand, varying in length from one species to the next—fourteen bases long in one hypotrich species and sixteen bases in another.[1] To identify similar repeat sequences in human telomeres would be the capstone in confirming this as a fundamentally conserved mechanism.

Blackburn left this task to others because she did not want to conduct research on human cells, though funding for such research was easier to obtain, since the NIH, the major source of funding for academic research, emphasized studies related to human health. During her stay at Yale University, when she and Sedat lived in a rented beach house, a colleague in John's department, Professor Paul Howard-Flanders, lived just down the street. Flanders and his wife, June, had taken Sedat and Blackburn under their wing and helped them to find their beach rental, and the two couples often went for an afternoon sail in Paul and June's small sailboat. On one of those sails, Paul had encouraged Blackburn to work on telomeres in humans. Although she recognized his logic, Blackburn did not take his advice. For one thing, at the time researchers couldn't get fast answers working with more complex human cells, and Blackburn, steeped in the élan of the MRC at Cambridge, favored searching for quick and elegant solutions to problems in molecular biology.

Blackburn and Greider had first named the enzyme *telomere terminal transferase* because terminal transferase enzymes could add any DNA to the end of a DNA strand, with no specific requirements about what kind of DNA they add to. But this new enzyme added a specific sequence (GGGGTT) to telomeric DNA, suggesting it belonged in an entirely novel class. The term *telomerase* (ta-LOM-er-aze) was coined in 1986 by Claire Wyman, a graduate student in Blackburn's lab, and adopted thereafter.[2] In Blackburn and Greider's initial experiments on telomerase, a discrepancy in their results had thrown up an intriguing hint not strong enough to include in their paper in *Cell*. When they had first employed micrococcal nuclease to create shorter restriction fragments for an assay, they had done so with a crude cell extract containing many proteins and nucleic acids. Though micrococcal nuclease destroys both RNA and DNA, they found evidence of telomerase activity. Knowing they would have to clean up the extract in order to begin purifying the enzyme, Greider now conducted column chromatography, which provided extensive purification of the enzyme. This time the cellular extract did not contain so many other nucleic acids for the micrococcal nuclease to cut, and telomerase unexpectedly ceased to work. If they had been lucky in getting results in the initial assay, Blackburn and Greider registered the possible cause of this discrepancy, alert to clues that the enzyme contained a nucleic acid component, unlike the DNA polymerase enzymes that built strands of DNA.

Blackburn and Greider could look to recent precedent in hypothesizing that an RNA component might be integral to the enzyme's functioning. Until the 1970s, biologists generally assumed that nucleic acids and proteins had distinct functions in the cell: DNA stored genetic information, RNA transferred that information, and proteins acted as catalysts, capable of speeding up chemical reactions without undergoing any change themselves. RNA was considered to be an exact copy of DNA, and any changes to it resulted from the action of protein enzymes. But in the 1970s and 1980s, researchers discovered that not only protein enzymes but ribonucleoproteins (RNPs) could carry out catalytic actions in the cell, including protein synthesis and mRNA splicing. Splicing was first proposed in 1977; when a stretch of RNA included tracts of nonessential intervening sequences copied from DNA, these introns were later cut out by small ribonucleoproteins, snRNPs. In 1981, Tom Cech and his collaborators, studying the ribosomal RNA (rRNA) in *Tetrahymena*, discovered that this rRNA spliced itself and behaved like a protein enzyme, making it a ribozyme.[3] Further eroding the assumption that only proteins had catalytic properties, in 1983 Sid Altman and Norman Pace discovered that an RNA component in the enzyme RNase P acted as a catalyst.[4] (Altman and Cech would later share the 1989 Nobel Prize in Chemistry for their discovery of the catalytic properties of RNA.)

To determine whether telomerase contained an RNA component, Blackburn and Greider conducted assays similar to those of their original experiment. They "pretreated" the cell extract by adding an enzyme, pancreatic ribonuclease (also called ribonuclease A, or RNase A) that cuts the phosphate backbone of RNA but does not cut the backbone of DNA, which differs chemically. If, as they anticipated, the enzyme relied on an RNA component to do its work, telomerase activity would be sharply reduced or eliminated in this assay. On the day in 1986 when they were conducting this assay, Cech, who'd been invited to Berkeley to give a lecture, visited the lab, and Blackburn introduced him to Greider.[5] Cech worked on the same organism, had an intense interest in

RNA, and only recently, working with Alan Zaug, had demonstrated that in the test tube the self-splicing rRNA in *Tetrahymena* could act as an enzyme, using an internal template.[6] Throughout the day, he checked back in at the lab to see if the results were in yet.

Telomerase activity ceased in the presence of RNase A, demonstrating that an RNA component was required for the enzyme's activity. In subsequent experiments, Blackburn and Greider used a number of controls to ensure that the ribonuclease was responsible for this, including an enzyme (DNase 1) that cut only DNA. They confirmed that the enzyme contained RNA, but couldn't say for sure that it was a template, though the precise and constant order of the telomeric sequence implied the enzyme had instructions for which triphosphate (a nucleotide plus a building block) to add next to the strand of DNA.

If Cech had provided the precedent for a catalytic action that relied on an RNA template, Blackburn and Greider's hypothesis still flew in the face of accepted dogma: a central tenet of molecular biology at the time held that in the essential cellular mechanisms of eukaryotes, enzymes copied DNA into RNA, and *never* the reverse. The only exceptions found had occurred in then obscure retroviruses, which copied RNA into DNA (hence the term "retro"; HIV later became the most notorious of these viruses), thanks to the action of reverse transcriptase enzymes. Such enzymes were thought to hark back to earlier, more primitive life-forms, and biologists assumed that these enzymes did not exist in more complex organisms, which evolved later, with protein polymerases taking over the role played by RNA in synthesizing the genetic material of the cell. Making Blackburn and Greider's hypothesis still more risky, reverse transcriptase typically copied thousands of nucleotides at a time in a viral genome. No reverse transcriptase had ever been found to copy only a short sequence of DNA.

Chemistry determines how any enzyme reads the DNA code. The DNA strand's phosphodiester backbone (a convenient slang term that collapses some beautifully complex chemistry) is held together by covalent bonds between its molecules. These bonds occur at specific sites on each sugar molecule in the backbone. To each of the five carbon atom positions of this pentagon-shaped molecule, scientists have assigned a number for identification, notated as 1′, 2′, 3′, 4′, and 5′. A nucleotide base chemically bonds with the molecule at 1′; positions 3′ and 5′, located on opposite sides of the pentagon, are the points at which a phosphate links it to the next sugar molecule, as beads might be strung lengthwise on a necklace. The DNA polymerase enzymes that construct the strand of DNA always add to the chain in just one direction, from the 5′ end toward the 3′ end, and bases are numbered according to their position (residue) along this continuum, counting from the 5′ end, analogous to reading words from left to right. (Complex mechanisms of DNA replication compensate for the ability of enzymes to add to a strand in only one direction.) When two strands of DNA align in reverse direction to form a chromosome, one strand is read from left to right, and the other from right to left.

At either end of the chromosome, on opposite strands, "extra" genetic material is dangling, offering a tempting opportunity for telomerase to

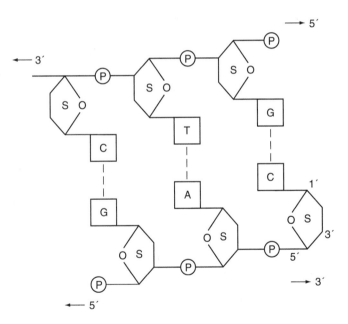

Figure 5.1
The phosphodiester backbone of the DNA strand. chemical bonds link the sugar molecules (S) of the DNA backbone as well as linking one base to its complementary partner. During replication, DNA polymerase enzymes add to the chain in just one direction, from the 5′ end toward the 3′ end. The two strands of DNA in the chromosome align in reverse direction. *Illustrator*: Alan Stonebraker.

go to work, synthesizing DNA in the same direction as polymerases do. But to perform its task, telomerase must possess special recognition properties, possibly derived from its RNA component. Since findings reported in their 1985 paper suggested the enzyme added to the 3′ overhang, the evidence in hand enabled Blackburn and Greider to ask increasingly specific questions: How did the enzyme in *Tetrahymena* cells recognize and extend even yeast telomeres? How did the enzyme identify the "starting point" at which to begin adding to telomeres?

Concurrently, Blackburn and her student Janis Shampay were exploring the curious fact of the heterogeneous lengths of telomeres in a typical cell, which Blackburn had first noted while working in Gall's lab at Yale. Shampay had again observed this variation in yeast in their experiments with Jack Szostak, but now Blackburn and Shampay could pursue the question of whether this variation resulted from the process of artificially inserting a plasmid into the cell—which essentially shocks the cell into accepting foreign genetic material—or occurred normally. In their proposed model, telomerase continually added to telomeres, thus compensating for the loss of genetic material during cell division when enzymes could not copy all the way out to the end of a DNA strand. In observing yeast cells (*Saccharomyces cervisiae,* or baker's yeast), Blackburn and Shampay could trace several daughter generations of a single cell and thus deduce what had happened to individual telomeres in living cells over time. They observed that a given telomere sometimes got longer and sometimes shorter, resulting in a heterogeneous mixture, exactly what they would expect if their model held true. Significantly, this variation in telomere length occurred within a fairly narrow range, suggesting that some mechanism reacted when telomeres became too short and also prevented them from growing too long.[7] These findings provided important confirmation of the results that Blackburn and Greider were obtaining in the test tube; a process so strictly regulated must contribute to the health of the cell. When the findings in both studies were integrated, they formed a coherent initial picture of how telomerase operated in cells.

In 1986, in the midst of this enormously productive work, at the age of thirty-eight, Blackburn discovered that she was pregnant. Immersed in their work, she and Sedat had never discussed having children. But she was immediately pleased: "When I became pregnant, I knew that this was an answer to terminal yuppiedom. Having a child would take me beyond the somewhat narcissistic preoccupations that were coming to take up most of our lives outside of work."[8] When Helen Wittmer congratulated Blackburn on her promotion from associate to full professor and asked what she would do with the extra salary, Blackburn answered that she would have plenty of use for it, and told her the news. She did not tell anyone else she was pregnant until after she received the results of chorionic villus sampling, a test for chromosomal defects that is conducted earlier in a pregnancy than amniocentesis.

Like many other women scientists of her generation, Blackburn had postponed having a child. The postdoctoral fellowships necessary to a career as a researcher required that scientists put in long hours for little pay and made few provisions for granting leaves of absence to women who chose to have children. The subsequent struggle to establish one's credentials as a researcher and gain tenure placed equally excessive demands on an academic scientist's time but coincided with a woman's peak childbearing years. While for male scientists marriage and family have little effect on participation in the labor force, these factors significantly decrease women's full-time participation, and married women with small children are less likely than single women to have a full-time career in science and engineering, although this gap has steadily declined since 1979.[9] As late as 1999, when five women became fellows of the Royal Society of London, one of these scientists, physiologist Frances Ashcroft, declared that women could not combine motherhood with research of the first rank, despite the fact that three of the other new female fellows had children.[10]

Ironically, if a woman interrupted her career in order to have children, she jeopardized her professional future, and if she did not interrupt her career, she still risked being seen as compromised in her commitment. In a by now established pattern, Blackburn denied the possibility that these impediments might apply to her, even as she rationalized that having been promoted, she had arrived on secure ground: "I had achieved enough to have a child." Far more confident of her professional abilities than she had been as a postdoctoral fellow at Yale, she never entertained the notion of interrupting her work. Her male colleagues would not regard fatherhood as an intrusion on their careers, and she could once again pass as one of the boys and dismiss any notion of curtailing her research: "There was never a question in my mind that I wouldn't continue or that my career would stop. I would just build in raising a child. I'd gone to a scientific conference in June, early in my pregnancy, and when Peter Bruns invited me to join him for a drink, I was pleased to tell him the reason I had to refuse his offer. I anticipated things would go on as they had—maybe I'd cut back my evening hours—and I'd just have to be much more organized."

Only in July 1986, when she was four months pregnant, did Blackburn belatedly announce her pregnancy to her colleagues, first to the members of her lab, who, she felt, "had the biggest stake in knowing" because of its potential effect on their research. On the very next day, she went into preterm labor. Hoping to avoid a miscarriage, she spent the remainder of her pregnancy on bed rest, but she immediately notified the chair of her department and offered to take a sabbatical, as if she had to compensate for this interruption in her work. The chair assured her this was not necessary and arranged instead for a leave of absence. For the next five months, she ran her lab from a distance and canceled travel to meetings and conferences. She felt bad when she first heard who had been chosen to replace her in giving a conference talk, but then realized she hadn't looked forward to the talk: "So much of my work had evolved into things that had become a burden. It was a relief not to have to pursue these peripheral aspects of a scientific career for a while."

In contrast, Blackburn was determined that however she had to accommodate her enforced bed rest, her experimental work would not suffer: "I had a totally functional lab, with excellent students and post-docs. I arranged for them to forward paperwork to John to bring home, kept a keyboard at a bedside table for sending e-mails—no laptops existed then—and had the members of the lab videotape their weekly group meeting. It was important to me to feel the work was going on, maybe not quite the same as if I had been there—people drifted a bit on their experimental courses. Toward the end of my pregnancy, Drena Larsen and I were writing up a paper, and I was reviewing the last draft while I was having contractions, saying, 'We've got to get this done!'"

Blackburn's son, Ben, was born, full-term, on December 5, 1986. To her surprise, Blackburn found that motherhood offered unexpected gains rather than a new handicap: "I felt like I had been in the faraway world of science and had reentered the human race, the world of normal pre-occupations. I got support from all sorts of people—even letters from distant colleagues. When I returned to work after Ben's birth, I felt a warmth I hadn't sensed before. My older colleagues, most of whom were fathers, now regarded me as a human being. They were pleased to talk to me about motherhood, as though this had forged a link with something they understood about women, and they were more at ease with me than they had been." Pregnancy and motherhood may also have made it possible for Blackburn to regard herself as a woman among other women. Habituated to disregarding gender in her professional life, while she was on bed rest Blackburn had turned to Zakian, who was sustaining a brilliant career while she raised two children. Blackburn recalled Zakian offering sympathy and insights borne of shared experience: "Ginger knew about having a difficult pregnancy and trying to be a scientist."

To help her daughter, Marcia Blackburn had come from Australia two months before Ben's birth, and she stayed for a month afterward. Blackburn joked that it required no less than four people to care for an infant: late at night, after Blackburn fed Ben, her mother burped him while John slept (his contribution as the designated "functioning brain"), and in the morning, Sonia Menjivar, whom Sedat and Blackburn had hired on the recommendation of a pediatrician friend, arrived to care for Ben during the workday. For Blackburn, it was essential to have a child-care provider whom she trusted: "I could go to work and forget for nine hours that I had a child. I felt privileged to have the resources to pay for a good person to care for Ben full-time." Yet her professional life changed dramatically in the wake of Ben's birth. Up until this time, Blackburn routinely worked at the lab from late morning until nine or ten at night and spent the weekend recovering from exhaustion—a schedule that would not work for someone with a young child. Blackburn cut back to an eight-hour day and carefully calculated her hours so she could get home to Ben on time, which meant she had to allow extra time to commute during rush hour instead of after it. She could regiment her hours only because as a senior researcher, she no longer had to carry out experiments herself. But working at the bench remained, for her, not a chore to be delegated to others but a necessary pleasure. Despite her increasing professional obligations and family demands, she contrived a means to keep her hand in, reserving the semester break at Christmas for doing hands-on work. A postdoc recalled that when Ben was young, Blackburn occasionally showed up at the lab in a sweater worn inside out, and she and her husband gave up outside pursuits, forgoing the sailboat on which they'd spent so many weekends with friends. Yet by the time Ben turned three, Blackburn would publish two more significant papers with Greider, confirming the uniqueness of telomerase and its central importance in the functioning of a cell.

When Blackburn returned to work, she reengaged with experiments that had continued in her absence. Greider was making steady progress in the study of the RNA template of telomerase. The experiments she conducted on *Tetrahymena* from 1985 to 1987 employed radioactive tags to distinguish newly added sequences from existing ones. Further experiments confirmed findings reported in the 1985 paper, in which the enzyme had "corrected" for the difference in the sequences of yeast and *Tetrahymena* telomeres when it began adding *Tetrahymena* telomeres to yeast. Supplied with a variety of DNA fragments, the enzyme wouldn't extend certain strands, only those that had the clusters of G bases characteristic of telomeres. In the test tube, provided with only a single strand of DNA, the enzyme still knew what to add in order to sustain a perfect tandem repeat sequence, depending on whatever base was last at the 3′ end of the strand. So, for example, if only one G base occurs at the end, the enzyme first adds GGGTT to complete the six-nucleotide sequence; if two Gs occur at the end, the enzyme adds GGTT; if three Gs occur, the enzyme adds GTT; and if four Gs occur, the enzyme adds TT. As shown in figure 5.2, the enzyme has primer recognition properties: it can "read" the existing sequence (its primer at the 3′ end of the DNA strand) and correctly add to it in order to form the uniform G_4T_2 (GGGGTT) sequence characteristic of *Tetrahymena*, a crucial clue to the existence of an RNA template and a promising suggestion that the enzyme could repair chromosomes whose ends had worn away.

GGGGTTGGGGTTggggttggggtt...

GGGGTTGGGGTTGgggttggggtt...

GGGGTTGGGGTTGGGgttggggtt...

← 5′   3′ →

Figure 5.2
Primer recognition properties of telomerase. Telomerase can begin adding to the 3′ end of a primer strand of DNA, sustaining a perfect tandem repeat sequence dependent on the enzyme's recognition of the sequence at the 3′ end of the strand. *Illustrator*: Alan Stonebraker.

In 1987, Blackburn and Greider published their second paper in *Cell*, demonstrating that telomerase contains an RNA component. To provide a broad base of evidence, they had used primers that mimicked the telomeric DNA in five organisms and demonstrated the activity of the enzyme in each case. They now clearly identified telomerase as an RNP, in which the RNA could contribute "precise recognition of nucleic acids" to an enzyme's functioning. This marked a clear shift in the proposed class to which this novel enzyme belonged. The paper points backward, providing detailed evidence for a number of possibilities posed in the previous paper, and it points forward, speculating on the nature of the enzyme and clearly indicating their next concern: identifying a likely candidate for the RNA component and determining whether it contained a template.

The "incidental" findings Blackburn and Greider reported show that they were laying a solid foundation for further characterization of this enzyme. They noted, for example, that it worked better on a DNA primer that already had three or more repeat sequences, and they speculated on the reasons for irregular repeat sequences in organisms such as *Saccharomyces cerevisiae* and *Dictyostelium*. Drawing an analogy between telomerase and the self-splicing intron ribozyme identified by Cech and Zaug, Blackburn and Greider considered whether the RNA component of telomerase might also have catalytic properties. And yet, though they clearly laid out mechanisms that telomerase had in common with a reverse transcriptase, they didn't assign it this classification.[11]

Overthrowing an accepted dogma, Blackburn and Greider had now demonstrated that RNA was copied back into DNA by an enzyme that was a normal part of normal cells, required for replication and renewal of the cell. Normally, reverse transcriptase enzymes, which copied RNA into DNA, were composed only of protein, but telomerase was a reverse transcriptase with the unusual properties of synthesizing short repeats and carrying its own internal RNA template for DNA synthesis, confirming the notion that it might be a relic from the time of the evolutionary transition from RNA to DNA genomes.[12]

Next, Blackburn and Greider attempted to identify the RNA sequence of telomerase. Eventually, this would enable them, through recombinant genetics, to clone the gene for the enzyme, making it possible to create deliberate mutations and thus conduct further experiments on its function. They had the luxury of not having to worry about competing with other labs to get results first, because as yet, no distinct field of molecular telomere research existed, though David Prescott was conducting important investigations into telomeres in other ciliates. As Blackburn and Greider forged ahead, in their wake other researchers would fill in the details, identifying the telomeric sequences of other species by relying on the ideas, conditions, and methods outlined by the two women in their 1985 and 1987 papers.

In order to isolate this particular RNA from all other RNA normally found in a cell and then purify adequate amounts for sequencing, Greider had to perform tedious enzymological and biochemical processes for fractionation and purification in an effort to cull an unimaginably tiny amount of material. (As a point of reference, the slight overhang of a dozen or so nucleotides at the end of a strand of DNA is one-ten-thousandth the size of a pinhead.) Because one could not predict in advance how to purify any protein or protein-RNA complex for the first time, the work had to proceed by trial and error. Blackburn emphasized that "a lot of what science is about is drudgery":

You're inspired by chasing an elusive goal, but you have to spend months on unrewarding trial and error. You do experiments that you *can* do—pick the low-hanging fruit—and in the mid-eighties the methods for identifying a scarce RNA were better than those for identifying a scarce protein, so we focused on finding RNA. We gathered a huge amount of data, but the picture only began to clear after a while. Almost unconsciously, you start thinking of connections. Always you're asking, how can I test this? What question can I ask that would destroy the hypothesis? You're applying syllogistic logic, but by itself logic can't solve the problem—my dog has four legs, my cat has four legs, therefore my dog is a cat. Because of the complexity of biology, things have not evolved by a logical process. You see this when you teach undergrads—on a test, if they don't know the facts, they'll try to deduce something, but even sound deductions are often wrong. You can't operate from first principles, as physicists can, but have to test logic against precedent and the data itself. Proof in the purest sense is something that comes in a messy way.

Blackburn and Greider tried various methods to purify the RNA in the enzyme, percolating the cellular extract through chemical resin in long glass cylindrical tubes, refining several hundred molecules from this step, and then repeating the process to further refine the sample (called a fraction) so that it would include only the RNA they sought. At this work, Greider wore not a white lab coat but a parka; because heat degrades protein, the fragile extracts of cells were kept in a "cold room"—essentially, a walk-in refrigerator. Eventually, Blackburn and Greider decided to add radioactive tags to RNA in the cellular extract and ran it out on sequencing gels just to visualize all the RNA. They assayed the various fractions for telomerase activity to see which RNA molecules tracked with that activity and then proceeded with further purification.

Blackburn and Greider found that two classes of RNA always tracked with enzymatic activity: transfer RNA, which they suspected was a contaminant but could not rule out, and another rare band, approximately 160 nucleotides in length, which looked like the best candidate. After performing more tedious purification, scaling up the size of their sample to extract the tiny fraction of this material that would be RNA, they began efforts to sequence this molecule in 1987. If they could not sequence it in its entirety, they hoped to sequence enough to locate the gene that coded for the RNA (because it was now easier to sequence DNA than RNA—"clone out" the gene) and then manipulate the gene to perform further experiments.

At the same time, Dorothy Shippen (formerly Shippen-Lentz) pursued a parallel track in the Blackburn lab, seeking evidence of telomerase activity in another organism. Shippen had arrived in the lab as a postdoctoral fellow in October 1987 after a careful search for a position. When she had first written to Blackburn to inquire, she received no reply, but she mustered her courage, called the lab, and spoke with Ed Orias, who was spending a sabbatical there. Orias explained that Blackburn was on bed rest and unavailable. When Shippen finally interviewed with Blackburn, she immediately felt this lab was the place for her: she had extensive experience with sequence analysis of RNA and was intrigued by Greider and Blackburn's work on identifying the RNA template of telomerase. Further, she found that "the people in the lab were very excited about being there. You can tell so much about the way a mentor runs a lab by the attitudes of the people there—upbeat, excited, nothing but wonderful things to say about Liz. She was very rigorous but very kind to me. It was important to me to be in a lab where there was a nurturing environment, and I wanted to work with a woman."[13]

At first Shippen was frustrated, because Greider had already staked out the territory of sequencing the RNA template in *Tetrahymena* telomerase. For six months, Shippen did "boring work, basic characterization of the *Tetrahymena* telomerase, nothing that could be called ground-breaking."[14] But then Blackburn found an opening that would allow Shippen to use her RNA expertise and expand the lab's investigations. She provided grant funding for Shippen to go to the lab of Larry Klobutcher, who worked on *Euplotes crassus*, a different protozoan, and Shippen learned how to cultivate this organism so that she could use it as a model system to study telomerase. *Euplotes crassus* possessed a hundred times more telomeres than *Tetrahymena,* and Prescott had shown that it had even smaller chromosomes, which meant its telomeres would constitute a proportionately larger percentage of a cell's DNA and thus provide potentially fertile ground for tracking telomerase. But as Gall had pointed out to Blackburn many years earlier, this organism was more difficult to grow and maintain in culture. On her return, Shippen cultivated the bright green algae that would provide a food source for the *Euplotes*, growing the algae in large watercooler bottles filled with seawater and set on the lab's windowsill under bright lights day and night. Once the algae were plentiful, she added *Euplotes* to the seawater to feed and multiply.

Shippen's research would eventually establish that telomerase functioned in other protozoans in the same way it did in *Tetrahymena*. But for each organism, deriving a process for extracting and studying the

RNA posed completely new problems. To attempt to purify the RNA, as Greider was doing for *Tetrahymena*, Shippen drew on Blackburn's resourcefulness and expertise as she devised a workable method: "All we knew about the *Euplotes* RNA was our prediction that it was going to contain the sequence AAAACCCC, complementary to the telomere repeat sequence of this organism. With so little information, there was no obvious way to go after the RNA. But Liz had the idea that we could find the *Euplotes* RNA template by employing a strategy called oligonucleotide-directed RNase H [ribonuclease H] cleavage. She got the idea from reading splicing literature—she knew Joan Steitz had succeeded with this approach for studying spliceosomal RNP complexes. So she said, let's try this." Much like Greider, Shippen found Blackburn to be scientifically optimistic and fully engaged with the research as it unfolded:

Liz didn't breathe down my neck for data, but she was so enthusiastic about any data you brought her. In those days, when you developed autoradiograms, you did so in tanks of chemical fixers and it sometimes took days to expose the film. Liz would come out of her office and say, "I smell fixer. Who has a result?" Whether the result was positive or negative, she never lost this incredible enthusiasm for what you were doing. She would love to be the one developing the film and seeing the data for the first time. She was so engaged in the process—it was not about trying to push the person. Even with stuff that didn't work out the way you hoped—no big results—you didn't sense disappointment in her. Instead, her response was, "Oh! *That's* the way it works."[15]

The collaboration between Blackburn and Greider continued past the time when Greider completed her thesis and accepted a prestigious postdoctoral fellowship at Cold Spring Harbor Laboratories. Typically, a lab would hold on to a graduate student's project, and as a postdoc, she would start over on new research. But Greider continued her efforts to clone the gene after her move. Determined to pursue what she'd begun, Greider even turned down an offer of a postdoctoral fellowship in Bruce Stillman's lab at Cold Spring Harbor and instead took a position as an independent postdoc.[16] The unusual arrangement between Blackburn and Greider acknowledged Greider's role in the research, and it provided a useful division of labor. At Cold Spring Harbor, Greider could consult with experts in RNP chemistry, and finding out more about the protein in telomerase would be another important step in defining its role in the cell. Blackburn continued working on the RNA gene and the effect of telomerase activity on the functioning of a cell. "It would have been easy for either of us to feel territorial or to fear that our equal contributions to our original collaboration might fail to be acknowledged," Blackburn said. "But I never felt any clashes with Carol. We read each other well, there was plenty to do, and our work diverged according to differing interests. So I didn't feel as if we competed with each other in pursuing the same questions. Carol declared an interest in proteins that led to her own career path. She later went on to work with human and mouse telomerase."

By April 1988, Blackburn and Greider had nailed down the DNA sequence of the gene that coded for telomerase and thus could piece together the sequence of the RNA template. Most of the 159-nucleotide sequence of the RNA component remained as mysterious as hieroglyphics—its function as yet unidentified—and searching for a recognizable sequence that complemented the telomeric sequence was analogous to scanning a word puzzle of 159 run-together letters, trying to discern a word containing the six letters they searched for, which could be configured in any number of ways, as *sail* might occur in *assail, sailor, sailed,* and so on. Once Blackburn and Greider found the sequence that could act as a template, it stood out as if boldfaced. The results of these 1988 experiments led Blackburn and Greider to propose as the template a 9-nucleotide sequence that ran from residue (position) 43 to 51 on the RNA strand:

5′ . . . CAACCCCAA . . . 3′

Since conventional practice dictates that DNA be notated so that it can be read from left to right (from the 5′ to the 3′ end), later they would habitually notate the sequence in the reverse direction, so that it could be diagrammed in relation to the DNA strand it complemented:

RNA   3′ . . .   AACCCCAAC . . .   5′
DNA   5′ . . .   TTGGGGTTG . . .   3′   →

This nucleotide sequence immediately suggested how the mechanism for adding on to telomeres worked, especially in light of precedent. It was already known that DNA polymerases, constructing a strand of DNA, required a specific prompt of from three to four nucleotides at the

3′ end of the strand in order to continue, and Blackburn and Greider naturally looked to precedent, assuming (correctly) that telomerase would work in a similar way. The RNA template aligns with the G-rich strand of DNA, three bases pairing with the last three bases at the 3′ end of the DNA strand, providing a "hook" that positions the template exactly for the next nucleotide to be inserted. The enzyme then adds the six-base sequence before it moves down along this new strand to begin over again (a process known as translocation). Figure 5.3 shows this

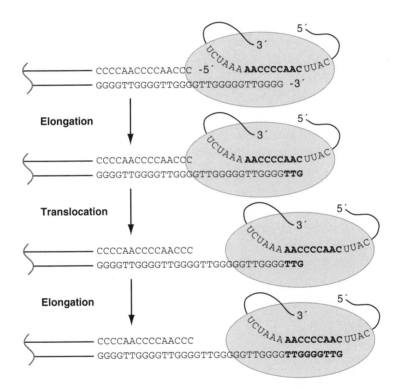

Figure 5.3
A model for the action of telomerase. The RNA template in telomerase provides primer recognition properties so that the enzyme can begin adding the correct nucleotide to the 3′ end of the DNA strand. Because the template contains more than one repeat sequence (in this case, AACCCC), it can align correctly in this recurring process. Reprinted by permission from Macmillan Publishers Ltd.: Carol W. Greider and Elizabeth H. Blackburn, "A Telomeric Sequence in the RNA of *Tetrahymena* Telomerase Required for Telomere Repeat Synthesis," *Nature* 337 (1989): 331–337, copyright 1989.

recurring process. The beauty of this template is that it contains more than one repeat sequence—enough extra to establish alignment as the enzyme moves along the strand of DNA to continue adding the repeat sequence. Telomerase can make tandem repeats—reposition and continue extending the DNA strand—because of these additional bases, and their existence in the template also explains how the enzyme knows where to line up along the 3′ end of the DNA strand, whatever its last three bases may be.[17]

In 1989, Greider and Blackburn proposed this model for how the template works. They extended the argument of the earlier papers by making a careful case for the uniqueness of telomerase, again referencing the self-splicing *Tetrahymena* intron ribozyme that Cech had studied for years. In contrast to other known DNA polymerases, "the *Tetrahymena* intron ribozyme and telomerase represented a new class of polymerase characterized by containing a template sequence that is an integral component of the polymerases." But like many RNPs, the *Tetrahymena* ribozyme used base pairing of its RNA components with other RNAs to determine reaction specificity, making telomerase "unique among eukaryotic RNPs in that it uses an RNA template for DNA synthesis." Fitting neatly within no single category, telomerase had some properties in common with polymerases, RNPs, *and* reverse transcriptases, though this last connection gets only a rather cagey nod at the end of the paper: "Telomerase enzymes with RNA components may be relics that have descended from an ancestral type of DNA polymerase whose action was based on an internal RNA template. In this model, the primitive DNA polymerase was replaced in evolution by the current protein-based DNA polymerases and reverse transcriptases which use exogenous templates, whereas telomerase retained the endogenous template."[18] Whether or not Greider and Blackburn were conscious of shaping a field in these three papers, they provided a broad base of evidence for their conclusions and articulated the questions that would follow from these findings. In the second and third papers, they carefully reiterated previously established evidence and filled in added detail, building out their argument "horizontally." Across the three papers, their findings are given still richer context by no less than eleven types of relevant experiments conducted in the Blackburn lab between 1981 and 1989.

This last paper would prove central to a decades-long effort to understand more completely the function of telomeres in the cell. By this time, telomeres had been sequenced in a wide variety of species; in their paper, Greider and Blackburn reported that telomerase activity had been identified in *Euplotes* by Shippen (citing as yet unpublished results[19]) and in *Oxytricha* by Alan Zahler and Prescott at the University of Colorado. In 1988, R. K. Moyzis and collaborators had identified the human telomeric sequence—a gratifyingly consistent pattern of AGGGTT repeats (resembling *Tetrahymena*'s GGGGTT repeats), and in 1989 Gregg Morin, a postdoc in Joan Steitz's lab at Yale, demonstrated that telomerase, as predicted, was also active in human cells. Using Greider and Blackburn's published reaction conditions, Morin found the exactly analogous telomerase activity in extracts of human cancer cells.

The importance of Blackburn and Greider's work was only fully established by this range of evidence that they had discovered a fundamentally conserved biological process. Even before publication of their third paper, Blackburn received an Eli Lilly Research Award in 1988, evidence that her discoveries were already recognized as significant. In what seems to be a characteristic understatement, Blackburn noted that this third paper "was readily accepted" by biologists and claimed she recalled little else about its reception. This verbal shrug is rooted in her scientific ethos, which places confidence in the work itself and regards any embellishment as self-aggrandizement, an intriguing reconfiguration of the notions of gentility and decorous modesty that characterized her upbringing.

In the wake of Greider and Blackburn's identification of the RNA template, Shippen and Blackburn confirmed the universality of the template's structure by identifying the nucleotide sequence of the template in the protozoan *Euplotes crassus*.[20] Telomeres in *Euplotes* have a C_4A_4 repeat sequence, and the RNA template contains this sequence plus seven additional bases, allowing for a longer alignment region, or hook. The entire sequence, with the alignment region boldfaced, thus reads **CCAAAAC** CCCAAAAAC. So where *Tetrahymena*'s template contains $1\frac{3}{6}$ repeats, the *Euplotes* template repeats $1\frac{7}{8}$ of the sequence; later, researchers discovered that for human telomeres, which have a six-nucleotide sequence, the templating domain sequence consists of $1\frac{5}{6}$ repeats.

Shippen had succeeded at identifying only the second RNA component of telomerase that had ever been isolated, providing crucial new
information; as she noted, "The second one tells you so much because
it shows you what's similar and different, how things have evolved. You
can compare the two to see what's important—what's fundamentally
conserved."[21] Blackburn gave Shippen credit for declaring at the outset
of her paper that telomerase was a reverse transcriptase: "I was joking
that I was so tired of starting off papers, 'Telomeres are,' and Dorothy
starts off this paper, 'Reverse transcriptases are.' Nice perspective
change." Like Shippen, Blackburn emphasized the critical importance of
this, or any, second finding:

Carol found the first evidence, and using the term *reverse transcriptase* didn't
change experimental thinking in the least. What it did change was the significance that could be claimed for both findings. You can extend the definition when
you see the second one. Dorothy said, isn't it interesting that an essential cellular mechanism is carried out by this unusual enzyme? Reverse transcriptase mechanisms were thought to be an oddity confined to viruses. By the nineties it was
known that some nonessential bits of the genome were copied by reverse transcriptases, usually in retroviruses that had come to live in the genome over evolution, but don't act like viruses anymore. Telomerase has in common with other
reverse transcriptases the ability to copy RNA into DNA, but the rest of its features differ. The definition has had to be broadened to include this specialized
reverse transcriptase.

Although this second finding by Shippen provided important confirmation, a culminating, formal proof still remained elusive. Results in the
test tube (in vitro) did not always bear out in living cells (in vivo). There
were unsettling precedents for such research results not panning out.
Years before, molecular biologist and biochemist Marian Grunberg-
Manago had conducted well-regarded research on the enzyme that
polymerizes RNA. Her candidate for this enzyme was convincingly
demonstrated in the test tube, but later studies on living cells showed
that the process worked differently and was carried out by another
enzyme, known as RNA polymerase.

Once Greider successfully cloned out the gene coding for telomerase,
researchers might conceivably exploit recombinant genetics to insert
altered copies of the gene into a living cell to test its properties. No one
had yet figured out how to engineer any gene in *Tetrahymena*; for

example, no method had been developed for inserting a plasmid vector, as Blackburn and Szostak had done in yeast. A technical process for inserting recombinant genes into *Tetrahymena* had yet to be worked out because the idiosyncrasies of each organism and the limits of the technology available at the time had stymied researchers. In a beautiful synchronicity, however, the other major strand of research in Blackburn's lab provided the key to solving this problem. How these two seemingly divergent strands of research cross-fertilized and informed each other exemplifies the syncretistic and opportunistic way in which science advances.

The other focus of research in the lab, lasting from the early 1980s to the mid-1990s, also grew out of Blackburn's work with rDNA minichromosomes in *Tetrahymena* (which code for the ribosomal RNA) and thus bore a seemingly tangential connection to her telomerase research. Blackburn collaborated with Orias at the University of California at Santa Barbara, taking advantage of a finding by Peter Bruns at Cornell University, who both collaborated and competed with Orias in the field of *Tetrahymena* genetics. Bruns had selected for *Tetrahymena* mutant cells resistant to antibiotics, and he suspected that the drug-resistance gene was located in the cell's rDNA minichromosomes. He approached Blackburn and asked her to sequence the entire gene. The collaboration was completed when Beth Spangler, a graduate student in Blackburn's lab, sequenced the rRNA gene in *Tetrahymena*'s rDNA minichromosomes, showing that Bruns's mutations lay in that gene and accounted for the drug-resistant mutants.

The study of ribosomal RNA, a highly specialized area of RNA research, explores how the genetic code is translated from RNA into the amino-acid sequences of proteins, a process involving the coordinated activity of mRNA and tRNA as well. In the factory of the ribosome, an elaborate process employs RNA as the actual catalytic enzyme (ribozyme), in coordination with proteins that help it to work accurately. Blackburn was especially interested in another advantage of the rDNA minichromosomes. They offered a unique system for understanding how chromosomes replicate themselves, because where most DNA is usually replicated in a strictly controlled way—completely and only once each time a cell divides—each of the approximately ten thousand rDNA minichromosomes in a nucleus replicates a *random* number of times per cell division until the total number of minichromosomes doubles.

Blackburn and Spangler visited the lab of Harry Noller at the University of California. Noller was a ribosomal RNA specialist who, in the days before much of this work could be computerized, attempted to create three-dimensional models of RNA based on sequencing information. The RNA molecule can fold over on itself to form helical structures, each like a hairpin loop, with complementary base pairs forming a bond (in contrast to DNA's partnership between bases on two separate strands). In those days, Noller drew diagrams in which he color-coded bases so he could look for sequences that mirrored each other and thus offered a clue to the molecule's elaborate three-dimensional structure. Noller had already identified how RNA folded in a few bacteria, but no one yet knew how the ribosomal RNA molecule folded in *Tetrahymena*. When Blackburn and Spangler brought their sequencing results to Noller, he had a new RNA sequence to play with, and they could learn from an expert the game of conjecturing how it might fold. Although this had no bearing on Blackburn's work with Orias, she couldn't resist the chance to check it out. The seeming digression illustrates an "outward rippling" movement of thought that counters the popular notion of life sciences proceeding by a series of focused investigations rather than a far more digressive, rambling, and often abruptly thwarted curiosity about how life works. Any idea or experimental finding, dropped like a stone into the sea of information about biological processes, might send out concentric ripples that at any point could overlap with another finding or hypothesis. Ultimately, this particular digression would illuminate Blackburn's later work on conserved structures in the RNA component of telomerase.

For his part, Orias considered their identification of the drug-resistance marker on the rDNA minichromosome a potential springboard for further investigations into the biological rules governing the genetic inheritance of this molecule. This drug-resistance marker now made it possible to select for other traits, through an elaborate, multiple-step process that would enable researchers to identify another crucial gene on the chromosome despite the impossibility of a more direct technical route. The researchers could chemically treat cells to force random mutations and then kill off all survivors except those carrying this drug-resistance marker, offering intriguing opportunities for comparing the mutant chromosomes to the original strain, or wild type. "Ed's genius was devising cunning genetic screens," Blackburn said. "In science, opportunism is critical. You plan research but you're always on the lookout for opportunities, because biology is so difficult to study—you're trying to breach a solid wall of complex, interrelated systems, and you try any little opening that you can. And we had to be scavengers in the days before technology advanced—save every little bit of string that might prove useful."

What proved useful was the fact that each nucleus contains two different allelic forms of rDNA minichromosomes, essentially two different versions of the same gene, analogous to a bag full of red and white marbles. During cell division, these two types of chromosomes randomly replicate until their number has doubled, essentially competing with each other. By accident, Wei-Jun Pan, a Chinese professor visiting in Blackburn's lab, had recently discovered that over several generations (cell divisions), the number of "white marbles" declines in relation to the number of "red marbles."

Because the drug-resistance marker occurs on the whites, it provided a tool for genetic selection. In Orias's lab, researchers chemically mutagenized the reds so there'd be random mutations and then mated these cells with cells containing white minichromosomes carrying the drug-resistance marker. (*Tetrahymena*'s capacity to reproduce asexually or by mating further aided this research.) Whites would win the replication competition in those cells in which reds had been disadvantageously mutated. When the culture was then treated with antibiotics, only the cells carrying minichromosomes with the drug-resistance marker could survive. That marker now became an indicator of mutation in the reds.

Blackburn's lab could study the mutant red rDNA minichromosomes to see what base changes they had undergone. If the mutations mapped on the rDNA, then which DNA bases had changed in comparison to the original strain? This would lead to identifying which conserved sequences on the minichromosome served as replication origins, because it was a good bet that this would be the key to their survival. The researchers already had a clue as to where to look for these replication origins, thanks to the work of cell biologists Gall and Truett, who had used electron microscopy to study rDNA minichromosomes in *Tetrahymena* cells during mitosis. Gall and Truett had made deductions and measurements based on replication bubbles they observed, and now molecular biologists could exploit this visual mapping of the molecule to pinpoint their target biochemically.

A postdoctoral fellow in Blackburn's lab, Drena Larsen, worked on this project to identify the DNA sequence critical for the function of the replication origin in the rDNA minichromosomes. Eventually, Blackburn's postdoc Jeff Kapler learned genetics from Orias so that after 1990, genes could be manipulated in Blackburn's lab. Few replication origins had been identified in eukaryotes, so just exploring this unknown territory was exciting. The "let's use it" mentality meant that once they had this knowledge, Blackburn and her collaborators would employ it again to explore further how replication is controlled. Since Challoner had already sequenced this region of the chromosome in several species, Larsen could superimpose her data over his to identify mutations in regions known to be conserved among several species and eventually map the replication origin sequence in *Tetrahymena*.[22] This experiment would further illuminate which DNA sequences on the rDNA minichromosome were conserved in different species of *Tetrahymena* and which were free to change, or drift, with evolution.

At just the right time, these findings crucially intersected with Blackburn's other major strand of research, telomerase. A graduate student in Blackburn's lab, Guo-Liang Yu, had been trying to devise a means to test the activity of telomerase in living cells, which would provide the formal proof of its existence. Once Larsen identified the replication origin sequence on the rDNA minichromosome, Blackburn and Yu seized the opportunity to use these results as a tool for their work on telomerase. Because the replication origin conferred the ability to replicate, they might now create a DNA fragment, ligate on to it a mutated telomerase RNA gene, grow it in a bacteria plasmid, and reinsert it into a *Tetrahyamena* cell. Blackburn's lab could therefore test for the activity of telomerase in a living cell and prove conclusively that telomerase contained an RNA template.

Yu was in the first wave of Chinese students who came to work in the United States, and in 1989, when protesters in Tiananmen Square were attacked by Chinese troops, everyone in the lab lived through that trauma with him. Because in China an afternoon nap (*shiu shi*) was a firmly entrenched custom, Yu managed the best he could in his new surroundings, putting his head down on his lab desk to rest every day after lunch. Developing new methods for transforming *Tetrahymena* using recombinant genes demanded creativity. Yu had to invent a whole new system for developing a plasmid vector that could carry a gene into the *Tetrahymena* cells and replicate itself. To achieve this was no slam dunk; the idiosyncrasies of *Euplotes crassus*, the organism Shippen was studying, have so far foiled all attempts to devise a method for inserting recombinant genes into the nucleus of the cell. Using an exquisitely fine, long glass needle, Yu employed a painstaking microinjection technique, recently developed by Meng-Chao Yao in Seattle, that was especially difficult to perform on *Tetrahymena*, which swam constantly and had to be immobilized in a Jell-O-like bath for the procedure to work. In order to have even a chance of success, the procedure also had to be undertaken at a time when cells were actively dividing, and in advance of more efficient methods, cells had to multiply much longer before it became possible to detect signs of change. Yu knew he'd succeeded when he looked at the cells after several generations and found the plasmid had made thousands of copies.

Now Blackburn and her colleagues could manipulate the telomerase RNA gene, recently cloned by Greider, and see how it behaved in a living cell. If mutations in the template were copied into telomeric DNA in the cell, then the enzyme worked the way Blackburn and Greider had predicted. To mutate the gene, Blackburn's lab partners ordered up synthetic DNA oligonucleotides (primers) that corresponded to the patch of gene, making small changes to the sequence, such as changing a single C base in the sequence to T. Each type of plasmid vector, containing a mutant sequence with just one base change to the template, should produce a corresponding change in the telomeric sequence synthesized in a living cell. In this experiment, wild-type cells were mixed with mutant types; if the model held true, naturally occurring telomeric GGGGTT repeats would be randomly mixed in with mutant variations.

These experiments, conducted in 1989 by Yu, with help from Laura Attardi and John Bradley in Blackburn's lab, met with success. (Later, Blackburn would remark that they had been "incredibly lucky" to get results, since some of the mutations they chose at random make it difficult for the enzyme to work at all.) The wild-type genes produced the familiar sequence G_4T_2, but a mutant gene carrying an extra A produced the sequence G_4T_3, and a mutant gene carrying an extra C produced the sequence G_5T_2.

Serendipitously, a change the researchers made to a particular C base in the sequence AACCCC (making it AACTCC instead) unexpectedly crippled the activity of the enzyme. It was as if all telomerase had been deleted from the cell. Over several generations, the telomeres in the *Tetrahymena* cells grew shorter and shorter until the cells ceased to multiply. Blackburn and her colleagues had conclusively demonstrated that telomerase acted as a reverse transcriptase enzyme in living cells, with the astounding bonus that when they altered C at residue (position) 48, the ceaselessly replicating *Tetrahymena* cells suddenly became mortal.[23] This experiment did not merely provide the confirming formal proof for the existence of telomerase but also made it clear that the enzyme's functioning was critical to the survival of cells.

6

Gold Rush

Up until now, Blackburn had managed both to succeed extravagantly and to avoid the more competitive and crowded fields of molecular biology. Her career so far had been marked by this paradox of her nature: a fiercely competitive drive to succeed and an equally ingrained instinct to avoid head-on confrontation. Even while still in high school, she battled to outperform another high-achieving girl (who eventually became an engineer), yet she did not acknowledge her competitive feelings until decades later. The thirst for glory in the sciences was nowhere more intense than at the MRC, where members of Sanger's lab and other labs competed to be the first to publish a DNA sequence—any sequence, so long as you got it first—yet Blackburn kept her distance from the scramble. When she worked in Gall's lab, she chose to pursue research in a related but clearly demarcated area (thanks to her unique training in Sanger's lab), which mitigated against direct competition with other lab members. During her twelve years at the University of California at Berkeley, telomere research was in its nascent stages, conducted by a small group of aficionados, and in this wide-open frontier, one could easily stake a claim in uncharted territory.

Like the early speculators in the California gold rush, biologists converged on telomere research in the early 1990s, after Blackburn and

Greider's findings on telomerase were confirmed in other species, including humans. Telomere research would soon become a thriving, highly specialized field of molecular biology—one that drew the rapacious interest of biotech companies as well. A sudden surge of interest in the same questions seems to be the way science operates—suddenly everyone leaps onto the bandwagon—yet Blackburn continued to try to put psychological distance between herself and the crowd: "I've never felt comfortable with this aspect of competition: the race. So I've always looked for a way to move ahead and stake our own turf as a lab, a way to move away from the pack."[1]

Blackburn consistently portrays herself as an outsider, operating on the fringes—the quiet colonial girl among the elite at the MRC, the person who crossed her fingers and hoped Gall could find room for her in his prestigious lab. This perception persisted despite her changing circumstances, another means of putting distance between one's self and the crowd. She had quickly won recognition for her work on telomeres and telomerase, receiving the 1988 Eli Lilly Research Award for Microbiology and Immunology from the American Society for Microbiology. This prize, sponsored by Eli Lilly Pharmaceuticals and reserved for scientists under age forty-five, was awarded at a banquet on a Sunday, the first night of the society's annual conference. More anxious to limit her time away from her toddler son than to bask in the limelight, Blackburn made plans to fly to Miami on the day of the ceremony. When her flight was canceled, she had to fly out on a different airline, competing for a seat with other displaced passengers desperate to make cruises that departed from Miami. Knowing she had a slim chance of arriving in time, Blackburn changed into her formal dress in the airplane's bathroom—a hasty preparation for her public appearance that seems emblematic of her diffidence. She barely made it to the awards ceremony, held at the Fontainebleau Hotel, and though she was too harried to enjoy the accolades, she ruefully noted that embossed on the back of the heavy, round bronze medal she'd received was an image of a man in a long lab coat. The next day, she gave a lecture at the conference, a standard obligation for award recipients. She just had time to enjoy the luxurious, beachfront hotel before her flight home that evening, and on impulse, she bought a swimsuit in the lobby shop. As she sat by the pool, enjoying a drink from the poolside bar, her illustrious fellow biologist Sid Altman wandered by in a dark suit, carrying a briefcase. Oblivious to his surroundings, he promptly sat down to chat about science with Blackburn.

Blackburn described the Eli Lilly Award as "a nice pat on the back"; given the low-key reception for the paper in which she and Greider first identified telomerase, the award assuaged both ambition and self-doubt: "I guess *somebody* thought it was important." Only two years later, Blackburn received the prestigious National Academy of Sciences Award for Molecular Biology. The same tournament mentality that emphasizes being the first to publish results also influences the granting of prizes in the sciences, typically judged by committees of scientists, including former recipients. This affirmation of status generates still more buzz, consequently increasing a scientist's chances of further awards.

Despite the gratification and prestige they bestow, prizes are also problematic. Often, award rules stipulate that only one person may be recognized, perpetuating the image of the lone scientist hero when so much of the research in molecular biology, as in other fields, is as delicately interconnected as the components of an ecosystem. Collaborators may learn from one another and part ways, competitors may spur each other on in ways that enhance their respective research, and in the best case, researchers eagerly share ideas, results, and techniques. Biological research today is also moving toward big-team science—as, for example, in the massive number of collaborators on the human genome project—a reality at odds with the emphasis on individual achievement. Two competitors, approaching the same problem differently, may both arrive at solutions more or less simultaneously, so that it is difficult, if not impossible, to award one of them first place and the other second. As a side effect of the tournament mentality, "junior" collaborators (postdocs or grad students) may fail to receive equal recognition for their role. Conscious that her work on telomerase was a collaboration between equal partners, whenever Blackburn is honored for her work on telomerase, she raises the question of whether the award should be shared with Greider and often insists on the point.

Blackburn was surprised at the relief she felt on receiving the second major award from the National Academy of Sciences, a clear signal that she no longer operated on the fringes of the scientific community. She had spent most of her career cultivating a disinterest in the trappings of success that was seemingly at odds with her ambition and with the vulnerability she often felt, so vividly expressed by her willingness to retreat during the stress of her first job search and by her "nervousness" when she and Greider first laid claim to the discovery of a novel enzyme. Recognition complicated her divided feelings about competition. It disrupted her reliable protective strategy of pretending she was not running neck and neck with the competition, and it troubled the clarity of her sharp distinction between personal and professional ego. In both cases, recognition impinged on her idealism about the work itself even as it rewarded her on the only terms that mattered to her.

Blackburn is intellectually tenacious ("very hard to argue with," to borrow Greider's words), yet she imparts anything but a fierce impression. Even her physical appearance is cheery and reassuring: she's apple-cheeked, with softly waved brown hair just barely tidied, and nearly always wears sensible shoes. A visitor to her lab might easily mistake this cozy sort of person for a fellow visitor, perhaps a teacher leading a field trip, chatting in her flutelike voice about the amazing goings-on in the lab. Playfulness, along with a facility for translating scientific findings into colloquial terms and familiar metaphors, enlivens her scientific reviews: "Telomeres: No End in Sight," "The End of the (DNA) Line," and "Telomerase: Dr. Jekyll or Mr. Hyde?" In talking about her work, Blackburn so consistently eschews anything that smacks of self-aggrandizement that it comes as a relief to catch a glimmer of ego in her humor; she refers to the researchers who flooded the telomere field in the 1990s as "latecomers," and her speech is laced with comically violent metaphors that translate aggression into harmless caricature.

Soon after Blackburn received the Eli Lilly Award, she made a significant decision about the future course of her career. The research in her lab was flourishing—the experiment conducted by Guo-Liang Yu had just turned up the fascinating, unanticipated phenomena of cells "going mortal"—and she could contemplate change from a strong position. For some time, she had not been completely satisfied at Berkeley. When she was pregnant with Ben, she had been asked by the department chair if her lab would move its microinjection setup (crucial to Yu's research) to another room to accommodate a new faculty member. Even though Blackburn recognized objectively that this imposed a reasonable collegial obligation, she was upset by this potential upheaval in a central area of the lab's research. "I felt like a mother tiger defending its young," she said, using a metaphor that was not coincidental. After becoming pregnant, she had discovered a wellspring of strong emotion that belied her public persona as someone who didn't look up from her research to bother with the peripheral matter of her actual surroundings. "Before Ben was born," she noted, "all my passions went into my work." She felt she had no choice but to accede to this workable, if unsettling, proposal: "I didn't realize you needed to negotiate to make it better." The practice that had guided her entire career—"just keep your head down and do good work"—could not ensure the continuity of her research.

Despite feeling that her work was valued at Berkeley, by 1988 Blackburn believed she didn't have the power or skill to insist on certain conditions. The planned move of her lab and department to another building on the Berkeley campus meant she would soon have to fight for a workable new setup, and the university was also reconfiguring its life sciences departments, so she would face change at the administrative level as well. The fate of the graduate students and postdoctoral fellows in her lab depended on her successful navigation of these issues, just as their fate would be tied to hers if she moved, since lab personnel usually accompanied the primary investigator to a new situation.

She applied for a position at Stanford University and also inquired at UCSF, where Sedat worked. Since the 1960s, the UCSF medical school had strongly recruited faculty in the life sciences; because molecular processes often held constant across species, research on a variety of organisms was conducted with a view to gaining greater understanding of human cells, with consequent implications for clinical research. J. Michael Bishop, then a renowned researcher at UCSF, sent a graceful and supportive letter in response to Blackburn's inquiry. Though campuses within the University of California system were not allowed to poach each other, administrators could hire faculty who had indicated a desire to move. Because Blackburn had also approached Stanford, UCSF had important leverage in bargaining for her: if they could not woo her, then the University of California system would lose her to Stanford. She had already interviewed with the Department of Biochemistry at Stanford and given a seminar there, and gossip had gotten back to her that Stanford planned to offer her the job.

On hearing that she might leave the university, a senior faculty member at Berkeley discussed with Blackburn the possibility of Sedat joining the faculty. Her colleague correctly understood that Blackburn wanted to avoid the eighteen-mile commute across the Bay Bridge, especially because of her young son, and proposed relocating her household to the East Bay as a possible alternative. But Blackburn had other doubts about her future at Berkeley. Committee work took up more and more of her time; she could "see the phenotype unrolling," and she wasn't sure she liked it. She wanted to devote her time to the excitement of research, and though she had come to enjoy teaching undergraduate classes, her teaching load and administrative tasks constituted a distraction from her primary focus.

Stanford and UCSF both made offers to Blackburn, and for the first time in her life, she could negotiate on her own terms as the two institutions vied for her. UCSF won out because it offered her a higher salary and the campus was just minutes from her home, while Palo Alto was a long drive south from San Francisco. At UCSF she would have fewer, less demanding teaching duties than at Berkeley, primarily delivering lectures in courses for dental and medical students ("science lite" by Blackburn's standards).

Blackburn described this recruitment period as "a time when all the stars are aligned," creating the best possible opportunity to negotiate for space and lab resources as well as salary. Offered a lab space that wasn't large enough, Blackburn turned it down, but she agreed to occupy a temporary lab while the university converted another space tailored to her research requirements. She had never really tested the Berkeley system, and "it was refreshing" when UCSF readily met her demand for an air-conditioned room for microinjections and agreed to provide window ledges where Shippen could situate the large water bottles in which she cultivated algae to feed *Euplotes crassus*. Fully aware that egregious demands could create resentment before a new faculty member took up her post, Blackburn facetiously asked if a wave might be painted on the copper roof that obscured the view from her office window.

By the time Blackburn arrived at UCSF in July 1990, the medical school had become a premier research institution, supporting the kind of pure research—curiosity driven—that Blackburn loved to do. UCSF had much of the momentum, scope, and scale in biological research as a program at Berkeley, with the distinct advantage that clinical activities produced healthy revenue to support research facilities. In addition, the NIH tended to favor medical schools and clinical research in its funding process. From the early 1970s to the 1990s, basic research like Blackburn's produced a watershed of new information that would ultimately swing the pendulum back from support for research on a variety of organisms to more focused studies of human health and disease. Rapid advances in knowledge, in tandem with advances in technology, now made it possible to study human cells with the necessary rigor, asking truly substantial intellectual questions, and a corresponding shift in the understanding of how to approach such questions also created new opportunities. The predilection for an elegant solution to problems in molecular biology was giving way to a renewed interest in the complexity of biological systems that required integrated approaches.

Increasingly, translational research bridged the gap between basic science and clinical research. The explosion of knowledge in molecular biology made partnerships between basic science researchers and clinical researchers advantageous, if not absolutely essential, since each could contribute highly specialized expertise. Yet this change also incurred hazards. Dave Gilley, a former postdoc in Blackburn's lab who today collaborates with clinical researchers studying human cancer cells, noted that it has become harder to get funding for research on simpler organisms: "People working on some of the classic model organisms like yeast have trouble getting money compared to people working on mammalian systems. As a junior person especially, you have to work on a mammalian system to survive. Human research is incredibly interesting, but some incredible things we've learned from other model systems get lost, even though they might be relevant to human health and human systems. A scientist who can't get funded can't keep the lights on in the lab."[2]

In this changing funding climate, moving to a medical school offered Blackburn an entrée into working with human cells in culture and conducting research that intersected with clinical studies. In preparation for the move, she had hired a lab technician, Cathy Strahl, who had experience working with human cells in culture. As a member of UCSF's Department of Microbiology and Immunology, Blackburn could now take advantage of an environment where her colleagues had extensive expertise in this field and plentiful resources were available. In labs just down the hall from hers, researchers were studying leukemia and lymphoma cells in culture, and almost immediately Blackburn began working with human cancer cells. Blackburn's shift to working with cancer cells also corresponded with the lab's overall shift to a focus on the processes in living organisms.

Blackburn's lab at UCSF was a warren of rooms, the largest being the lab. Blackburn also had a separate office and rooms dedicated to performing microscopy, handling radioactive chemicals, and storing fragile cell extracts at cooler temperatures. Arranged in closely packed rows, or bays, the benches in the lab were Formica counters of the height and depth of a typical kitchen counter, complete with a sink. Shelves above the benches stored frequently used chemical solutions, and shelves below held supplies such as test tubes and pipettes. Ceiling-high shelves above the desks were crammed with archived lab notebooks and supplies purchased in bulk, so that this cramped space verged on the claustrophobic. Each person in the lab was assigned a bench and a desk, and desks were usually as cluttered and off-limits to intruders as a teenager's bedroom. No one dared to touch anything on someone else's desk or bench without explicit permission. Limited space at the benches precluded any personal touches, but desks were decorated with photos or cartoons, and photos of group outings and baby pictures adorned the lab's refrigerator. Socially, the lab was a casual workplace—there were "office" romances, and lab members occasionally brought their babies for a visit.

Blackburn deliberately kept her lab at UCSF the same size as her lab at Berkeley: about a dozen people, including a lab technician with experimental expertise, a lab assistant to maintain equipment, a technician with experimental expertise, graduate students, and postdoctoral fellows. This size fit the cultural norm at UCSF, and "cottage industry" science suited Blackburn because it allowed for close interaction among lab members. The apprenticeship system, in which graduate students and postdoctoral fellows often perform drudgery in the service of larger intellectual aims, rests on the premise that for this long service they will one day be rewarded with research positions of their own. But by the early 1990s, the NIH had contracted, and the exponential growth in the field of biology from the 1950s to the 1970s had slowed, so that a highly qualified field of candidates competed for fewer jobs. Where postdoctoral fellowships had typically lasted two to three years, now these same people might work four to five years as fellows in a lab before they found academic or research positions, a shift that reflected economics as well as the requirement for increasingly specialized knowledge.

In the past, those postdocs who went on to teaching posts at small liberal arts colleges rather than research institutions might have been regarded as less successful, but now at least some primary investigators strove to provide graduate students and postdocs with opportunities to teach undergraduate courses. Shrinking public funds directly affected the future chances of these young scientists even if they did manage to land a job at a research institution, and though the NIH budget would double between 1998 and 2003, a young principal investigator continued to face steep odds when applying for a first research grant. Blackburn recalled that when she started out as an assistant professor at Berkeley, funding was comparatively plentiful, but "today an assistant professor has to fight every inch of the way to get her first grant, and the grant will likely go through one or two revisions before it is funded." This problem has become so severe that in December 2005, the NIH announced a pilot project aimed at increasing the chances for first-time grant applicants to succeed, including awarding research grants that would induce prospective employers to hire these applicants.[3]

As Blackburn observed, "It takes courage and persistence to succeed as a researcher, and you have no guarantee of success even if you do everything you're advised to do." Over time, she had found her feet as a mentor, despite the lack of any training for this role. Early in her career at Berkeley, she had felt hampered by a lack of experience and came to doubt the "tough approach" that characterized the academic sciences, which treated apprenticeship in a lab as a trial by fire:

The nature of science is to be critical. So is the culture. Scientists don't give strokes. It's rare for people to realize, as Joe Gall did, that members of their lab need praise, though the culture has begun to change. One of my first students at Berkeley was floundering, but she was attempting difficult work, and I hadn't given her praise for this. Flying home from a conference, we began talking. I started praising what she had done, which I had never done before. Suddenly, we could communicate a lot better. This was a big lesson for me: praise and appreciation really matter. It's important to appreciate that students have done a lot of work, even when it doesn't pan out with glamorous results. Luck plays a role in success, and sometimes the nature of a project means you won't meet with easy success.

Where some labs would not accept those who didn't obtain fellowship funding on their own, Blackburn found it hard to plan her research in light of these contingencies. She hired postdoctoral fellows without such a precondition, providing initial funding from her own grants, presuming they would succeed in obtaining their own grants on the second try, if not the first. This policy reflected her growing confidence in the reputation of her lab. She also developed an atmosphere of collaboration in her lab that would ensure a continuity of expertise, passed on from one "generation" of lab members to the next, despite the constant turnover in personnel. For example, one postdoc after another had passed along hard-won skill in devising successful reactions for yeast telomerase activity in the test tube. In balance with such efforts at continuity, Blackburn believed turnover in personnel was essentially healthy for the lab: "You want turnover because people need to move on with their careers. Also, a big resource of the lab is brains, and you don't want things to become static."

Though the size of her lab remained the same, Blackburn's higher profile and the widening scope of her research engaged her in an increasingly sophisticated managerial role. In a *Science* editorial calling for more thorough training of postdocs, Cech and Enriqueta Bond noted that each primary investigator "runs something rather like a small business, and they do it with little or no training in what it takes to run a business."[4] A 2005 survey of 6,083 university researchers found that they spent 42 percent of their time on administrative tasks, including filling out grant progress reports and managing finances.[5] An established primary investigator must function as an intellectual manager as well as a business manager. Primarily, the primary investigator nurtures talent, providing postdocs and graduate students with intellectual capital for their experiments. But the job also demands skill at balancing a ledger, savvy at negotiating the politics of funding and publication, and a taste for civic-minded volunteerism that consumes what little free time remains. Each of these often conflicting responsibilities has to be synthesized with the others; mentoring must somehow translate to the only hard evidence that counts, publication in reputable peer-reviewed journals, and to ensure future funding, the lab must churn out that product with extraordinary efficiency.

Peer review, an essential mechanism for grant funding and publication in the sciences, is not immune to conflicts of interest and political jockeying, but it is constructed so that it tends to sustain equilibrium. The NIH, with a budget in the $20 billion range annually, provides grants to independent research and also to its own "intramural" research institutions. At the NIH, 150 standing peer-review panels, or study sections, review applications in closely related research areas. Like many midcareer scientists, Blackburn served on NIH panels, and such voluntary service constitutes an expected civic duty for a scientist.

Typical grant applications to the NIH, the product of exhaustive effort, include a description of the proposed research, a detailed budget, a detailed research plan indicating the aims, background, and significance of the research, preliminary studies (or a progress report for a continuing grant), research design and methods, and more. There are three review cycles, with each round requiring about nine months to complete. For grants submitted in October, for example, the study section meets the following February, which allows time for three panel members to read each grant application and offer a critique, written anonymously. After discussion, the panel assigns a priority score and a percentile ranking to each grant to rate its relative value and determine if it falls within the funding range. The study section must recommend which grants to fund by a triage process, since usually only one-quarter (or less) of all grant applications receive funding, and the staff of an NIH institute or an advisory body then reviews these recommendations for relevance to the agency's objectives and funding availability. By the time an applicant learns a grant has been rejected, the deadline for the next funding cycle has already passed, so that revising and resubmitting a grant application may take as long as fifteen months. The 2005 NIH pilot program, aimed at increasing the success rate of young investigators, attempts to improve this turnaround time and may one day be expanded for all applicants.[6]

For first-time grant recipients, the success ratio is one in five, compared to a ratio of one in three for all applicants.[7] (Since 2003, however, budget restraints have lowered the funding rate for grants to below 19 percent.)[8] Favorable reviews of grant applications are based on the quality of the science, but the chances of approval increase if the primary investigator has visibility in the field, so once again reputation perpetuates itself. Conversely, members of a study section often compete directly with those whose grants they rank—a potentially strong conflict of interest. Many NIH rules mitigate against favoritism or prejudice; when an application by a committee member's colleague or former postdoc is reviewed, that committee member leaves the room. But a further, effective pressure reinforces impartiality. If one's competitor has applied for a grant, science is so interconnected that other committee members recognize a potential source of bias, whether or not it is intentional.

Because the majority of grant applications cannot be funded, subtle pressure is exerted on committee members to seek faults that can help to winnow the field, and in this atmosphere proposals that range widely or seem unusually ambitious can be dismissed as fishing expeditions. Grant applications must present strong scientific evidence for the success of proposed research, and they are scrupulously analyzed for airtight logic and linearly derived experiments based on a clear and specific hypothesis, building on prior results. Having done service on a study section, Blackburn perceived a significant flaw in this system: "The process favors grants that cannot be faulted—research that seems guaranteed to succeed. The exciting papers you publish aren't risk-averse, or the work wouldn't have been done." Yet she quickly added that a researcher can finesse this liability: "Countering this tendency to support more cautious research is the fact that the grant itself is not a contract." While a grant may need to be worded adroitly so that it fits within semantic parameters, the research lab can pursue questions beyond the concrete, careful aims laid out in the grant itself.

By the early 1990s, Blackburn had established a successful track record at obtaining grants from the NIH. Consistent with her approach to experiments, she tackles grant writing with a ferocious thoroughness that betrays self-doubt: "The desire to overdo it and prove myself has never gone away, and in grant applications it still takes the form of my trying at first to include every type of evidence of our productivity and prowess as a lab and of demonstrating my ability to conceive of great research to be undertaken." Since the late 1980s, her annual budget, drawing on a number of grants, has climbed within range of $750,000. Only once was a grant application denied. Since the application provided for a line of research tangential to her main focus, Blackburn did not pursue the option of submitting a revised grant in the next funding cycle and discontinued the work, knowing that a postdoc who was about to move on from the lab planned to continue the research elsewhere.

Later in her career, however, a grant application she'd prepared for a new line of research was assigned a percentile score of 22.10 by a study section, which meant the grant was unlikely to be funded. By this time, her grant applications to the NIH consistently scored as low as the .2 percentile, so she was shocked by this ranking and further rankled when she read the peer review itself. She well understood the implicit code in these evaluations; grants were typically rated as outstanding, excellent, and very good, but to be deemed good spelled disaster. The peer review claimed "a lack of preliminary data demonstrating feasibility," which Blackburn regarded as a formulaic and inaccurate criticism, and commented on the "extensive proposal," which she translated as "the proposed research has a broader aim than desirable."

Rather than submit a revised grant application, Blackburn took the rare step of appealing to a program officer at the National Cancer Institute. At this point in her career, Blackburn did not hesitate to presume. She felt her established credentials in the field warranted a chance to rebut the review rather than to accept defeat and submit a revised grant proposal, delaying funding for at least a year. And she had recently picked up a valuable lesson while at an NIH-sponsored research retreat for intramural programs. Blackburn had attended a workshop in which the director of an intramural NIH institute pointed out that when a male scientist at the institute fell out of the funding range, he'd collar the director and argue his case. In the director's experience, women did not take this step, with consequences for their careers as well as success in getting funding. Blackburn's rebuttal to the peer review, limited to five pages, guardedly enumerated oversights and errors in the review, in particular focusing on extensive preliminary results that supported the feasibility of this experiment. She succeeded in her appeal, though the project was eventually funded for less money than she had sought.

As part of the cycle in which publication leads to funding leads to publication, the peer-review process for scientific journals shares some of the strengths and weaknesses of the grant-funding process. Most research scientists review submitted papers for journals in their field, evaluating the validity of experimental methods and conclusions. Established scientists also serve on editorial boards of journals, as Blackburn has frequently done, and together these voluntary tasks comprise a time-consuming but critically important responsibility for members of the scientific community.

Once again, researchers who compete directly with one another may be in a position to evaluate one another's work, and in this case, with the prize of publication—and prompt publication at that—at stake. Reviewers are asked to recuse themselves from reviewing papers by their fiercest competitors, but often vested interest, masquerading as curiosity, gets the better of reviewers, making the ethical territory more ambiguous. This confidential process, which provides for anonymous reviews, enables reviewers to be completely frank yet offers room to be unfair without being held accountable. The process can lead to paranoia; Blackburn reported that authors sometimes call editors to declare, "I know my enemy so-and-so reviewed this paper." In many cases, the editor knows (but does not divulge) that this is the case. Yet Blackburn also spoke of this as a matter corrected by "self-policing. If you always reject papers by competitors in your field, you are less likely to be asked to review again. The fact of the matter is, papers are improved by this process. More minds reflect on the problem, the results, and their implications." As a further counterweight to biased reviews, an editor, who has the final say on publication, can request additional reviews if he or she questions the initial ones. After papers have been reviewed, they might be rejected, or else the author might be asked to conduct further experiments or revise the paper before publication. It is so unusual to have a paper accepted without any requests for revision that Blackburn and her colleagues were once caught short when a paper based on Spangler's work on rDNA minichromosomes was instantly accepted and published with a typo that had inadvertently been incorporated into a genetic sequence described in the paper.

A single scientific paper may constitute the outcome of several years of experimental investigations, and the importance of getting primary credit is so inflated in the scientific community that the order in which authors' names are listed is sometimes hotly contested. When a primary investigator is working with a junior researcher, the battle over a prime position is not a fair fight. In this instance too, Blackburn could find cause for optimism, arguing that science cannot afford an emphasis on getting there first as a sole priority:

> Everyone understands this is a tournament—who gets published first. On the other hand, you are trying to train people in your lab in the best practices in science. Though it matters—up to a point—if you publish first, the good paper is the one that will prevail. It will be cited more often, another criterion of the success of a paper, though this can sometimes be a function of herd mentality. So there's tension between getting a paper out first and wanting to do the best work. There is never a time when it's perfect to publish. Everything you do raises more questions. So there's a judgment call—be complete, be sure the paper can withstand scrutiny by expert reviewers who are also your competitors.

In private written comments, though, Blackburn revealed that beneath this stance of cheerful and constructive objectivity, she has always felt devastated by the experience of having a paper rejected or faulted:

> The celebratory moment of completing a paper and submitting it is quickly followed by suspense and insecurity. Who knows what knives are out there, under the cloak of anonymity? When the reviews come back, there is first the decision letter from the editor to read, and if the paper has been sent to a high-profile, competitive journal, it is rarely a straightforward, enthusiastic acceptance letter. Usually it is a qualified letter, suggesting that more could be done. Sometimes it is an outright rejection. This is accompanied by a knot of tension in the stomach and a feeling of instant depression. This is always very personal—our hearts and souls go into this work. Days and nights are consumed, other things in life are neglected, and our decision to do all this has been predicated on the premise that this work was worth giving up all this for, and indeed worth doing in its own right.
>
> When the editor has sent a letter of rejection, the next step, reading the reviews, is even more painful and fraught. First there comes the quick scan, with occasional recognition of the falsity or accuracy of a comment. The review may be objective about the science and the suitability of the paper for the journal, though

human nature being what it is, I think it is inevitably colored by positive or negative inclinations on the part of the reviewer. Then comes the attempt to sum up the situation. Hopeful and worth rebutting, or not?

At that point I stop. The emotional toll would be too high at this early stage. I need to assimilate it, vent a bit to whoever will listen, and also feel rising in me the will to fight, because whatever we decide to do, we are now fighting for the paper. I have never published a bad paper—meaning an untrue paper or a paper that had sloppy science in it. I have probably published some overly dense papers. But usually I can be proud of this research, and proud of the lab members who have done all this work. Yet still, after all these years, I suffer over bad reviews. I am also assailed by a sense of unfairness and a sense that someone is deliberately being mean. This feeling contrasts so strongly now with the success I have come to experience.

There is no question that we will give up on the paper. The work has been done, it is good, lab people's careers and all their effort must not be put in jeopardy. In theory, I am the least vulnerable lab member, so I have to reassure co-authors that the work is good and it will be published in a respectable journal sooner or later. But there is work to be done. I carefully reread the editor's letter and the reviews to see if critical comments might be valid, or if they result from misreading or a wish to nitpick, or if we need to revise the paper to state the concepts more clearly. If a reviewer did not understand or like the paper, others will likely have the same reaction. Much of this process remains the same as when I was a more junior researcher. But then I tended to blame myself more for the shortcomings of the paper.[9]

Today, in an atmosphere of increasing competition for limited resources, prestigious journals reject a higher proportion of good papers, and reviewers are more inclined to list all the reasons why a paper shouldn't be published. Blackburn reported an incident in which a reviewer congratulated a colleague on a forthcoming publication he'd reviewed, only to discover the paper had been rejected by the journal. The editor had excised portions of the review that would justify publication of the paper, sending the author only the negative comments. The objectivity of peer-reviewed journals can also be undercut by the impact factor, a privately compiled citation index that ranks journals according to the average number of citations per article. Journals have been known to fudge their numbers and even to lean on authors to add to their papers citations from the journal to which they're submitting. Critics of "impact-factor fever" suggest that it can skew the course of science, with researchers tailoring their investigations to the kind of questions most likely to be of interest to high-impact journals, particularly *Nature*, *Cell*, and *Science*.[10] Partly in reaction to the politics, in the 1990s

J. Michael Bishop and others formed the *Journal of Molecular Biology of the Cell*, and Blackburn served as an associate editor.

Blackburn revealed that she struggles to sustain an idealistic stance toward her profession, a sense that competing forces ultimately balance out: "We all have the same information, but each person can perceive it in her own way. I would rather be upbeat and positive and see the good. I'd be miserable if my life were dominated by uneasiness about the politics involved, and I refuse to let that happen." Consistently, Blackburn minimizes any intimation that a research career may be susceptible to the vicissitudes of politics, but this is by no means a consequence of timidity. In an interview at her home, when asked about the impact factor, she called to her husband in the next room to ask if he knew what it was. "No," came the reply. "What's an impact factor?" Blackburn said, "I don't even really know what the impact factor is either. I hate it. Real scientists don't pay attention to this crap. It's the bottom-feeders who talk about the impact factor."

Up until this point, when Blackburn gave talks on her research at conferences, she was usually slated for a late session of miscellany. But other scientists had begun to register that telomere research promised enormous rewards and held great medical potential. Both Cech and Altman, who had worked on the catalytic properties of RNA, had a keen interest in a new enzyme that incorporated an RNA template. Cech now did extensive telomere research, studying the ciliate *Oxytricha*, and Zakian studied telomeres in the same ciliate and yeast, and their work confirmed that Blackburn's research results held true for different species. Paralleling the discovery of cell senescence in Blackburn's lab, Vicki Lundblad, while a postdoctoral fellow in Szostak's lab, had discovered a similar phenomenon in yeast: a mutation that shortened the telomeres of yeast cells resulted in shorter and shorter telomeres with every cell division (a phenomenon that came to be known as "ever shorter telomeres," or EST), precipitating the death of the cell line. In 1989, Lundblad and Szostak proposed that this process in yeast resulted from mutations to a gene, named *EST1*, speculating that the gene encoded a protein component or regulator of telomerase.[11] (Many years later, it was shown to be the latter.)

Taken together, these experimental results suggested that in many organisms the attrition of telomeres contributed to cell senescence, in which cells cease to multiply and then gradually die off. In other words, this evidence showed that telomeres and telomerase ensure the cell's capacity to replicate normally, even though in many cell types, telomerase exists only in limited amounts. As telomeres wear down with each cell division, telomerase valiantly tries to repair the damage and replenish the sequence, but in most cases the race is ultimately lost, with more material being eroded than the enzyme can supply. This makes telomeres still more vulnerable, because only when an adequate amount of repeat sequences is present can telomeres bind the special telomeric proteins that help protect the chromosome end. In an analogy popular with researchers in the field, telomeres, coated by these special proteins, function like the plastic cap on a shoelace, keeping the precious DNA code from "fraying" with every cell division.[12]

Interest in telomeres and telomerase intensified as findings on human cells confirmed earlier observations on single-celled organisms, but with qualifications. Unlike single-celled organisms, which can divide indefinitely, normal human cells in culture usually divide a certain number of times and then hit a preset limit, known as the Hayflick limit (for its discoverer, Leonard Hayflick), at which they stop reproducing and die off. In 1986, Howard Cooke and his collaborators had reported that telomeres were shorter in adult human somatic cells than in reproductive cells (eggs and sperms), with the suggestion that telomeres might shorten as a human ages.[13] In 1990, a year after Gregg Morin had isolated the activity of telomerase in a human cancer-cell line, Greider collaborated with Cal Harley, who had shown her an obscure study published in the 1970s by A. M. Olovnikov, a Soviet scientist. Olovnikov had proposed that the Hayflick limit was reached when cells stopped dividing because their chromosomes grew too short.[14] Studying human somatic cells, Greider and Harley tested this hypothesis and observed that telomeres grew shorter with each cell division.[15]

Armed with this knowledge, researchers could speculate that cells arrived at the Hayflick limit when their telomeres dwindled to some critical threshold that triggered a crisis. Shampay's observations of the variation in telomere length in yeast, confirmed by research on other organisms, further suggested that for each species, there exists some optimal mean length for telomeres in healthy cells. Although telomerase had been shown to sustain the length of telomeres in a number of single-celled organisms, the enzyme appeared to be inactive in typical adult human cells. Researchers speculated that the gene for telomerase might be turned off in human cells, with the exception of stem cells, specialized self-replenishing cells (such as immune-system cells), and germ-line cells (eggs and sperm) in which the enzyme maintains the telomeres. (Not until the end of the 1990s would sensitive, accurate assays detect the enzyme's presence, in tiny amounts, in all other human cells.)

Dramatic but puzzling findings on cancer cells complicated the emerging picture of how telomerase worked in human beings. In 1990, Titia de Lange, working in the lab of Harold Varmus at UCSF, reported that telomere shortening was implicated in human cancer, and a research group in Edinburgh, led by Nicholas D. Hasties, found that telomeres in human tumors were shorter than telomeres in the surrounding tissue.[16] In 1992, in a large collaborative group that reflected the changing times, Greider, Chris Counter, Harley, Silvia Bacchetti, and others found that telomerase was active in cancer cells in culture, even though the cells' telomeres were shorter.[17] Further investigation provided a possible explanation. In a precancerous cell, genomic instability triggers an alarm that signals the cell to stop reproducing. In this hypothetical model, a cell that ignores this alarm signal initially suffers a drastic shortening of telomeres, a crisis that enables some survivor cells to activate telomerase and maintain these stunted telomeres indefinitely.[18] In 1994, Counter, Bacchetti, Harley, and collaborators demonstrated that telomerase was active not only in cancer cells sustained in culture but in ovarian tumors in the human body. Eventually, a number of studies confirmed that telomerase was active in 90 percent of human cancers.[19]

On the door of her office, Greider keeps a graph on which she plots the number of citations for papers on telomerase by year. The field began to coalesce in the late 1980s, yet the numbers remained relatively flat—at most several hundred annually—until 1996, when studies linking cancer and telomerase began to appear; citations climbed to about three thousand by 2002, and six thousand by 2005.[20] But as yet, no one had been able to clone the gene for the RNA component of human telomerase, and no one had isolated the gene for the protein component of telomerase in any organism. Through recombinant genetics, researchers could exploit this gene to answer many questions about how the enzyme operated in a living cell, as Blackburn was doing with mutations of the gene encoding for the RNA component of *Tetrahymena* telomerase. The potential therapeutic implications of answering these questions are profound and far-reaching. The difference between the telomerase level in cancer cells and that of normal human cells may make it possible to target cancer cells with a telomerase-inhibiting drug that will have minimal toxic effect on normal cells, a crucial problem in cancer therapy. As early as 1991, Blackburn suggested in a review that "telomere synthesis could be a target for selective drug action" to combat disease—for instance, to kill a parasite while not harming human somatic cells in which telomerase "may normally be low or absent."[21] Conversely, the aging process might be countered by sustaining the maintenance of telomeres in cells.

The race was on to clone the human genes that encoded for the RNA and protein components of telomerase. To find the gene for any eukaryotic organism might fairly quickly lead to finding it for all, thanks to DNA-sequencing computer programs that rapidly processed a monumental database of gene sequences, searching for analogous patterns. Blackburn opted out of this race; she had already ceded her "interest" in the protein component to Greider when Greider moved on to Cold Spring Harbor Laboratories, and she had found a gene she could exploit to create mutations in *Tetrahymena*.

Suddenly, telomeres were sexy. Biotech companies joined the fray, since recombinant genetics and other technological advances now made it possible to explore the medical implications of the research. By the early 1990s, biotechnology companies such as Genentech, Biogen, and Sugen had sprung up to capitalize on the profit potential of the new science, and stock market interest in biotechnology was soaring. The founder of a major biotech company once explained to Blackburn that the point of a biotechnology company was to arrive at the moment where it could offer an initial public offering; the more the company could hype the promise of an as yet imagined product, the higher the cost of shares bought by the public. Economic incentives drove the aggressive hype produced by many biotech companies; Blackburn regarded the profit promises of the biotech companies as being more akin to speculating in pork belly futures than investing in the stock market. As with many other scientific advances, the response to biomedical research followed a cycle in which the first flush of enthusiasm for new discoveries gave way to disappointment when the promising new technology failed to produce a magic bullet. In turn, disappointment generated reservations about problems in treating human diseases and legitimate fears over the ethical dilemmas posed by the new technology. The same recombinant genetic techniques that produced synthetic insulin for diabetics and interferon for cancer treatment have also given rise to bioengineered crops, which have prompted protests and boycotts.

Oddly, the seemingly improbable implications of exploiting telomerase to cure or counter the aging process first compelled the interest of the biotechnology sector. While working in the lab of Woody Wright and Jerry Shay in Texas, Michael West, who had earned a PhD in cell biology and had a consuming interest in the problem of aging, wondered whether telomerase might prove to be a magic elixir that could replenish the supply of telomeres and thus ward off the aging process. West tirelessly pitched this notion, and although he admittedly "got kicked out of a lot of conference rooms," by 1992 he had attracted venture capital and recruited outstanding researchers to the staff and scientific advisory board of his new biotech company, Geron. Avidly touting the promises of telomere research, West touched a ready chord in a nation of graying baby boomers, earning his company the dubious distinction of achieving "the highest buzz-to-equity ratio in biotech history."[22]

The media, like the public, has an appetite for singular events and dramatic discoveries, which often runs counter to the more speculative and qualified truths that science produces in fits and starts. Worse, once an appealing idea takes hold, it becomes difficult to dislodge. Early on in telomere research, a study reporting that human skin cells usually divide eighty to ninety times in culture whereas cells taken from a seventy-year-old man might divide only twenty to thirty times, was viewed as evidence that the maintenance of telomeres might be implicated in the aging process. This "fact" (which even Blackburn and Greider cited in a 1996 *Scientific American* article) later proved to be an artifact, yet the correction has never entered the mainstream press and is only occasionally quoted in newer reviews.[23] As scientists raced to clone the human genes for the components of telomerase, mainstream press accounts, hyping the headline-making novelty of this new chase, repeatedly noted that no one had yet isolated the actual molecule of telomerase, though of course Blackburn and Greider had partially accomplished this when they cloned out the gene for its essential RNA component.

Geron, with a vested interest in the profitability of research on telomeres, promoted the seductive image of telomeres as a mitotic clock ticking down and determining the life span of a whole human being, not just a cell—an image that came to dominate press reports on the research and captivated the imagination of the general public. In the early 1990s, researchers had studied telomeres in short-lived organisms such as roundworms, which live three weeks, and also in mice, which live for two years, even when they are fed and kept disease-free in laboratories. Every organism ages as a result of at least two processes: an increasing susceptibility to disease and this mysterious "preset" limit to a life span (analogous to the Hayflick limit for individual cells), which varies for different organisms. But to conclude that shortening telomeres announce the ticking of the mitotic clock glosses over significant evidence, confusing cellular immortality with that of the entire organism. Laboratory-bred mice, for example, die of old age even though their telomeres are five times longer than human ones, so aging can't simply be a function of short telomeres. A far more likely theory, since substantiated, posits that healthy telomeres play a role in the renewal or replenishing of tissues, which significantly affects the disease susceptibility that comes with old age. Because in the human body telomerase is active mainly in self-renewing cells like those of the immune system, to generalize from simpler organisms to long-lived species such as humans further oversimplifies the actual biological processes.

As commercial possibilities emerged in a new direction, Geron simply switched horses. When it became clear that telomerase was active in many types of cancer cells, the potential for targeting telomerase in order to treat cancer prompted an about-face on the question of how active telomerase was in normal cells.[24] If normal somatic cells had no telomerase activity and cancer cells had extensive telomerase activity, then potentially profitable cancer therapy targeting telomerase would likely have few toxic side effects. This gave a biotech company a vested interest in countering new evidence that telomerase was active (even if minimally) in normal human cells.

On the other hand, if researchers were to discover types of cancer cells that did not have active telomerase, it would at the very least qualify claims for manipulating telomerase levels to treat cancer, which would also be bad for business. In the mid-1990s, John Murnane, a researcher at UCSF, found a precancerous line of immortal human cells that had no telomerase. He showed his results to Blackburn, who was already investigating a similar fascinating exception. Doing a second postdoc in Blackburn's lab, Lundblad had pursued a curious phenomenon that had cropped up in her work with Szostak. The *est1* mutation was normally lethal for yeast, yet a small fraction of cells continued to grow. Lundblad found that these cells, which she termed "survivor cells," patched telomeres on to the ends of the chromosomes in a process that involved recombination.[25] In Blackburn's lab, Lundblad and later Mike McEachern showed that this backup process sometimes maintained telomeres when telomerase did not function normally; though this process was not as effective as the activity of the enzyme, beggars could not be choosers.

Geron researchers tested Murnane's cell line in an unsuccessful attempt to detect telomerase, and Blackburn's lab also did an assay with the same result. Morin, then at the University of California at Davis, tried the same assay and also could not find telomerase activity. Blackburn speculated that Murnane's cell line was also a survivor cell line, and further investigation by various labs demonstrated that these cancer cells had striking similarities with yeast survivors. In contrast, for several years Geron never, in any public company statements, acknowledged the findings on this and similar cell lines.

The energetic and well-funded participation of biotech companies in cutting-edge biological research can prove dangerous, because unexpected and novel findings that present a commercial liability are exactly the findings that drive pure science. Furthermore, a company's (natural) interest in profit requires proprietary rights over research, which runs counter to the open exchange of ideas necessary to science. Eager to share in any windfall, universities commonly enter into licensing agreements, or "sponsored projects," with corporations when research seems close to generating a profitable product. Such sponsored projects often entail a material transfer agreement that preserves a company's commercial prerogative over any results—and potentially any *future* discoveries that in any way rely on the patented material. Proprietary rights are dangerously murky in molecular biology, which produces not new inventions but intellectual property, such as gene sequences, with significant medical and social consequences. When such research tools are patented, a company might waive the cost of access for academic scientists, but its commercial interest in any resulting product impedes ready access and can curtail further exploration. As early as 1998, an NIH-sponsored study deemed this a "serious threat" to scientific progress.[26]

Paradoxically, biotech companies also support academic scientists with whom they compete, because pure research essentially functions as a form of research and development for these companies. (In its ongoing quest to find a cure for aging, Geron would soon take a commercial interest in therapeutic procedures derived from stem cells and would champion public funding for this research.) But corporate funding always comes with strings attached, as Geron's entry into the telomere field demonstrated. First, researchers employed by Geron or working in partnership with the company had hitched their wagon to a star and thus were vulnerable to bias. Greider, who had served on Geron's scientific advisory board and collaborated with its scientists, ultimately severed her connection to the company after Geron scientists took issue with an article of hers that called into question any direct correlation between telomere length and aging. When told by Harley, then a Geron scientist, that she wasn't supportive enough of the telomere hypothesis, Greider replied succinctly: "Cal, you don't support a hypothesis. You *test* a hypothesis."[27]

Second, Geron's staunch protection of patent rights granted it the power to restrain academic research that encroached on its patents. For instance, the race to clone the human gene for the protein component of telomerase involved several research labs, including Greider's and Cech's. Greider had been collaborating with Harley to clone this gene, and when he moved his lab to Geron, they were suddenly competing with independent researchers at the company. Greider defended her territory, threatening to withdraw from the collaboration. According to Stephen S. Hall, whose *Merchants of Immortality* documents the quest to find a cure for aging, the tension spilled over into an "in-house turf war," even as so many researchers joined the hunt that "it wasn't a race; it was a mosh pit."[28] Greider continued her efforts to isolate the human gene for the RNA component, but in 1993 Geron researchers beat her to it. Not long after, the company unsuccessfully attempted to prevent a graduate student in Greider's lab from including the DNA sequence of this gene in her doctoral dissertation because free access to these ideas might interfere with the company's pending patent application.[29]

In the mid-1990s a number of labs, including Greider's and Cech's, attempted to isolate the protein component of telomerase. In 1996, Lundblad collaborated with Joachim Lingner and Cech to identify and sequence the gene for the protein component in yeast and *Euplotes aediculatus*, and once they published their results in March 1997, it was only a matter of months before researchers sequenced the human gene for the protein component of telomerase. The race to clone this gene ended in August in a dead heat between Cech's lab at the University of Colorado at Boulder and a team in Robert Weinberg's lab at the Whitehead Institute, led by Christopher Counter. But the Cech group was first (by only a week) to publish its results, and since Geron had entered into a licensing agreement with the University of Colorado, the company promptly filed a patent on the work. Scientists who wanted to conduct further research on the gene claimed they had difficulty getting the DNA sample from Geron or found that too many restricting conditions were imposed on their research.[30] (To clone the gene independently required sophisticated and time-consuming, if straightforward, recombinant methods.) As Hall reports, Weinberg, who subsequently made his lab's independently cloned sample of the gene available to anyone who asked, "went so far as to suggest that Geron's stance on intellectual property may have impeded the speed of research on diseases like cancer. 'I happen to know,' he said, 'that a number of companies have shied away from thinking about developing telomerase inhibitors, for fear that they may run afoul of the Geron patent and the very aggressive policy of Geron in warding off anyone who might come within light-years of their terrain.' "[31] Weinberg's remarks reflect a striking difference between academic and corporate research; a 2005 survey by the American Association for the Advancement of Science found that 76 percent of industry scientists reported difficulty gaining access to patented technologies, as compared with only 35 percent of academic researchers who met difficulties that affected their research.[32] This difference is in part due to the fact that researchers seek patented technologies primarily within their own sector, with industry relying more on licensing, which restricts or delays access, and academia relying more on the traditional, open sharing of results through publication or informal contact.[33]

Blackburn had tried to keep clear of the media-hyped quest to clone the human genes for the protein and RNA components of telomerase (which are, intriguingly, located on different chromosomes). In Hall's account of this scientific tournament, he portrays Blackburn as a "merry personality," who humorously referred to biotech funding for academic research as "money for jam."[34] Hall also speculates on the reasons why Blackburn, who carefully frames any scientific claims, may have been a disappointing media source: "Scientists who championed the view that shortened telomeres 'caused' aging were courted and quoted by science journalists; those, like Blackburn, who expressed a more complicated, equivocal, or nuanced view tended to see their opinions appended, if at all, to the subterranean plumbing feeding the 'fountain of youth' stories."[35]

It would be easy not to register the subtext of Blackburn's reference to "money for jam." The phrase refers to getting money to do exactly as you please, and a strong will is very much a part of this "merry" personality. Blackburn didn't want anyone else telling her what to do in her lab, and that meant a reluctance to be courted by a biotech company or to partake in a large collaboration, at least for the moment. Yet she was hardly sitting on the sidelines. The work that led to isolating the gene for the protein component of telomerase depended on assays developed in her lab, first with Greider and later with Marita Cohn, working with yeast. Blackburn was continuing enormously productive research on the RNA component of telomerase, and her work with McEachern and Lundblad seemed to clear up the mystery of scientific findings that had found recombination to play a role in telomere synthesis. Since telomerase most likely played some role in the proliferation of cancer cells, Blackburn had an intrinsic interest in how a cell became cancerous, so she also pursued her work on human cancer cells. Yu's research had an interesting, if still seemingly tangential, bearing on cancer cells' proclivity to proliferate ceaselessly: cell division in *Tetrahymena* cells required telomerase activity, and inactivating the enzyme resulted in cell senescence, a potential angle of approach to cancer therapy.

Blackburn came across another opportunity she could not turn down. In the days before the development of protease inhibitors, clinical studies on reverse transcriptase inhibitors, such as AZT, focused on how these drugs might be used to treat AIDS. Since telomerase was a reverse transcriptase enzyme, she might study how these reverse transcriptase inhibitors suppressed the action of telomerase in the cell. Lab tech Strahl had demonstrated that AZT could shorten telomeres in *Tetrahymena*, and Blackburn hoped the drug could be employed to create drug-resistant cells, making it technically possible to select for mutated telomerase genes. (In a similar way, drug-resistance markers had enabled her lab to identify the replication origin in *Tetrahymena* rDNA.) Blackburn was interested not only in how AZT might affect telomeres in human cancer cells but also how it might affect telomeres in normal human cells, which could have implications for AIDS treatment.

Blackburn never started a biotechnology company or invested in one, though she was certainly wooed. One day in the early 1990s, Strahl reported to Blackburn, "There's a guy in a *suit* in the lab." The visitor was Michael West, who expressed interest in funding Blackburn's research in exchange for a share in any commercially viable product. After some discussion, West requested that Blackburn write up two pages on a current research project. Serendipitously, her lab had begun studying *Candida albicans*, a fungus with interesting telomeres. A few years before, McEachern, having just earned a PhD, had planned to look for a job in industry but wanted to hand off to a telomere researcher an interesting project he'd begun on *Candida albicans*. He made an appointment to talk to Blackburn, and, he reported, "she agreed it was a good project and said I should be the one who did it."[36] More commonly known as thrush, *Candida albicans* afflicts patients with AIDS, flourishing in the throat and mouth so that difficulties with eating further sap a patient's strength. In the report she drew up for West, Blackburn proposed developing a therapy that might genetically alter the telomeres of *Candida albicans* so the cells could not thrive; because its telomeric sequence closely resembled that of humans, researchers would first have to pinpoint a variant to target.

Geron offered to provide $440,000 to sponsor this research, but Blackburn acquired reservations after talking with other academic researchers who had worked in partnership with biotech companies. As an added temptation, company lawyers at biotech companies often encouraged researchers to buy stock in their companies, assuring them that many collaborators availed themselves of this opportunity. But one did so at the potential cost of objectivity, and Blackburn was also wary of the proprietary interest of a corporation in research results, which demanded secrecy. Where did the corporation's rights over research results end? She felt it made no sense to demarcate intellectual property in a basic science lab such as hers, where information spills from bench to bench in free flow, and she doubted whether she could take on such a burly partner. Not long after West approached her, she got a taste of what corporate partnership might entail when she and Greider were asked in 1995 to write a report on telomeres and telomerase for *Scientific American,* an important opportunity to reach a larger audience and a tip of the hat to the significance of their research. Blackburn recalled that "a biotech guy tried to muscle in on the authorship, not above throwing an elbow against these two little ladies. Carol was handling our communications with the journal, and she had to strenuously resist his efforts to share the credit."

A corporate partnership was simply a poor temperamental fit for Blackburn, who had transferred her childhood notions of gentility— growing up in a home where it was considered beneath one's dignity to be too concerned about money—to the conduct of her work. She had an instinctive resistance to thinking in terms of profit: "It just wasn't a motivation for me. I used the good old avoidance technique—I can thrive without having to do this. I don't want my life ruined by getting entangled with the commercial sector at a time when there is no product."

While she might not appear fierce enough to survive the mosh pit, Blackburn could summon an impressive toughness of a different order. When a staff member of UCSF's Contracts and Grants Office called Blackburn to ascertain whether she'd declined Geron's offer, he mentioned that the university did not usually turn down $440,000 grants. Blackburn answered, "In this case, you do."[37]

Yet this conversation did entangle her with Geron, if only tangentially. UCSF administrators determined that Blackburn might make a patentable discovery that could produce profits for the institution and therefore had to be disclosed. The university negotiated with Geron and, rather than compete with the company, agreed to include Geron and some members of Blackburn's lab on a patent. Geron paid the patent lawyers' fees, but no treatment ever came from this research. Eventually, Blackburn also agreed to work as a consultant to Geron. She didn't have any hesitation about sharing experimental results, and she would not need to sacrifice her objectivity if she treated her occasional consultations with Geron researchers as a variation on the weekly meetings in her lab, in which participants questioned and critiqued the soundness of each other's experimental work. But rather than treat consultation fees—two thousand dollars per visit—as personal income, she used them as unencumbered research funds.

Though she recognizes that biotech companies play a crucial, often constructive role in developing important medical therapies, Blackburn retains her reservations about research partnership with a biotech company. The big-money stakes in biotechnology create a constant tension between the academic mission of a university and its commercial drive, just as the utilitarian goals of for-profit enterprises often run at cross-purposes to the aims of pure research. A researcher who desperately needs the funding or is dazzled by the potential for profit can easily succumb to the plentiful opportunities in this arena, potentially at the cost of scientific integrity. Blackburn believes that the only big-money force that can obviate this pressure is NIH funding for pure science:

For this reason, if there's to be any kind of culture of science as an activity that produces accurate truth, it's a good investment for government to fund pure research, not profit driven, to keep it honest. This is clear in the field of stem-cell research, because potential profit depends on which of two types of stem cells—embryonic or adult—holds the most promise for medical therapy, and scientists affiliated with companies are under pressure to assert their product is better and trash the other. The truth will not take care of itself in the commercial sector—at least, not until a lot of time has elapsed. Truth derived from evidence-based reasoning is the bedrock of scientific progress. Ultimately, money is saved if research is done right and not based on speculation about future profits.

7 | Entering the Fray

Along with Blackburn's prominence would come increasing obligations and pressure to reconceive her essential notion of herself as "an inward-looking researcher."[1] She was elected as a fellow of the Royal Society of London in 1992 and as a foreign associate residing in the United States of the National Academy of Sciences in 1993. Membership in these societies marked her as an elite, one of few women in this select company, a trend that continues to this day. (Nineteen female members were elected to the National Academy of Sciences in spring 2005, the largest single group of women to that date, bringing their total number to only 187 of the academy's 2,059 members.[2] In the same year, the extremely selective Royal Society elected 44 fellows and 6 foreign members, of whom only 2 were women.)[3]

Blackburn had grown accustomed to being in the minority as a researcher in a seemingly arcane field and as one among a tiny handful of female speakers at many scientific conferences. For much of the 1980s, telomere research was typically subsumed under another topic at a scientific conference, but by the early 1990s entire sessions of a conference might be devoted to it. Blackburn's assessment of her career at this time— she was "quite pleased" that her work was well received and grateful for an "easy collegiality with my colleagues, male and female"—

euphemistically sidesteps the fact that she had achieved a reputation as "the queen of telomeres."[4] Between 1990 and 1996, she frequently chaired sessions and cochaired three major scientific conferences, two of which were devoted entirely to telomere research. She also authored twenty-three invited reviews, chapters, and articles on the field and served on the editorial boards of five peer-reviewed journals, including *Science* and *Molecular and Cellular Biology*. Her expertise meant she played a central role not only as a productive researcher but as someone who wielded enormous influence on the direction of the field.

Blackburn began to see that her prominence made a difference not only because she was proof that women could succeed but also because it gave her the opportunity to shape a more inclusive scientific community. She traced the evolution of her thinking on this issue in relation to her role as a conference organizer. In 1986, newly pregnant with her son, she had cochaired a Gordon Conference on Nucleic Acids, which had required two years of advance planning. When Blackburn and her cochair, Christine Guthrie, chose speakers, they didn't pay much attention to the number of men and women they invited, though they made a conscious effort to include a junior woman on the roster. Blackburn recalled that she and Guthrie "were very much in the guys' world. The question of gender was always there as a subtext, but I was far more immersed in seeking speakers who were doing exciting science." Seven years later, when Blackburn organized the 1994 UCLA/Keystone Symposium on Nucleic Acids with Douglass Forbes (a woman), the cochairs explicitly discussed the representation of women among the invited speakers and made inclusion a priority. As she organized the 1994 Banbury Center Conference on Telomeres, the first conference ever devoted entirely to telomeres, Blackburn realized to her delight that "I didn't even have to worry whether women would be represented." Telomere research had become very competitive—Calvin Harley would attend the conference as a representative of Geron, which still avidly promoted the promise of telomerase—and attendance at this conference was limited to only thirty-five primary investigators, but most of these were women, an astonishing departure from the norm in the sciences. Though it may have been by chance that the central research in the early days was conducted by women—Greider, Blackburn, Lundblad, de Lange, and Zakian—as a result the field was characterized by a rare gender parity.

When Blackburn was first hired at UCSF in 1990, she was asked if she'd be willing to take over as chair of her department, and she agreed, provided that Leon Levintow, who was planning to retire, would stay on as chair for a few more years until she found her feet. On the face of it, chairing a department offers an opportunity to exercise power and influence within an institution, yet this is not necessarily a step up for a research scientist. Unlike clinical departments at medical schools, basic science departments generate no income from medical services, which often makes a department utterly dependent on a dean's good will and puts chairs in the position of justifying requests for resources in relation to benefits to the clinical mission. Fortunately, at UCSF, department chairs in the basic sciences spent far less time begging and pleading for necessary funds, partly because they were housed in the School of Medicine, by far the "big gorilla" at the university, which has much smaller schools of pharmacy, nursing, and dentistry. If Blackburn could expect few difficulties with this aspect of her duties as chair, taking up the position was still an unusual step for a researcher whose all-consuming work was vital to her field. Typically, the institutional power of academic research scientists derives from national recognition in a given field; Blackburn's own department gained clout from the presence of J. Michael Bishop and Harold Varmus, then the only Nobel laureates at UCSF. Research stars are not usually expected to prove their value to the institution in other ways; as Carol Gross, a longtime colleague of Blackburn's in the department, put it, "Why should she have to do this? She's one of the most brilliant scientists of our generation. She invented a field."[5]

Blackburn may not have evaluated herself in this light at the time she was hired, and serving as chair meant honoring a commitment and fulfilling a civic duty. Earlier in her career she had been unwilling to devote energy anywhere else than to her research; when Daniell was denied tenure at Berkeley, Blackburn drew the conclusion that Daniell had been penalized for not sustaining the impression of a "monotheistic" focus on science. Blackburn hadn't required a tough-as-nails professional exterior in order to thrive in the competitive arena of research science, so she presumed her competence would transfer easily to a mere administrative role. She also felt confident that she could rely on "good relations" with her colleagues. Yet as she prepared to take over as department chair, Blackburn harbored doubts, especially as a relative newcomer to UCSF, which has a somewhat insular culture. A survey conducted several years after Blackburn served as chair found that faculty members did not see the university as doing a "good" or "excellent" job of welcoming new faculty, and women faculty ranked the university lower in this respect than did their male colleagues.[6] According to Gross, "Almost everyone at UCSF began here at the start of their careers, as an assistant professor, and worked their way up. UCSF tends to build its own stars rather than hire stars. So Liz was not only a woman but also an outsider."[7]

For the first time in her life, Blackburn avidly sought a woman's perspective on her professional life. She consulted Dee Bainton, MD, a cell biologist in the department of pathology at UCSF, and Zena Werb, another cell biologist and cancer researcher at UCSF, and also sought advice from Lucille Shapiro, chair of the Department of Developmental Biology at Stanford. If Blackburn took on the role of chair, she didn't want the shame of doing the job badly, and she felt women would give her more honest replies about the pressures and obligations of serving as chair while trying to sustain her research: "I was at home dealing with men or women in science, but this was something where I felt I'd get more honest, realistic answers from women. Women faculty would realize that gender was an issue. If I went to male colleagues, I figured they'd say, 'It's great for you to serve as chair! Don't worry about it.'" Yet none of the colleagues Blackburn consulted explicitly warned her about sexism or suggested strategies for contending with it. In the sciences, successful women presumably had already adapted to their minority status and pushed this fact of life to the periphery in order to succeed.

Despite her concern to protect her research, Blackburn wanted to serve as chair because she understood the need for more women administrators at UCSF. Although the university tried to improve its numbers, in the mid-1990s, UCSF could count only one woman department chair in the School of Medicine, with no women serving as division chiefs, only two as deans or senior administrators (outside the School of Medicine), and four as associate or assistant deans. This paucity of women in high-level positions was similar to that at most U.S. medical schools.[8] In 1996, of these 125 medical schools, only 8 could count twelve or more women administrators at any level; 11 did not have any women chairs or division chiefs; and 7 had no women serving as deans.[9] This disparity reflected endemic inequities. According to the Association of American Medical Colleges, even though by 1996 women made up 42 percent of medical students, only 10 percent of full professors and 19 percent of associate professors were women.[10] The salary gap between men and women of equal rank has remained fairly constant for full-time medical school faculty (differing by roughly 27 percent in 1993), with one survey suggesting that the concentration of women in less lucrative clinical specialties (pediatrics and internal medicine) contributes to this disparity.[11]

When Blackburn attended her first medical school meeting as a chair, she rode up in the elevator with Virginia Ernster, chair of the Department of Epidemiology and Biostatistics, who told Blackburn that she was about to step down as chair of her department, citing the needs of her teenage children. This made Blackburn the sole female department chair, and she felt acutely self-conscious at the next administrative meeting, as if she had to start all over again to prove herself: "It felt like a time warp back to the 1970s, when I had often been the only woman scientist present in a room." For all of Blackburn's tenure as chair, from 1993 to 1999, she was the only woman serving as a department chair in the School of Medicine, though women chaired departments in the pharmacy, dentistry, and nursing schools. (After Blackburn's tenure ended, Nancy Ascher, who became the chair of surgery, was the only woman chair in the School of Medicine.)

Though Blackburn had carefully established conditions to preserve her work while she served as chair, she betrayed little concern for other dimensions of her changed status. The dean of the School of Medicine, Joe Martin, was an MD who had conducted research in neuroscience; as dean, he had more direct financial power than even the chancellor. Blackburn bluntly told Martin, "I can do more for UCSF if my research program stays strong," and sought guarantees for the necessary resources, including a salaried postdoc who would oversee the continuity of work in the lab. She asked for little in terms of space and office requirements, even though such perks visibly communicated status. The culture of research scientists in which she was firmly grounded looked down on all things administrative—"You've gone to the dark side if you become an administrator," she said—and she regarded the five-year appointment (a common arrangement in science departments) as temporary.

Behavior that might be read as unambitious and naive often reflected only that Blackburn's ambition was intently focused *elsewhere*. This ambition was shaped in part by the cultural ambivalence of lab rats toward bureaucrats—not unlike the disdain of creative artists for the marketers and managers in their field of work. Blackburn admitted, "I started off unambitious because I wanted my research to be the major thing in my work life. I didn't ask for things such as renovations to the department's labs. I figured things were pretty good and we already had many of the necessary resources. I always chafed at the time I devoted to being a chair; I felt it was taking time from research. I never felt research took time from chairing." Disowning any particular ambition in this regard can seem like another form of protective coloration—modest claims ruffle fewer feathers and minimize the risk of being judged a failure—but it also reflects how deeply Blackburn had imbibed the notion that to be careerist or "worldly" somehow diminished the science. She speaks just as diffidently about her scientific track record, where she really has no worries about success.

Preparing for her first meeting with Haile Debas, who had succeeded Martin as the dean of the School of Medicine, Blackburn realized she was entering a new culture; a revenue-producing medical school had much in common with the corporate world. Immediately, she understood that her wardrobe wasn't up to snuff. In the lab culture, you dressed casually to demonstrate commitment to your work; if you dressed up as a woman you weren't serious and would look like a lowly administrator. For years, Blackburn had left her shoulder-length brown hair unstyled, and she wore wire-rimmed glasses that might have survived from her own days as a graduate student. Fearing she'd be "stomped on" by scientists if she wore suits but risk offending administrators if she didn't, she settled on wearing jackets with pants on days she had to meet with administrators.

Conflicted identity made Blackburn reluctant to assume the trappings of her position in other ways as well. When Blackburn first moved to UCSF, she had helped design her new lab and chosen an office for its proximity to her lab. The office, which she described as "small but functional," was approximately nine by eleven feet—a jail cell is six by nine feet—and stuffed with filing cabinets. Once she became chair, this office proved inconvenient; she had to go back and forth between the department's administrative staff office and her lab. Yet when Varmus left the faculty and his large office became vacant, Blackburn allowed the office to be reassigned to another member of her department; this colleague had requested the office because it was near his lab, and he was also considering a job offer from another university at the time. Early in Blackburn's tenure as chair, Harry Noller, whose work on determining the helical structures in RNA had inspired Blackburn's work on the structure of telomerase RNA, visited UCSF and stopped by her office. Noller wryly remarked, "I notice your office follows UCSF custom: the more important the person, the smaller the office."

Gender issues intersected with the lab-rat culture's suspicion of the "pretension" of an impressive office, compounding Blackburn's hesitation to insist on a more functional setup: "I just didn't know how to have the dean's office expect that I would demand an adequate office. If a guy said, 'I have to have a larger office,' it would happen. But unless I really pushed it, it wouldn't happen for me. And I felt, this business is below me. Science is more important. But I was selling myself short."

Though Blackburn had unhesitatingly reworked her lab schedule to suit her new priorities as a mother, once she became chair, she felt more pressure to hide the difficulties of arranging for child care. The School of Medicine's administrative and executive committee meetings were often slated early in the day, at 7:30 or 8:00 a.m., but Sonia Menjivar, Blackburn's child-care provider, arrived for work at 9:00 a.m. In order to make the meetings, Blackburn had to negotiate with her husband or ask Sonia to come early. Even then, her problems were not "solved." If her workday started at 7:30 a.m., it would still end at around 7:00 p.m., and she did not want a daily routine of eleven hours away from her son. Once she arrived in the higher administrative ranks, Blackburn had to start from scratch to overcome self-consciousness and embarrassment about the realities of juggling her family life with her career: "It never occurred to me to say this is really awkward because of my child care, so you have to change the meeting time. Culturally, it would have made me appear 'not normal'—a weak chair. I felt I would not be taken seriously if I was seen to make compromises for family."

Whenever department chairs had to attend an evening event or a leadership retreat, usually scheduled on weekends, Blackburn once again felt conflicted. She made it to annual, off-site leadership meetings focused on significant issues, such as plans to build a new campus; often, John and Ben would accompany her, and John would look after their son while she attended meetings. But a number of retreats were aimed primarily at developing collegiality, featuring social events such as dinner dances, and Blackburn, who already traveled frequently to scientific conferences, didn't feel she could add even more trips away from Ben. She never went to a single one of these meetings, which meant she effectively withdrew from an arena in which expansive social contact influenced the way power was wielded at UCSF.

Fortunately, the actual administrative tasks of a department chair were made easier by sound institutional policies at UCSF. The six basic sciences departments shared graduate programs, an unusual federalism that had been pioneered by the university. Because similar research might be conducted by different departments, graduate students did not apply to specific departments but chose labs based on their research interests, eliminating departmental competition for the best students. Other aspects of the basic sciences were also run on a federal model that emphasized common goals; for example, the departments combined to teach required courses for PhD students. As a result, departments did not vie overtly with each other for influence and resources, in marked contrast to Blackburn's experience at Berkeley. Within the Department of Microbiology and Immunology, a careful balance had been struck between the department's obligation to teach basic science to medical students and the desire of most faculty to conduct research—often a point of contention at other institutions.

The department chair had the final responsibility on financial matters for a department of 140 students, faculty, and staff. So that she could carry out her fiscal responsibilities as chair, Blackburn received tutorials from the management services officer for the Department of Microbiology and Immunology at UCSF, Pat Clausen. Each department had its own officer, whose expertise was vital to managing funding that ranged from $10 to $20 million a year. (As a whole, UCSF has a $350 million annual research budget.)[12] Blackburn learned from Clausen about funding mechanisms and conducting business within a byzantine academic bureaucracy. For instance, faculty salaries, calculated by a university formula, were paid from a combination of sources, including both the state of California and foundations that had awarded individual grants, and the money had to be routed through complex channels within the university, according to a logic that, as Blackburn put it, "would not have been deducible from first principles." Blackburn also had final jurisdiction over the space assigned to the department, and only delicate negotiating skills could avert turf wars. For example, when a new faculty member needed more lab space, Blackburn recognized that a senior faculty member had more than necessary, but the status-sensitive implications of space allocation meant she couldn't make a simple transfer. Given her own priorities with respect to research, she saw an exchange of space as necessary, but she appealed to altruism and softened the blow by offering the senior person a nicer office.

Blackburn's department was something of a forced marriage between microbiologists, whose primary interest lay in molecular research, and immunologists, whose interest in diseases made for strong ties with clinical research but diverged from those of microbiologists. Faculty members in the department were equally distributed between both specialties, and many immunologists had joint appointments in other departments, such as medicine. (Underscoring the many ways in which research boundaries differed from those of departments, Blackburn herself had insisted on a joint appointment in the Department of Biochemistry and Biophysics because her central research connected vitally with biochemistry, though the vacant post for which she'd been hired happened to fall within another department.)

Given the diversity of the Department of Microbiology and Immunology, Blackburn focused on strengthening collegiality within her department, especially in ways directly relevant to research. She chose three vice chairs, each of whom represented the interests of the different kinds of researchers in the department. Vice Chair Gross, one of only two other women in the department, became a close friend and colleague. When Gross first moved to UCSF, her lab had to be redesigned, exactly the kind of detail work to which Blackburn devoted conscientious attention. Gross reported that "Liz spent an enormous amount of time overseeing this. She helped me when I came and after my husband died, a very difficult time."[13] Having been forced to learn the hard way when she first began teaching at Berkeley, Blackburn also initiated a mentoring program for the four assistant professors she hired. She instituted weekly meetings at which department faculty presented their research and inaugurated similar meetings, open to everyone, for postdocs, who had no other organized public forum for presenting their work and would thus gain much-needed practice as well as a chance to engage with colleagues beyond the specialized world of their own labs. Collegiality hadn't been a particular concern of Blackburn's male predecessors; when she hosted a dinner for the department faculty at her home, one colleague commented that the previous chair had never organized a similar social event.

Blackburn's lack of interest in campus politics meant that she didn't expend a great deal of energy promoting the department to the institution, didn't perceive opportunities that could be leveraged in relation to this, and didn't necessarily register political conflicts within the department. Though Blackburn felt "relationships within the department were good," Gross reported that "some people didn't feel she was doing a good job," but added that Blackburn operated under a handicap: "She was playing catch-up, coping with inherited problems. She didn't have the skill set of an administrator but had to learn on the job." According to Gross, Blackburn "had great ideas—she's one of the most far-sighted people I know," but did not always know how to strategize to achieve her goals: "Liz tends not to perceive the undercurrents in social relationships or see why a good idea can't be immediately carried out. Often, skill at this is one of the ways women get along in competitive environments. That makes it doubly hard for her. It's hard for a woman to operate the way a man does. Maybe she could have succeeded better as a chair if she'd used the 'woman's card,' but she doesn't have it to play."[14] Gross's comment might be translated to read: what a man could get away with in this situation, a woman could not.

Like many scientists, Blackburn segregates the intellectual work of problem solving from personal fears and loyalties. If this offers another way for a woman to get along in a competitive environment, how it is manifested depends on complex qualities of character. On intellectual turf Blackburn does not negotiate her views except in the face of hard evidence, seemingly out of sync with her otherwise tentative manner. Polite and soft-spoken, she has a singular capacity for "tuning out" her own ego and is surprisingly willing to give ground when only her personal feelings are involved or when she perceives the other person has a larger stake in the outcome than she does. Dave Gilley, a postdoc in her lab in the mid-1990s, suggested she might easily be read as a pushover: "In this very, very competitive environment at UCSF—kind of a bizarre environment—where you have the type of people who rush to be first on the elevator, some were not respectful to her, and this wouldn't be tolerated by others at the institution. She was incredibly tolerant, I think to a fault. But she is so clear about herself that it is not a challenge to her."[15] Blackburn's track record as a mentor demonstrates sensitivity to particular kinds of social nuance, yet her indifference to status symbols underscores how easily she might discount political nuance, at least in relation to *campus* politics. That she felt divided about this—and about what it meant to do her job well—is demonstrated by the value she placed on getting good offices for other faculty while declaring she didn't care about her own.

During her tenure as chair, Blackburn was forced to acquire a more attuned sense of the political, particularly in relation to gender barriers. As the sole woman chair in the medical school, she was sought out even by faculty women she had never met. Women came to her office to explain a situation that had left them feeling helpless and ask her advice. A number of these women worked in clinical departments, which had a separate pecking order from the basic sciences, and Blackburn felt "out of my depth" when confronted by these reports of persistent problems with discrimination. How useful was her own example of putting on blinders and focusing on the work?

Despite institutional goodwill, a credibility gap existed between enlightened policies and actual practice. Blackburn had joined the

Chancellor's Advisory Council on Women soon after she became chair of her department, and while she served on the council, the university conducted an internal survey on women's salaries. At first glance, the statistics were deceptive: for the most part, men and women of equal rank had equivalent salaries. But a gender gap existed in the time it took for men and women to rise a given step in the promotional ladder. This gender gap was more pronounced in the clinical sciences than in the basic sciences, though in the basic sciences women constituted a smaller percentage of faculty, despite their equal numbers among graduate students and postdoctoral fellows. As a result of this survey, UCSF promptly adjusted salaries for women to eliminate discrepancies between faculty of equal rank. But the more intangible problem of the slower rate of advancement for women could not be so neatly targeted by a single action.

Slower rates of advancement have complex causes at every level of the hiring and promotion process and have persisted in academic sciences despite the fact that overt discrimination is illegal and institutional barricades are erected to prevent it. To Blackburn's knowledge, within the last dozen years, fewer women than men have applied for assistant professor positions at UCSF. Faculty search committees winnow finalists from among this de facto limited field of candidates and recommend the best candidate to chairs, who merely consult on the process and persuade deans to provide the necessary resources. As in many other departments, all too often the list of finalists produced by search committees in Blackburn's department included no women. Yet she hesitated to challenge the search committees because their autonomy constituted an essential aspect of academic freedom. In retrospect, she came to question her choice: "My wish to work with the boys was strong. In one instance, another female faculty member wanted to advocate for a good woman candidate, and I followed up on this, but when it seemed too hard, I did not pursue it. I was trying not to make waves by sticking my neck out in a way that would clearly be motivated by the fact that this was a woman candidate. I capitulated too easily on that. Ostensibly we were making the best possible hires in the field. Why was the very best person each time turning out to be a man?"

Despite genuine efforts to implement nondiscriminatory hiring policies, the unconscious bias of predominantly male search committees often means that they will assess women candidates as less strong, at every rank from postdoctoral fellow to administrator. Such bias has been noted anecdotally but has also been measured quantitatively. A 1997 Swedish study analyzed the peer-review scores for applicants for postdoctoral fellowships at the Swedish Medical Research Council and concluded that "peer reviewers cannot judge scientific merit independent of gender." Quantitative analysis demonstrated that a woman applying for a fellowship had to be two and a half times more productive than a man to rate the same scientific competence score from reviewers; women's success rate in obtaining fellowships was half that for men.[16] The researchers who conducted this study also cited several other studies showing that female evaluators tended to be more objective than their male counterparts in assessing the scientific competence of women.[17] In another recent study, neuroscientist Rhea Steinpreis randomly selected professors to evaluate 250 academic résumés, asking them to rate the candidate as a job applicant. Half of the identical résumés were sent by "Karen Miller" and half by "Brian Miller." While two-thirds of the professors would hire Brian, fewer than half found Karen an adequate candidate.[18]

A hidden catch-22 impedes institutional efforts to ensure women's advancement: at higher ranks, women have the opportunity to influence the judgment of search committees, but so long as their qualifications are consistently undervalued, few women can rise to this level. And once they get there, they might hesitate, as Blackburn did, to challenge the discriminatory consequences of practices bolstered by tradition and by values such as academic freedom, to which these women also owe a deeply felt allegiance. Finally, the dearth of women applying for positions at a prestigious research institution such as UCSF stems from a subtle accrual of factors that create the perception of a hostile environment, from women's accurate anticipation that their chances are lower at elite institutions to the reputation of clinical medicine as a sexist bastion. A dismaying lag time exists between policy initiatives that counter discrimination and actual practices that achieve this goal; Blackburn noted in a 2005 interview that a UCSF faculty search conducted that year once again produced an exclusively male list of finalists.

Among the women whom Blackburn tried to assist as they contended with perceived discrimination was Sally Blower, a mathematical and evolutionary biologist hired in 1995 in a soft-money position at UCSF. At the time Blower was hired, her husband, Nelson Freimer, was a key member of UCSF's human genetics program, and the university did not want to risk losing him if his wife found a job elsewhere. As an adjunct associate professor, Blower had to raise her own salary from research grants, and she also did not enjoy the same status and prerogatives as tenure-track faculty.

From the start, Blower's new position at UCSF posed delicate problems: her work didn't fit neatly with other research in her department, and a scientist with an international reputation might well chafe at the low status and uncertainty of her position. During her five years at UCSF, with space on the Parnassus Street campus at a premium, her office space was relocated four times, and at one juncture, she was forced to work at home.[19] Blower complained and assertively negotiated for better working conditions, with both her department chair and Debas, the dean of the medical school.

In an effort to help, Ira Herskowitz, a senior faculty member in the Department of Biochemistry and Biophysics, asked Blackburn if she would speak to Blower. When Blackburn and Blower met, Blower reported that she felt the series of moves had been politically motivated and gender biased. When Blackburn suggested she raise the issue with members of her department, Blower said she couldn't, because an influential member of the department wanted to share research credit for work that Blower had done, and the department chair was supportive of this male colleague. Wanting Blower to feel less alone, Blackburn and Gross took her out to lunch. "We liked her," Blackburn recalled. "She was funny, witty, acerbic. I admired her for standing up for herself and being willing to be unpopular over what she felt was an injustice."

Already considering the possibility of inviting Blower to join her own department, Blackburn asked her to give a research seminar, and her talk impressed Blackburn as scrupulously thorough, scientifically substantive, and well presented. A position in the Department of Microbiology and Immunology was created for Blower, but although a move might temporarily solve her space problems, the faculty had agreed to offer only another soft-money position. Blower was provided with an office and nearby lab benches, the only space available for her team of four post-docs, with the unnerving proviso that if these benches were needed for research, she might again be forced to relocate.

Blower eventually lost patience with what she perceived as discriminatory treatment, and she and her husband left the university to take positions at the University of California at Los Angeles. In 2000, after she secured the new post, Blower sent an e-mail to a British medical journal and several colleagues, charging that she had been victimized at UCSF by a network of "Senior Boys" and accusing two male colleagues of misconduct. She reported that when she discussed potential office space with Lee Goldman, chair of the Department of Medicine, he told her she would be on probation for a year while they "dated," and she would need to make him "happy" in order to stay on.[20] Blower also claimed that when she had tried to pursue her complaints through formal channels, she had been told by "an official in the Dean's office (in the School of Medicine) that the Chairs can do anything they like with me & that the Dean's office will back them up."[21]

After Blower's e-mail became public, J. Michael Bishop, then the chancellor of UCSF, authorized an investigation into her charges. Zina Mirsky, associate dean of the School of Nursing, conducted a monthlong investigation and found no evidence of wrongdoing or sexism. Debas claimed the university administration had made extraordinary efforts to accommodate Blower and later noted, "Most of what we did, we did to retain her husband."[22] The facts might be read as neutral: deans at UCSF did not typically interfere with a chair's final authority over space. Not proof but only inference might link Blower's eviction from her office to her clashes with powerful colleagues, leaving open the question of whether sexism played a role.

As a star asking for the star treatment accorded to others, did Blower elicit responses that had more to do with gender than with the nature of her demands? Blackburn, who like everyone else cannot be sure of the facts in this case, worried that labeling women is a time-tested method for evading the issue: "I felt sympathy for Sally because of the temptation to say, 'Oh, she's a difficult woman.' We have to get beyond such stereotypes and find out why she's being a 'difficult woman.'"

Making a formal or public charge of discrimination means sacrificing other kinds of negotiation and compromise in order to achieve one's goals, and at medical research facilities, the consequent bad publicity will almost certainly undermine efforts to recruit women faculty. (Blower's charges were reported in *Science* and *Nature Medicine* as well as in local newspapers.) Whistle-blowing demands another significant sacrifice: it consumes time and energy that might otherwise be devoted to one's career. Although Blackburn tried to be helpful to Blower, she acknowledged that "Sally wasn't very impressed by my efforts. She said publicly that I was really interested in my research and not particularly interested in the predicament of women. Yes, there was some truth to that. I didn't drop everything else to go after what she felt were really deep-seated problems of sexism at the university. My conscience could only be completely clear if I'd gone to fight the larger injustice." If there are costs for whistle-blowers, clearly one consequence of compromise and political trepidation is guilt, as Blackburn's reflections on her interactions with Blower reveal: "Here was someone prepared to make waves. I had hidden behind my protective veil of rationalization and denial to avoid dreaded confrontations that in my quailing heart I feared would endanger my standing among scientists and colleagues. I felt she would be right to see me as pretty much a lackey of the system that she felt had failed her so badly."

Late in her tenure as department chair, Blackburn would again confront the dilemma of how to respond to perceived discrimination and experience firsthand the self-doubt that makes so many women hesitate to voice accusations. Blackburn's term as chair had gone well for four years, and throughout that time she had "trusted in the general goodwill of the other chairs and deans." But in her fifth year as chair, UCSF embarked on a huge expansion project that would alleviate the space shortage that had exacerbated the Blower controversy. Comprehensive planning for the new Mission Bay campus also generated tension as administrators determined which departments would relocate from the foggy Parnassus Street campus to a new site situated in the sunny part of the city.

At a projected cost of $1.5 billion, the Mission Bay campus would double the research space of UCSF and provide an academic home for

9,100 faculty, scholars, staff, and students, making it the largest urban development in San Francisco since the creation of Golden Gate Park. Complete plans included housing, stores, office space, and a community center. Located just south of downtown, between the San Francisco Bay and Interstate 280, the campus would take up forty-three acres, and it would be devoted to the basic sciences, including affiliated institutes (the Gladstone Institutes and the California Institute for Quantitative Biomedical Research) and two new research buildings (Genentech Hall and the Arthur and Toni Rembe Rock Hall). The first research building to be occupied, Genentech Hall, would be built at a cost of $223 million. Of this total, $50 million was provided by Genentech, but this was not mere charity: the funds were part of the $150 million settlement of a lawsuit UCSF brought against Genentech for patent infringement of its human growth hormone products.[23]

Genentech Hall would house research groups in structural and chemical biology, cellular, molecular, and developmental biology, genetics, and developmental neuroscience. Its 385,000-square feet accommodated fifty-four spacious labs, patterned after the university's existing "interactive research environment," which organized labs into neighborhoods that fostered collegiality.[24] In advance of the move, slated for January 2003, many of the meetings of department chairs focused on divvying up the new space and resources available at Mission Bay. Even in comparatively minor areas, the division of space was a hot-button issue on the crowded campus, and the chairs might spend hours discussing the disposition of a broom closet or a desirable office—exactly the aspects of status to which Blackburn turned a blind eye.

Blackburn could not afford genteel aloofness when she had an obligation to protect the interests of her department, including space requirements. And a more serious issue was on the table: the move threatened to split the Department of Microbiology and Immunology. The microbiologists, who felt their research was allied with biochemistry and biophysics, wanted to move to Mission Bay, but the immunologists wanted to remain at the Parnassus campus, close to clinical programs. In planning meetings, Blackburn explored various solutions, though to her dismay it became clear that the department would probably be split geographically, an administrative nightmare that might lead to its disintegration.

Blackburn had left Berkeley before she would have had to conduct territorial combat over space allocation in a new science building. She did not want to preside over the demise of her department but could not reconcile the interests of different factions, and she didn't know how to hold her own in the intense, aggressive negotiations taking place at UCSF. The personal traits that worked in her favor as a researcher—religious scrupulosity with regard to deriving decisions from evidence and exacting logic—also made it harder for her to factor into her reasoning the force of influence, ego, and political alliances. Her disregard for rank and prerogative translated to a liability in this administrative context. "My empire-building genes are not very good," Blackburn admitted, and this might in part be due to the acutely fine distinction she drew between "grabbing the goodies for the department rather than serving the good of the department." Blackburn felt bullied in the wrangling over space, uncertain about the rules of this competitive game among her all-male cohort, and distressed by the tension of these confrontations. If she didn't have a "woman's card" to play, she also didn't have an alpha male's thirst for combat.

In the same year that she struggled to contend with these tense negotiations, Blackburn would receive international awards in recognition of her research: the Gairdner Foundation Award, the Australia Prize, and le Grand Prix Charles-Leopold Mayer Award, a French prize accompanied by forty thousand dollars. She would be elected as president of the American Society for Cell Biology (ASCB). Several times, at administrative meetings, Debas had confidently asserted that one day she would win a Nobel Prize, and in 1999 Blackburn would be informed that she had been nominated for this prize. Yet simply to contend with colleagues of equal rank, for the sake of debating space allocations, Blackburn was cramming like a college student. She diligently combed the pages of *Games Your Mother Never Taught You*, gleaning advice predicated on the notion that women must learn to play the games that men play in order to get ahead.

Blackburn presided over a department in the midst of an identity crisis and deprived of an important advocate after Varmus left to become director of the NIH. Painfully conscious of her dwindling power and increasingly uncomfortable with her role in planning for the move to Mission

Bay, Blackburn wanted out: "I wondered whether I should think about leaving. Evasion. But it wasn't evasion. I think you can negotiate if you are threatening to leave." This reaction echoed her response years before when she had felt powerless at Berkeley. Happy at UCSF and content in her role as a department chair, Blackburn had previously dismissed an attractive offer to direct a biomedical research institute in Australia. Now, toward the end of 1998, she received an offer of a job as director of another institute in Australia, with the promise of a job for her husband at a nearby institution. This time, she accepted an invitation to visit. Flattered by the offer of a prestigious post, Blackburn, still an Australian citizen, retained strong ties to her native country, making relocation even more tempting. As negotiations progressed, Blackburn and Sedat made a second visit to the institute. They had just bought a Lexus—the money from her French prize made a luxury car affordable—and when they returned from this visit, Sedat looked into having the steering wheel reinstalled on the right, so the car could be driven in Australia.

This possibility on the horizon provided Blackburn not only with a literal reason to renegotiate her position at UCSF but also with external validation of her worth. That Blackburn, so eminent in her field, still could not see herself as powerful speaks volumes about the difficulties that continue to beset women as they enter the higher ranks of academia. She couldn't imagine a negotiating strategy other than threatening to leave, and she was willing, finally, to test her own star status within the UCSF system. Yet she had mixed feelings about capitalizing on the job offer as a negotiating tool: "I liked and respected Haile Debas and didn't want to play a cynical game. Yet I also didn't want to become absorbed by petty politics at a time when I was passionately invested in my research and confident that it was going well."

Ultimately, Debas negotiated with Blackburn to keep her at UCSF. She set out her conditions: faculty members in her department should be able to choose to join a department most suited to their research interests. (This was her only condition that went unfulfilled: the department would remain intact even though it was physically split, with the immunologists staying at the Parnassus campus, home to immunologists from other departments.) Blackburn, who had been hired with a joint appointment in the Department of Biochemistry and Biophysics, wanted to move to this department, where she felt her work properly belonged. Having spent more than five years as a chair, she wanted to step down; indirectly, this meant she could retreat from the tense negotiations over space allocation at Mission Bay. Throughout her tenure as chair, Blackburn felt she had managed to make sure other department faculty had good offices, but the office assigned to her was inconveniently located and woefully inadequate. Last but not least, she obtained a promise from Debas for a decent office.

Three years before she ended her term as a department chair, Blackburn was approached by a nominating committee from the ASCB, in which she had long been active. She readily agreed to run for president. With a membership of about eleven thousand in 2005, the ASCB focuses on disseminating research through scientific meetings and its own journal, training students and young investigators, and promoting the value of biomedical research to Congress and the public. Gall, Blackburn's mentor at Yale, helped to found the society in 1960, and past presidents included Mary-Lou Pardue, Susan Gerbi, and J. Michael Bishop.[25] Women are wholly integrated into the society; the Women in Cell Biology Committee published and distributes for free *Career Advice for Life Scientists*, two volumes aimed primarily at young investigators. From 1980 to 2005, ten of the ASCB's twenty-five presidents (40 percent) were women, an unusually high proportion for a scientific society.[26] According to Blackburn, society members joke that if a man ran for office against a woman, he wouldn't stand much chance of election, and she attributes the society's inclusiveness to the influence of Gall, who'd been such an effective and generous mentor of women scientists.

   Blackburn cited strong personal attachment as the reason for her willingness to serve as ASCB president: "When I was just starting out at Berkeley, I went to ASCB meetings to present my research and felt welcomed. Other conferences, like the Gordon conference, had a more edgy, competitive feel, but this society was particularly inclusive of junior faculty. There was an atmosphere of supportiveness and warm interest— a real sense of community. This was a whole unknown aspect of science to me." Though she portrayed her willingness to serve as paying a debt of gratitude to the ASCB, Blackburn was asked to run for office because she was already a highly visible and effective leader in the scientific community, which offers an instructive contrast with her conflicted approach to campus politics.

As president elect of the ASCB, Blackburn served for a year on the organization's executive committee before she began her one-year term in December 1997. When she took office, Blackburn provided policy oversight for publication of the society's journals, annual meetings attended by ten thousand members, and congressional education and lobbying efforts. But she felt the ASCB "was already a well-run ship. I didn't launch any big initiatives but chose to devote time to two crucial areas, early career issues and federal funding." She worked closely with Executive Director Elizabeth Marincola, a skilled strategist who had served as deputy director of the Office of Policy Analysis and Congressional Education in the National Institute of Mental Health before coming to the ASCB in 1991.

Marincola found Blackburn to be unusually willing to give of her time and attention: "She's so much in the present. So many busy, important people, when you're with them, they're literally or figuratively looking at their watch, stressed out, worrying about other burdens in their lives, which is completely understandable. When you're with Liz, it's unbelievable, because she clears the mental space to be with you. It's like you're sitting talking with your sister." Marincola's observations also shed light on the way Blackburn negotiated her authority: "Liz is very modest, which is one of the reasons people like her. Nobody doesn't like her. She's a very genuine person. On the other hand, she's not a pushover." To illustrate, Marincola reported an incident that took place several years later, in December 2005, not long after she left the organization. At the society's forty-fifth annual meeting, when she was honored for her service, Marincola delivered an address:

After I gave this speech, a society member came up to me to comment on something I'd said about threats to science, such as the intelligent design movement, and this member in effect defended intelligent design and challenged me and wasn't letting up. He was determined to make his point with me at a quite inappropriate time (it was my moment of recognition) and he wanted to engage me in a debate. I'm not a scientist, and he wanted to pull rank, if you will, by citing scientific evidence. And Liz witnessed this happening and quietly moved in and said, "Actually, you're wrong in your arguments and let me tell you why." I think he knew who she was and was embarrassed to be called on his specious arguments by such an important scientist. She was very polite, very appropriate, but she clearly was protecting me and decisively putting this guy in his place. If she doesn't agree with you, she'll let you know why. She is never personally rude or nasty.[27]

As ASCB president, Blackburn became increasingly involved in efforts to lobby Congress on behalf of the scientific community. In 1989 the ASCB, along with the Genetics Society of America and the Society for Neuroscience, had been instrumental in forming the Joint Steering Committee for Public Policy (JSC). On behalf of more than forty-nine thousand researchers, the JSC advocated for federal funding for basic biomedical research, with the corollary aim of educating Congress on scientific issues.[28] Marincola, who also served as executive director of the JSC, defined it as "a prominent group of researchers supported by a coalition of scientific societies, a very special group able to do what none of the remarkable members could have done individually."[29] Blackburn served on the board of directors, which today includes Bishop, Gross, and Keith Yamamoto (all colleagues at UCSF), Paul Berg, Gerald D. Fischbach, Maxine Singer, Susan Taylor, Peter von Hippel, Terry Orr-Weaver, and Varmus, the current chair.[30] According to Marincola, "Liz helped found the caucus with these other prominent scientists. She herself has worked with Congress through the caucus. She never declines to do something like that. She says yes with her typical enthusiasm, as if she has nothing to do except these sorts of things."[31]

Marincola described Blackburn as a "critical member" of the JSC, but Blackburn claimed the JSC was "a fully formed entity before I became ASCB president. They had prominent scientists in this group. They didn't need me." It's hard to say whether Blackburn is being exactingly precise or "modest," scrupulously avoiding any hint of self-aggrandizement. If Marincola's report of Blackburn's generosity seems inconsistent with her persistent effort to protect her research from other encroachments on her time, the issues at stake were intensely personal for Blackburn along an axis on which she felt no inner conflict. The JSC was launched in response to congressional stinginess in funding biomedical research (primarily through the NIH), and Blackburn viewed publicly funded research as a means to ensure intellectual freedom and preserve science from a profit-driven, utilitarian ethos. The creation of the JSC marked a cultural shift in scientists' traditional abstention from politics, and Marincola reported that vigorous debate preceded this decision:

When we first started getting into congressional advocacy, there was a vocal segment of our membership who said scientists should not be involved in advocacy, period, because it cheapens what we do. We should be above politics. It's a dirty, immoral business. It took a lot of persistent argument for people like Liz and others, whom the members trusted and respected as scientists, to help them understand we are at the public trough, we owe the public our time and explanations and gratitude because we can't take public funding for granted. If we do, it will disappear. Now that seems so obvious—of course you have to explain NIH funding is money well spent—but at the time public funding was expected, and literally any interaction with a member of Congress was considered inappropriate for a scientist. They should stay in their lab.

Marincola emphasized that only scientists could have persuaded their peers to change this cultural stance: "Many old-school scientists believe nobody is as smart as they are except other scientists. The willingness of people like Liz to persuade their colleagues that we needed to engage our government leaders is so important because people like me couldn't do it on our own. Liz's power of persuasion comes not just from her personality but from her respectability as a scientist."[32] Given a cause that she perceived as fully compatible with her notion of scientific integrity, Blackburn betrayed none of the reluctance for politicking that had characterized her tenure as a department chair.

Since its inception, the JSC has made presentations to Congress, advised the Congressional Biomedical Research Caucus (a bipartisan group of congressional representatives), and served other branches of government. According to Marincola, the JSC acquired a reputation as "the premier caucus in Congress for the quality of its presentations and its staying power. Every month for fourteen or fifteen years, we've sponsored monthly briefings on the Hill, inviting prominent scientists who gave twenty-minute talks on their research and then answered questions. A lot of congressional staffers tell us this is how they learn about science."[33]

While Blackburn was president, the ASCB continued its advocacy and educational efforts in an atmosphere of optimism about public funding for scientific research. Thanks in part to the efforts of the JSC, Congress had steadily increased NIH funding so that it doubled during the 1990s.

Increasingly, Marincola reported, lobbying efforts came to encompass controversial debates on scientific policy: "We became very involved in issues beyond funding—research cloning and teaching of evolution in schools, any policy issues that bore on our members' ability to practice our craft. In the Clinton administration this was easier because the administration was generally sympathetic, very supportive of the NIH in general, and worked hard to get good appropriations. Since then it has been much tougher in terms of funding and policy issues and in terms of ideological differences between what our members care about and the priorities of the Bush administration."[34]

Inevitably, participation in the legislative process led scientific lobbying organizations into the realm of political controversy. In the wake of the cloning of the sheep Dolly in Great Britain in 1996, fears about human cloning had contributed to repeated efforts in Congress to pass bans on cloning that encompassed stem-cell research. The ASCB, ahead of other professional organizations, early on had advocated for funding for this research. Accuracy in defining scientific fact was and continues to be a part of the controversy, and Blackburn gained "familiarity with the ways of Washington" that would prove useful when she later served on the President's Council on Bioethics and deliberated national policy on stem-cell research.

Often the ASCB contributes to the political process simply by providing scientific expertise to legislators and striving to ensure that scientific findings are accurately represented in congressional debates. For a pending bill that would distribute research funds provided by a huge tobacco company settlement with the government, the ASCB had politely offered to help write the legislation to ensure the money went to legitimate, peer-reviewed research, and Blackburn met with Senator Ted Kennedy to consult on the wording of the bill. Once again advocating for accuracy rather than a specific policy decision, the JSC responded when NASA bolstered its case for funding for a space station by citing biomedical research that could be done only in space. Not all of NASA's claims were true, and the steering committee emphatically countered the agency's inaccuracies. Perhaps to woo the opposition, NASA invited Blackburn to watch a launch of the space shuttle in Florida from a VIP viewing site. But since no children under sixteen were allowed, she couldn't bring her twelve-year-old son, who would have adored seeing the launch from close up as much as she would have. If she felt uncertain about whether conscience barred her from accepting the offer, her son's anticipated reaction settled the issue: "Ben would have killed me if I went without him, and I would have heard about it for years."

As the publisher of *Molecular Biology of the Cell*, a peer-reviewed scientific journal, the ASCB also had a role to play in the growing debate over open access to scientific literature. The unrestricted sharing of research results within the scientific community can run at cross-purposes to the proprietary interests of scientific publishers, including a number of scientific societies, which often charge high subscription fees for their journals. Open access to data had significantly advanced the pace of scientific discovery when gene sequences were posted in a public data bank (GenBank), and the Internet provided a vehicle for disseminating scientific data almost instantaneously, holding out the promise of a genuinely global scientific community. The ASCB might have opposed the shift to open access, as many journal publishers did, but while she was the society's president, Blackburn supported efforts to make the society's journals more readily accessible, and as Marincola noted, she did so for a particular reason: "Liz cared very much about access to the scientific literature and journals to which scientists contributed. She felt strongly that research funded with public dollars shouldn't be restricted because of exorbitant subscription prices. She made sure the ASCB offered its own journals on generous terms and supported efforts to persuade scientific societies and publishers they should find a way to protect their own financial interests without unreasonably limiting access to publicly funded research."[35]

The ASCB staff debated how to balance the financial interests of the journal with the benefits of open access, and Blackburn credited David Botstein, editor of *Molecular Biology of the Cell*, for making the journal content freely available. With the proviso of a time lapse between print publication and online publication, the ASCB's journal became one of the first periodicals included in PubMed Central, an electronic publishing site developed by Varmus in 2000, when he was director of the NIH. In the same year, the nonprofit Public Library of Science (PLoS) was founded by scientists and physicians with the aim of providing a public online library archive of scientific and medical research, and Marincola served on its board of directors.[36] The nonprofit's first online journal, *Plos Biology*, launched in October 2003, publishes current research and, to cover the costs of publication, charges the authors of accepted, peer-reviewed papers a fee, waived in cases where authors cannot afford it.

Blackburn relished her term as president of the ASCB as a chance to effect change in the world beyond the lab: "Scientists need to function in the larger world. We had to speak up for scientific research if we wanted continued support for it." She had long since ceased to be a purist, entirely inward looking. Yet despite her success in this national forum, she would soon feel overwhelmed by departmental politics at UCSF. Immersed in pursuing a new line of cancer-related research and preparing to move her lab to UCSF's new Mission Bay facility, she faced another battle. On the strength of the dean's promise to her, Blackburn expected an office setup suitable for her work needs. But now some of this space was being delegated to others.

This time around, however, Blackburn reconsidered her pattern of strategic retreat. Denial had been a successful coping mechanism for many years; as Blackburn acknowledged in a published interview, "I was oblivious for a long time. . . . If I had stopped and thought about it, I would have felt so vulnerable to it."[37] But she had begun to register how costly this strategy was: "Maybe it was necessary for the earliest women in science to charge ahead, blinkers on—at a time when women needed to publish twice as much to be seen as equals in grant competition. Maybe it took that. But it was not a sustainable situation and not one I wanted to perpetuate."

Forced to contend once more with confrontational negotiations over space, Blackburn did so with an accumulated burden of anger. Again, she felt untutored—she simply didn't know how to do this. Self-doubt flooded her: she was a woman taking things personally when such encounters fell within the rules of war, as men understood them; she was so easily dismantled by the simple strategy of direct, aggressive opposition. Yet she was furious as she'd been only rarely in her life—when some students in the first class she taught treated her roughly in evaluations, when her colleague Ellen Daniell was fired from Berkeley. Her anger gave rise to many sleepless nights, "lying awake, upset, unable to focus on my research." In contrast to "my failure to use my clout to make a fuss when I was chair," Blackburn demanded the office space she'd been promised and fought hard—and successfully—to win this particular turf war. Yet she questioned the degree to which this was a victory. She had not turned to women colleagues for advice and support—"For fear of seeming weak, I had not availed myself nearly enough of this strength"—and she saw that acting alone was not enough to combat "the hidden costs of discrimination."

During this period in her life, three pressures had converged to reshape Blackburn's disdain for the accoutrements of power, each of which elicited a different response. Her experience as a department chair left her distaste for institutional politics intact, and she responded as she had so often in the past. After she stopped chairing her department, she retreated from public life at UCSF. Campus politics had forced her to confront gender politics, and she'd finally reacted by using her clout to gain a specific objective. But this did not dispel her anger or retrospective regret. She would not become a crusader or a whistle-blower, but she would speak out on women's issues whenever circumstances permitted her to do so in a way that didn't engage her in direct conflict— in interviews, on panels, and in speeches. Her involvement in scientific policymaking at the national level was astonishingly unclouded and untroubled in contrast to these other experiences, largely because in this context, she did not have to parse the hidden motives of others in order to get a clear line of sight on what she valued. When she next entered the public arena, it would be on terms that tested her integrity as a scientist, something from which she could not retreat.

8

An Interlocking System

In biology, no lone hero in a garret can come up with the kind of sudden, brilliant insight that lights up the sky in mathematics or physics and instantly offer conclusive proof. Knowledge is instead accumulated and tested in fits and starts, and progress is chaotic and contested. New findings that do not fit neatly within current assumptions may or may not immediately suggest how an old model might be revised or a new one created, and seemingly conflicting evidence can make many models provisional for a long time. For example, although Blackburn and Szostak's provocative experiment on yeast helped Blackburn to deduce the existence of telomerase, respected biologist Zakian insisted that no incontrovertible evidence existed for the activity of telomerase in yeast until 1995, when Marita Cohn in Blackburn's lab provided conclusive confirmation of its existence by conducting biochemical assays like those Greider had done to demonstrate the existence of telomerase in *Tetrahymena*. When the evidence supporting hypotheses is suggestive rather than conclusive, further experimentation can seem tenuous at best as it builds on a series of inferences that *may* have been proved. As Blackburn put it, "The work proceeds in boring little increments that don't look exciting to outsiders and often don't at first appear to promise any breakthroughs."[1]

The collaboration between the labs of Cech and Lundblad in 1997 to clone the gene for the protein component of telomerase (called TERT, for telomerase reverse transcriptase protein) illustrates both the daily reality of painstaking trial and error and the interdependence of different approaches to the same problem, by now a hallmark of molecular biology. After completing a second postdoctoral fellowship at the Blackburn lab, Lundblad had taken a genetics approach to the search for the protein component, trying to create more yeast mutants like *est1* in the hope that a random mutation would lead to identification of the correct gene. In Cech's lab, Joachim Lingner took a biochemical approach to the same problem, breaking open the cell to fractionate different classes of molecules and enzymes. Lingner used the activity assay developed by Greider and Blackburn, and adapted by Shippen, to isolate telomerase from a related protozoan, *Euplotes aediculatus*, and he faced many of the same problems in trying to isolate the protein component of telomerase: sparse quantities of the substance and extreme fragility, since enzymes could so easily become degraded. (The difficulty of tracking the correct protein helps to explain why Greider and Collins, using the same biochemical approach at about the same time, incorrectly identified two other proteins that associated with telomerase as the core component.)

Lundblad's genetic approach to the problem presented equal difficulties. A long-standing hypothesis held that EST1, the gene Lundblad had identified while working in Szostak's lab, coded for a subunit of telomerase, and a number of people were attacking this biochemically. (This hunch would turn out to be wrong: the gene coded for Est1, a protein different from the core component of telomerase.) Lundblad, newly arrived at Baylor University to head her own lab, decided to use a brute-force genetic screen to identify strains of yeast incapable of replicating their telomeres. Just as Lingner faced challenges in isolating the protein biochemically, Lundblad anticipated having difficulty distinguishing among mutant yeast colonies, because strains with genes that encoded for defective proteins merely associated with the telomere might present the same phenotype as a defective gene for this core component. The screen was performed by a couple of undergraduates, a technician, and Lundblad, and it took a labor-intensive eighteen months just to identify likely candidates. After subjecting cultures of yeast to chemicals that would create mutations in the DNA, the researchers plated out a million colonies of yeast that bore a spectrum of different mutations in the six thousand genes in the yeast genome.

Just a few people had to pick by hand thirty-five thousand good candidates from the total of a million colonies, identifying mutant colonies by their appearance and scooping up the yeast with toothpicks to insert them into microsized individual test tubes for the next stage of the experiment. As Lundblad wryly noted, "Plating out a million colonies of yeast is not that hard to do—with a medium on plastic Petri dishes, it takes a couple weeks to get colonies—but what we did to analyze those colonies was *work*." She had to go to great lengths to persuade her colleagues to keep at this task:

> About halfway through the experiment, the undergrads and lab tech said, "We can't do it anymore. It's too much hunching over a Petri dish with sterile toothpicks, too painful, and we're burning out." We were doing the work in sets of a quarter of a million colonies. I knew we were hitting gold, so we made a deal: for the last awful part of the experiment, an undergrad hired a dozen of his friends to come to the lab on a weekend, and they were paid one hundred dollars and as much pizza as they could eat all weekend long. And they picked fifteen thousand colonies for the next stage of the experiment, with music blasting as they did this mind-numbingly boring work.[2]

Conducting further analysis to see which genes among six thousand had mutated, Lundblad and her colleagues found twenty colonies that were incapable of replicating their telomeres, and Lundblad knew one of these must be the mutant for which she was searching: "When I knew the experiment was working, I had a hard time sleeping at night. The two undergrads and tech were pretty excited too. A lot of bench science is technically very redundant and meticulous, so you have to be pretty excited and then confident about what the outcome is going to be. Delayed gratification is a huge component of biomedical research." Ultimately, the screen led Lundblad to identify three proteins associated with telomerase, one of which turned out to be TERT. Another of these mutant strains, picked out by one of the undergraduates, had an unusual alteration of Cdc13, a protein that had not yet been associated with telomerase regulation. A combination of dogged persistence, a gambler's sense of possibility, and alertness had led Lundblad to the discovery of a protein that played an important role in protecting telomeres: "Just one of those million colonies treated by a mutagen had an unusual alteration of Cdc13, a classic needle in a haystack, and this single colony opened up ten years of research. We got it because we kept pushing the genetic screen."[3]

While Lundblad was deducing the amino-acid sequence for the protein component by reading the DNA sequence of the yeast gene she had mapped, Lingner was looking directly at the amino-acid sequence in *Euplotes* and working backward to get the gene. After Cech and Lundblad compared notes during a break at a scientific conference, they decided to share their results. Each had a partial but complementary picture of the gene. Lundblad had the gene sequence, but she needed biochemical tests to prove it coded for the protein component. While Cech and Lingner knew for sure they'd located the protein component, they didn't know which genes had to be knocked out to make it inoperative. Looked at cold, the amino-acid sequences might not have been decipherable, but the knowledge that telomerase is a reverse transcriptase, derived from the work of Greider and Shippen in Blackburn's lab, cued Lundblad and Lingner to look for a sequence that resembled those of other reverse transcriptases. These three different approaches could be triangulated—Lundblad and Lingner could compare the genes for the amino-acid complex in ciliates and yeast from a now-biased viewpoint. Viral reverse transcriptases, much studied, have characteristic patches of amino-acid sequences, so identifying similar patterns in each of these new sequences provided the final clues both researchers needed.

To confirm that they had isolated the correct gene, Lundblad and Lingner mutated amino acids at the proposed active site of the enzyme to see if enzyme activity would be destroyed. Work by Blackburn's lab supplied another essential tool; her postdoc John Prescott had further refined the enzyme activity assay for yeast telomerase developed by Cohn into a painstaking but reliable protocol. Not long after Blackburn reported this at a scientific conference, Cech phoned Blackburn to ask for the exact sequence of the DNA primer that Prescott had found to be the most successful, and on Blackburn's advice, Lingner used this same primer. These experiments demonstrated that Lundblad and Lingner had identified the correct gene: cells with these mutated amino acids had an "ever shorter telomeres" phenotype, the null phenotype expected if telomerase was inactive.[4]

Though Blackburn had stayed out of the race to clone the gene for TERT, the seminal work by her lab provided the necessary underpinnings for the successful effort. The story of telomere research, especially in its early days, is so interconnected, and Blackburn's work so central to it, that the research in her lab offers a useful locus for understanding an increasingly accurate and complex picture of how telomeres and telomerase function in the cell. Though mere numbers do not tell a complete story, a 2004 *Cell* review of telomere research by Cech roughly mirrors the field, citing twelve papers and reviews by Cech or his lab, nine by the Blackburn lab, and four or more papers by Greider, de Lange, Lingner, and Collins. Overall, of the seventy-three papers and reviews cited, fifteen were authored by the Blackburn lab or former members of her lab.[5] A more narrowly focused chapter on telomerase in the 1998 edition of *RNA World*, written by Blackburn, cited ninety-four scientific papers and reviews; of these, thirty-nine had been produced by her lab.[6]

Blackburn's first love is basic science, in which exploration of nooks and crannies can cumulatively produce a new leap forward, and if she has a distinct fondness for striking out on her own, she will also persist at the same experimental questions for years. Through the 1990s, Blackburn was fascinated by the RNA component of telomerase. Shippen and Greider had by then identified telomerase RNA sequences in two species, but when their results were compared, the sequences beyond the template still read as indecipherable—no guesses could be made yet as to their function. A natural next step, well-known from research on other RNAs, was to look at species more closely related to each other. As species evolve over time, their DNA sequence (and consequently the RNA sequence transcribed from it) undergoes random changes due to evolutionary drift, but evolution selects for those sequences that improve fitness and don't undermine the functioning of the organism. Blackburn was interested in looking at the sequences in the telomerase RNA component in different organisms to determine which regions remained the same, because this would provide the first clue to interpreting its role in the enzyme's functioning. Research on ribosomal RNA had progressed precisely because of similarly diligent comparisons of extensive sequences across species.

Compared with ribosomal RNA, telomerase RNA sequences have diverged rapidly among eukaryotes, so sequencing a wide variety of species took time and a diversity of methods that Blackburn summed up as "drudgery," nothing like the elegant universal method that was the preferred approach in molecular biology. As with efforts to clone the gene for TERT, Geron's intense interest in commercial applications for research contributed substantial resources to the task. The difficulty in accumulating this information is underscored by the fact that even today, far less is known about the RNA component of telomerase in plants than in other eukaryotes, because the gene has yet to be cloned out and the RNA component diverges so greatly from one group of eukaryotes to another.

Sequencing the RNA component in closely related species promised to provide clues that would enable researchers to compare the sequences of vastly different species; by homology, conserved sequences essential to the functioning of the enzyme could be identified. By the early 1990s a postdoc in Blackburn's lab, Dan Romero, had sequenced the RNA component of telomerase in a number of relatives of *Tetrahymena* (fellow protozoans), and comparison showed patches of conserved sequence.[7] Romero was faced with two analytic tasks: to identify the actual sequence of nucleotides that can be "read" like a word and to identify how this sequence causes the RNA molecule to fold in a particular way. Much like a computer program that processes complex functions on the basis of simple patterns of zeros and the numeral one, the deceptively simple arrangement of just four bases into patterns that comprise the RNA sequence (its primary structure) allows the RNA molecule to fold in the shape of helices (its secondary structure). To understand this structure would be to understand one day the overall architecture of the whole RNA (its tertiary structure) and how it is built in with the protein component of telomerase (its quaternary structure) so that the enzyme can do its job. Thanks to the work of scientists such as Noller, more was now known about the three-dimensional structure of RNA molecules and how sequence patterns offered clues to their helical structure. Blackburn's visit to Noller's lab in the late 1980s to share her rDNA

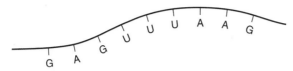

Primary structure of RNA

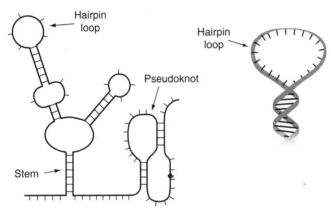

Secondary structure of RNA

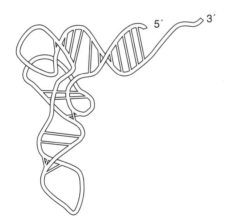

Tertiary structure of RNA

Figure 8.1
Primary, secondary, and tertiary structure of RNA. The primary structure of RNA is determined by the sequence of bases in the strand. The secondary structure is dependent on patterns that "mirror" each other, allowing for bases to bond with one another so that the strand folds in the shape of helices. How the entire RNA molecule folds in space is termed the tertiary structure (shown here for tRNA). *Illustrator*: Alan Stonebraker.

sequence, made largely to satisfy curiosity and speculate, would have bearing on her research now. As she noted, "Nothing ever goes away in science. That's why you're never rude to anybody."

Using color coding as a visual aid in discerning patterns, Noller had been able to identify complementary sequences in different places on the ribosomal RNA of several bacterial species. Because the RNA strand can flex in space, portions of it can bond together, forming a helix. Analyzing the sequence of RNA, researchers could identify patterns of base changes on one side of a helix, such as an A to a G, and find a complementary change on the other side (a U to a C); if such a compensatory change occurs in several species, it cannot be by chance but must confirm that this pairing forms a helix. This empirical observation could then be tested by physical methods—looking at the chemical and spectroscopic properties of a specific RNA. Where two portions of the molecule are brought together and held in the helix, a chemical that reacts to one side simultaneously produces changes that affect the other, demonstrating their proximity. When evidence from both sequencing and chemical analysis concurs, the helix is considered a "proven helix."

Before the protein component of telomerase had even been sequenced in any organism, Blackburn and other researchers had identified conserved sequences in the RNA component that determined how the molecule folded, though they did not yet know why it folded in this way. Among the first six telomerase RNAs identified by Romero, only half of the sequence was the same. Throughout the 1990s, Blackburn's lab sequenced a wide variety of telomerase RNAs from budding yeasts and protozoa, distantly related eukaryotes. Mike McEachern sequenced the first yeast telomerase RNA in *Kluyveromyces lactis*. Like thrush, this budding yeast has a small number of chromosomes (only six) with long repeat sequences in the telomeres, making it analogous to a high-yielding crop when it comes to harvesting the telomeric DNA. Elsewhere, at about the same time, through genetic screening, Dan Gottschling identified and sequenced the telomerase RNA for baker's yeast. These budding yeast RNAs were a shock—over 1,000-nucleotides long (compared to only 159 in *Tetrahymena*). The size of the RNA component of telomerase, as it turns out, can range from 142 bases in a ciliate to at least 1.3 kilobases in budding yeasts.[8]

Harold Stewart Blackburn and Marcia Constance Jack in June 1945.

The Blackburn children in 1952. From left to right: Katherine, Liz, Barbara, and John.

Liz Blackburn seated before a scale for a high-school photo in 1963.

Liz Blackburn and John Sedat in a classroom at Yale University, taken when Blackburn worked in the Gall lab.

Peter Challoner, San-San Chiou, and Liz Blackburn, 1979.

Members of Blackburn's lab at Berkeley in 1986. From left to right: Eric Henderson, Drena Dobbs-Larsen, Jim Forney, Carol Greider, Marietta Dunaway, Claire Wyman, Elizabeth Spangler, Judy Orias, Janis Shampay, Nicole Calakos, and Ed Orias (on a sabbatical year in the lab).

Liz Blackburn with
her son, Ben.

Liz Blackburn signing the Academy Register as a newly elected member of the
National Academy of Sciences, as James Wyngaarden watches, in 1993.

Joe Gall, Liz Blackburn, and Carol Greider, at Brandeis University, on the occasion of the 1999 Rosenstiel Award, shared by Greider and Blackburn. *Photo:* © Andrew Brilliant/Brilliant Pictures Inc.

Liz Blackburn and Fred Sanger in Cambridge, England, at the April 2003 celebratory meeting, "DNA: 50 Years of the Double Helix." *Photo:* Edward Ziff.

Members of Blackburn's lab at the 2003 Cold Spring Harbor Meeting on Telomeres and Telomerase. From left: Hinh Ly, Lifeng Xu, Liz Blackburn, Jue Lin, unidentified conference participant, Dan Levy, and Dana Smith.

Taking a boat ride on a canal in Amsterdam, Blackburn celebrates her 2004 A. H. Heineken Award for Medicine with the appropriate brew. Seated at the very right is Andrew Fire, recipient of the Heineken Award for Biochemistry and Biophysics the same year. *Photo:* © Capital Photos.

Blackburn and Elissa Epel, her collaborator in the stress study of women care-givers of chronically ill children, in a photo that appeared in the *New York Times* on November 30, 2004. *Photo:* Peter DaSilva/The New York Times/ Redux.

2004 family portrait of John Sedat, Ben Sedat, and Liz Blackburn.

Blackburn at the 2005 Franklin Institute awards ceremony, at which she received the Benjamin Franklin Medal for Life Sciences. Shown with the Benjamin Franklin medal laureate for electrical engineering Andrew J. Viterbi (left), and Dennis M. Windt, President and CEO, the Franklin Institute, Philadelphia. *Photo:* Rose Marie Riley, *City Suburban News.*

Elizabeth Blackburn in 2006. *Photo:* Elisabeth Fall.

Eventually, more than 30 ciliate RNA sequences were analyzed; the sequences differed by more than 80 percent. For some time, the length and sequence diversity of these telomerase RNAs made it hard to discern any common features within or across species. Yet despite this rapid divergence in primary structure, secondary structures betrayed telling similarities.[9] By 2001, Blackburn and her collaborators began to see some conserved elements in the telomerase RNAs of closer relatives of *Kluyveromyces lactis*, although these RNA sequences were so long that the researchers couldn't figure out if or how the molecule's folding structures would be related to those of ciliates and humans. By 2000, Greider's lab had sequenced the 400- to 500-nucleotide-long RNA component of telomerase in many vertebrate animals, from sharp-nosed sharks to human beings. A consistent core structure characterized this RNA in vertebrates and protozoans, though the vertebrate RNA had extra bells and whistles that didn't directly affect the enzyme's functioning but apparently influenced the stability of telomerase in the cell.

From 2003 to 2004, Blackburn devoted a yearlong sabbatical exclusively to research, collaborating with Yehuda (Dudy) Tzfati, a former postdoc in her lab who had moved to the Hebrew University in Jerusalem. When Tzfati hit on a way to write down on paper a similar folding structure for the RNA sequences of the two groups of budding yeast (relatives of *K. lactis* and of baker's yeast), Jue Lin and Hinh Ly, postdocs in Blackburn's lab, tested this hypothesis and confirmed it. Blackburn then tried to generate a unified theory of a common core structure for all telomerase RNAs. Many of the elements in Blackburn's universal model began to fall into place when she finally saw conserved structures in yeast that paralleled those thought to be confined to just one group of eukaryotes, such as ciliates or vertebrates. In vertebrate RNA a certain helix appeared not to fit the pattern that Tzfati had diagrammed, a frustrating oddity, since Greider had demonstrated that this helix was essential for enzyme activity and therefore likely to be conserved. Looking at the structures of these various telomerase RNAs, Blackburn realized they would all fall into a common pattern if the troublesome helix was simply repositioned to match the position of an equivalent proven helix in the RNA of yeasts and ciliates. If so, this helix would line up with another proven helix found in all telomerase RNAs. Clearly, all now *did* fit the same pattern.

In 2004, Blackburn synthesized from all this documented evidence a model for a universal structure for all telomerase RNA.[10] The sequences of these RNA components share a markedly conserved core structure, with relatively even spacing between different secondary structural elements largely conserved—an important hint that the RNA component of telomerase had functions other than providing a template for adding repeat sequences to the telomere.[11] Despite their nucleotide sequence differences, these RNAs can fold into similar overall shapes. A structure known as a pseudoknot, in which the RNA strand curves back toward a loop so that it looks as if it might thread itself through the loop to create a knot, is common to these RNAs. The template is flanked by a pseudoknot on either side, and these in turn are flanked by an enclosing helix.

Molecular genetics yielded another key way to identify and understand the properties of the RNA component of telomerase. Once again, what was learned from sequencing would be triangulated with what could be learned from experiments with mutation and biochemical analysis. This line of research in Blackburn's lab had begun in 1988 when Yu's experiments with mutations to the RNA template caused cell senescence in *Tetrahymena*. Lab members experimented with making a variety of changes to bases in the entire RNA sequence, not just within the template. The fastest way to determine the possible function of any stretch of the sequence was to manipulate the genotype and then see what happened in the cell—the same "shot in the dark" approach that characterized Lundblad's genetics work. The researchers had no map to suggest which changes in the telomerase RNA sequence might be

Figure 8.2 ▶
Conserved sequences in the RNA component of telomerase. These models of the secondary structure of telomerase RNA depict conserved sequences across species. Structures that are not yet known or vary among species within a group are shown by dashed lines. Adapted by permission from Jiunn-Liang Chen and Carol W. Greider, "Telomerase RNA Structure and Function: Implications for Dyskeratosis Congenita," *Trends in Biochemical Sciences* 29 (April 2004): 183–192, copyright 2004, with permission from Elsevier. *Illustrator*: Alan Stonebraker.

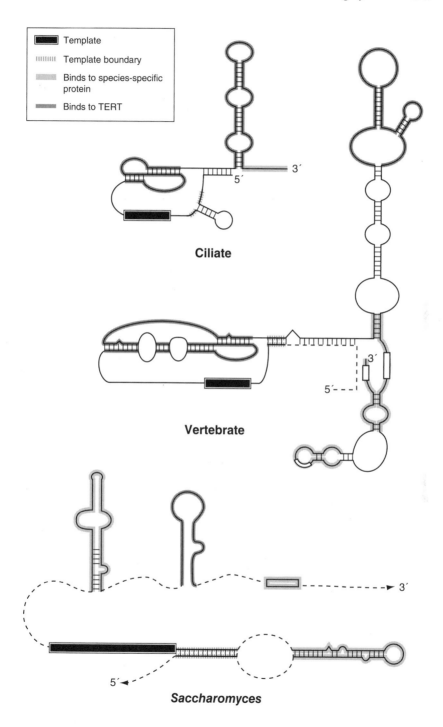

Template

Template boundary

Binds to species-specific protein

Binds to TERT

Ciliate

Vertebrate

Saccharomyces

crucial, so they couldn't predict the results for any specific experimental mutation of the gene. Blackburn reported that many changes had no apparent consequences for the cell, but then suddenly a change to a single base would have drastic consequences: "There's a scream, and a body falls down. This mayhem makes us very happy, especially when we're going in with one little dagger, not throwing a bomb into the room."

Blackburn and her collaborators were getting a lot of "hits" that produced violent results, another approach to the problem of speculating which patches of the sequence played a critical role. Those instances in which a tiny change had dramatic results for the cell began to suggest a pattern. Subsequently, other labs, including those led by Greider and her former postdocs Kathy Collins, Chantel Autexier, and Maria Blasco, showed similar results in experiments with mutant forms of the RNA component in human and mouse telomerase. They, too, would suddenly get an unexpectedly dramatic hit; sometimes it was understood why, and sometimes not.

For Blackburn, the most exciting results came from the hits that suggested the RNA component mattered as much as the protein component in enabling the enzyme to work. Members of her lab published papers on the results of these small changes to the RNA sequence, and experimental work undertaken from 1990 to 1998 showed that specific but small changes could stop the enzyme from working, which should not have been possible if the RNA functioned solely as a template for synthesizing new telomeric DNA. Minute changes made outside as well as inside the template region of the RNA really mattered—in some cases, the consequence was as if no enzyme existed in the cell. Now researchers could begin to assign functions to the regions of the RNA component. In humans and yeast, the base-paired RNA elements that form boundaries to the template enable the enzyme to terminate each cycle of reverse transcription.[12] The RNA also wraps the TERT into a shape necessary for the enzyme to carry out its job, providing binding sites for ancillary protein parts of telomerase. Working together, the labs of Blackburn and Cech found that in several yeast species, a bulged stem on the RNA component binds Est1, a protein essential for activating telomerase at the telomere.[13]

The miracle of enzymes lies at the heart of how biology works. Every enzyme targets a specific chemical or biological molecule (its substrate) and performs a transforming chemical reaction on it at a speed so devastatingly fast that it both awes and chagrins chemists, who cannot engineer chemical reactions even remotely as efficient. An enzyme can work so quickly because its protein component enfolds its substrate, which sits within a cleft on the protein. Certain amino acids have intimate contact with the substrate, thus enabling reactions far more efficient and accurate than would occur if the molecule were exposed to the same chemicals in solution. To determine whether a particular amino acid in the polypeptide chain is vital to the enzyme's work, a researcher can experimentally alter that amino acid in the chain, analogous to changing a single base in a DNA or RNA sequence. The consequences of such a mutation often cannot be predicted; as with the base changes Blackburn made to the RNA component of telomerase, sometimes a single change produces a hit that proves the critical role of a particular amino acid. But even such a dramatic result can be uncertain: the amino acid may be located in the active site of the enzyme or far away from the active site, since its flexible energy can travel to the active site and alter its action.

Cech and Altman had already shown that RNA could perform certain catalytic functions, posing a fundamental question for telomere researchers: Is the chemical reaction of telomerase actually carried out by the RNA or protein component of the enzyme? This possibility had intrigued Greider and Blackburn from the moment they became convinced that telomerase contained an RNA component. Telomerase would turn out to be a singular subset of reverse transcriptases. Reverse transcriptases are themselves a subset of DNA polymerases, about which a great deal is known. But knowledge about the behavior of these two other classes of enzymes did not automatically apply to telomerase, which is also a ribonucleoprotein (RNP). The protein component of the enzyme, which also has a three-dimensional folded structure (a polypeptide chain instead of a polynucleotide chain) may be made up of hundreds of thousands of amino acids. By itself the sequence of the gene for the protein component of telomerase could not illuminate how the protein folded, and only a portion of the *TERT* gene corresponds to the motifs identified in other reverse transcriptases. The much longer sequence of *TERT* suggests it may have other functions.

Researchers grappled with the delicate and complicated task of identifying how not one but two component structures were elaborately folded within the enzyme. In viral reverse transcriptases, certain changes made to particular patches of the amino-acid sequence located in the enzyme's active site had already been shown to block the enzyme from working. In identifying the *TERT* gene, Lundblad, Lingner, and Cech had mutated amino acids in TERT and clearly indicated that the protein, not the RNA component, of telomerase acts as the catalyst for the chemical reaction that joins a triphosphate (nucleotide plus a building block) to the telomeric strand. Other researchers found similar evidence in yeasts and ciliates, and after Geron isolated the human *TERT (hTERT)*, it was also shown to function in this way. Though Cech and Altman's earlier work had demonstrated that RNA *could* perform catalytic actions, the RNA in telomerase did not perform this task.

By the 1990s, Blackburn was stating definitively that telomerase RNA did not merely provide a scaffold for displaying the RNA template but enabled the enzyme to carry out its enzymatic action, polymerizing DNA.[14] The functioning of DNA polymerases, the enzymes that use the DNA strand itself as a template during cell division, could provide many clues to the tasks set for the "sequence-specific" telomerase. Yet unlike conventional DNA polymerases or reverse transcriptases, telomerase recognizes a specific primer, working best with a single strand overhang of G-rich DNA. A template-directed enzyme must make not just any chemical bond but the correct bond, discriminating between the right nucleotide to join to the strand and the three wrong ones within a nucleus bathed in all four of these essential building blocks. If the template contains C at the position at which the next nucleotide should be added to the DNA strand, then the enzyme will polymerize only the complementary base, G.

To function correctly, the enzyme must carry out a series of essential steps even before catalysis can occur: it must first align with (recognize) the correct triphosphate, complementary to the base within the RNA template (just as DNA polymerases match triphosphates to the base on the DNA strand); if by chance the enzyme joins the wrong triphosphate to the strand, it must somehow recognize the mistake and not elongate the strand further; it must then avoid getting stalled and copying the same base over and over; and finally, it must continue along the DNA strand in order to join the next correct triphosphate, moving it into the active site in order to continue rather than simply floating away from the DNA strand. Special protein components in conventional polymerases ensure this last process so that the enzyme doesn't simply fall off the DNA strand it is copying.

In Blackburn's lab, Dave Gilley had examined her favorite mutation to the RNA template: the simple, seemingly minor C to U mutation at position 48 that had caused *Tetrahymena* cells to die off in Yu's 1988 experiment, with their telomeres growing shorter and shorter, as if the mutant enzyme had been unable to add the expected telomeric DNA sequence. Gilley tried to determine how this mutation might affect any of the essential steps in a polymerase's function. He found something startling: as a consequence of this mutation, the enzyme was barely able to stagger along the DNA strand; it fell off before completing the template. In addition, the enzyme became blind to the difference between the bases G and A, so that before it fell off, 50 percent of the time it might add G and 50 percent of the time it might add A. Whether or not the enzyme made the correct choice, it still fell off before completing its task.[15] These experimental results provide conclusive proof that the RNA component of telomerase plays an essential role in the enzyme's ability to discriminate between one base and another as well as to continue copying the template. No one has ever found RNA to function in this way in any enzyme except for telomerase.

In conjunction with Cech's and Lundblad's findings, the work in Blackburn's lab cleared up the question of "who does what" in carrying out the enzyme's chemical reactions: subcontracting takes place to complete the enzyme's task of polymerizing the right DNA sequence, with *both* components carrying out different parts of the enzyme's work. The RNA component makes some crucial "decisions," and the protein component performs the actual chemical reaction. Researchers are still exploring the mechanics of this cooperative endeavor and examining how the actual three-dimensional structure of the molecule makes it possible.

For some time, Blackburn's curiosity had been aroused by the question of how the cell regulated telomere length. While at Yale she had wondered about the variation in telomere length in individual chromosomes in *Tetrahymena*, and her study of yeast with Shampay in 1988 found that telomere length varied within a fairly narrow range, suggesting that some mechanism in the cell regulated this process. In that study, Shampay and Blackburn noted that individual telomeres changed by small increments as cells divided. If nothing regulated this process, after several generations some telomeres in the cells would get continuously longer and some much shorter—a tremendous degree of variation at odds with their experimental findings.

Homeostasis, this tendency to sustain equilibrium, is a characteristic of all living things; every cell needs to keep its environment stable in order to reproduce itself. Yet if a cell—or an organism—is to respond resiliently to challenges, then many interacting components must reinforce each other, or the change to one component would throw the whole out of equilibrium. These dual requirements for stability and flexibility help to explain the complexity of living organisms. This natural characteristic of living organisms runs counter to molecular biologists' interest in elegant solutions; though a first breakthrough in scientific understanding might depend on a blinkered focus on a single principle, this emphasis has its limitations. A researcher might identify a simple principle—telomere length preserves the integrity of the chromosome—but discover on closer analysis that the process is subject to many interrelated factors.

Since the 1988 study by Blackburn and Shampay, researchers' observations of telomeres and telomerase showed that in microcosm, this system was also homeostatic. In further explorations of what happened when various components of this system were disturbed (through mutation, for example), researchers found that a complex of interactive components, not just a single determining factor, often supported centrally important processes, such as regulation of telomere length. By now it was known that proteins functioned in close association to regulate the activity of telomerase. Along with TERT, the core component of the RNP, other proteins associated with telomerase (ancillary proteins) enable it to do its job in the setting of the cell, and still other proteins (telomere structural proteins) bind to the telomere itself. Proper telomere maintenance depends on the interaction of all three, as demonstrated by Lundblad's work on the Est proteins in yeast. Est1 binds to the RNP, forming a bridge to Est4 (Cdc13), which binds to telomeric DNA, and Est3 binds to TERT (Est2) in ways that remain mysterious. Mutations that alter any of these proteins result in an Est phenotype. (Nomenclature varies for different specialties; for yeast, a distinction is made between wild-type and mutant genes—*EST* and *est*, with the product denoted as Est; but otherwise, acronyms are usually italicized for genes but not for proteins—*TERT*, TERT.)

A number of labs, including Blackburn's, now began in earnest to investigate the role played by the telomere structural proteins. Several studies had already provided clues that these proteins form a cap that shields the telomere from enzymes that would degrade it, thus regulating telomere length. While she was still working in Gall's lab, Blackburn recognized that telomere structural proteins differed from the chromatin on the rest of the DNA strand and so must perform a specialized role, and while at Berkeley she began using biochemical approaches to try to identify the telomeric proteins of *Tetrahymena*. Efforts in her lab persisted without success until the mid-1990s, when her grad student Phil Cohen found an unidentified protein that could specifically bind telomeric DNA from that ciliate. Yeast genetics provided a quicker answer to the question; by 1990, Judith Berman's and Roger Kornberg's labs each independently showed that a protein called Rap1 binds to yeast telomeric DNA in the test tube, and work from Zakian's laboratory offered the first evidence that Rap1 regulated telomere length in yeast.

Rap1 provided a prototype for telomeric protein in other species, although which proteins play primary roles in the carefully orchestrated interactions that protectively cap telomeres varies from one species to the next. Though a second-string player in human cells, Rap1 has a central role in yeast. Identifying its function led to insights into fundamental principles that remained the same for other organisms irrespective of whether cells normally had telomerase activity or not. Such discoveries also illuminate the shifting that takes place over the course of evolution; the same net result—a protective cap—might be wrought by any configuration of proteins, any combination of Lego pieces that could be fit together to build a fortress of the same shape.

The process of puzzling out how a complex of proteins works to protect the telomere is complicated on two levels: the technical difficulty of purifying and correctly identifying the scarce quantity of proteins involved and the conceptual difficulty of sorting out who does what in an interactive complex of proteins that bond not only with the telomere but with each other. To further complicate matters, different configurations of protein bind to the single-stranded overhang (the substrate of telomerase) than to the double strand of telomeric DNA. Titia de Lange's investigation of the human telomeric protein complex, known as shelterin, highlights the technical and conceptual difficulties of understanding these proteins.

In 1985, de Lange began working on telomere DNA in Varmus's lab because she thought it might bear an interesting relationship to genome instability in cancer cells, her primary interest at the time. When de Lange moved on to her own lab at Rockefeller University, she began looking for DNA-binding proteins specific to the telomeric sequence. Working with an extract from the nuclei of human cells, de Lange could rely only on her own hands and the help of a technician. But because these proteins were so difficult to purify, it soon became clear that she would need more than a thousand liters of cultured cells in order to succeed. So de Lange gambled the entire twenty thousand dollar supply budget in her NIH grant to purchase cultured cells in the quantity she needed, though if she succeeded in purifying a protein, it might not even be the right one. As de Lange recalled, "Even people in my own lab didn't believe in it, and I can't blame them. But you have to believe in your intuition that something will work."[16]

After de Lange derived the sequence of a few peptides of one protein, she still didn't have enough information to clone the gene (necessary to test its effects experimentally) and managed to do so only when another researcher deposited in a public database a short sequence of a gene that turned out to match the sequence of the protein de Lange's lab had purified. With this additional information, she isolated the gene in late 1994. The effort of years was merely a preamble for experiments in de Lange's lab to determine if this protein affected telomere maintenance. As it turns out, the way telomerase maintains telomere length in humans *is* affected by this protein, which de Lange named Telomere Repeat-binding Factor, or TRF1. As de Lange reported, "Having one protein in hand broke open the field. Because now we could ask, are there proteins that interact with this protein? Are there other similar proteins? We discovered a relative of this protein, TRF2, which looks much like it. Then we found interacting partners of these proteins that bound to the telomere, a complex of six telomere-specific proteins, some of which we discovered, some of which were discovered by others. We were able to begin to ask: How does this complex regulate telomere length and protect telomeres?"[17]

Blackburn's lab was looking at the same question—how telomeric proteins influence telomere maintenance—in yeast. In the complex of proteins in which Rap1 plays a part, it engages other proteins, such as the Rif1 and Rif2 (Rap-interacting factor) proteins, in ways that are still mysterious. After David Shore discovered Rif1 and Rif2, Susan Gasser and Eric Gilson found that every time Rap1 binds, like a C-shaped clamp, to the DNA strand, it puts a bend in the strand, thus influencing the actual architecture of the chromosomal end. From 1994 to 1995 McEachern, working with yeast cells in Blackburn's lab, found that certain changes in the RNA template caused a change in the DNA sequence added to telomeres by telomerase. Whenever a change destroyed the binding-recognition site for the crucial protein Rap1, it had two consequences: when Rap1 was just barely bound, some telomeres quickly grew twenty to fifty times longer than normal and some were greatly degraded, thanks to the action of enzymes that normally don't have access to telomeres; alarm bells went off in the cell, and the cell stopped replicating. Furthermore, the more seriously damaged mutants correlated with a decline in the ability of the protective Rap1 to bind to the mutant DNA sequence, so that it looked like Rap1 performed a critical role in telomere homeostasis.

Happenstance figured in McEachern's pursuit of these experiments. When he first came to the Blackburn lab, he had been stymied in efforts to clone out the gene for the RNA component of telomerase in *Candida albicans*, despite his familiarity with this organism. But once he had success cloning the analogous gene in *K. lactis*, he could manipulate the gene to create mutations, at first in an effort to show that he had correctly identified the RNA gene. In the process, he created a class of mutations that affected length regulation. McEachern termed one of these mutations "the first example of the delayed elongation class of mutants," in which "the cells initially produced short telomeres but were relatively normal. After I grew the things for weeks—I knew from my previous work in genetics that you had to wait and see—the telomeres suddenly got very long. This mutant was defective at length regulation; you had to replace most of the telomere with mutant repeats before getting defects. Until then, the internal repeats kept the telomere regulated."[18]

Further work enabled McEachern to identify a particular mutation that interfered with the ability of Rap1 to bind to the telomere, and this offered a new way to understand this homeostatic system.[19] McEachern observed that the implications of this evidence were very much a matter of debate: "At the time it was unclear in some people's minds whether or not telomeres were really going to be regulated in some sense. The model was passive—telomerase just replaced a sequence—but didn't incorporate the idea of negative regulation, which this experiment suggested."[20] The mere presence of telomerase in the cell did not guarantee its access to the telomere, and McEachern's study, soon corroborated, hinted that Rap1 and analogous proteins in other organisms could inhibit the enzyme's access to the telomere. In an experiment that clarified how the enzyme knew when to go to work, David Shore and his collaborators found that the number of times Rap1 bound to the telomere, which correlates with its length, regulated telomerase activity, a mechanism known as protein counting.[21]

In experiments that paralleled McEachern's, another postdoc in Blackburn's lab, Anat Krauskopf, also found that the proteins that coat the telomeres mutually reinforce each other. She made two tiny but different genetic changes that affected telomeres in yeast. One small deletion, which clipped off the very tail end of Rap1 (the C-terminal 28 amino acids), caused a minor lengthening of telomeres but had no deleterious effect on cell growth. The other change, an RNA-template mutation, resulted in the addition of mutant DNA repeats. By itself, this particular change barely affected the capacity of Rap1 and other telomeric proteins to bind to these newly created tips of the telomeres, causing only a modest elongation of the telomeres over many cell divisions. A cell with either mutation divided and grew normally to all outward appearances. But when these two mutations were combined in double-mutant yeast cells, catastrophe resulted.

Events unfolded in these double-mutant cells in an especially revealing way. Krauskopf spread the cells out on a nutrient agar plate so that each colony that grew from an individual cell was distinct. A healthy cell generates a sizable colony within a few days; if the original cell cannot divide at all or only a few times before its progeny lose this capacity, it produces either no colony or a small one with ragged edges, with the most recently dividing cells pushed out to the periphery of the colony. Krauskopf's double-mutant yeast, first spread on the agar plate, started off growing many healthy colonies, about the same number as normal cells might produce. The telomeres of these cells were merely slightly longer than normal, and their lengths all fell within a tight distribution range.

But twenty or twenty-five generations later, the colonies looked dramatically different from those of normal cells: some normal, large and round, and a fraction quite small. The small colonies also contained fewer cells, and their telomeres now fell into a bimodal distribution, either nearly normal or longer than typical. Though most of the telomeres were still apparently closely regulated in length, Krauskopf could identify a subset of telomeres whose length had increased enormously. The DNA of these longer chromosomes was also greatly degraded, which showed up on the gel used for separation as a streaking of these molecules—a sign that DNA has become partially single-stranded.

At the next few serial replatings of the cell on to agar plates, done successively some tens of generations apart, the cells became perceptibly and progressively worse at producing colonies. After a couple more replatings, most colonies had become small and sparse, with the percentage of well-regulated telomeres lessening each time. The telomeric DNA had now become mostly a mixture of very long and very short, partially degraded DNA molecules. With brilliant insight, Krauskopf realized that she was witnessing the whole population of telomeres converting, apparently telomere by telomere, from capped to uncapped—a situation analogous to popcorn popping, one kernel at a time, from one state to another entirely different one, with no transitional stage between. Any given telomere started off nearly normal and well regulated, and at some point abruptly became deregulated.

This study, published in 1998, provided crucial evidence for a two-state model of telomeres, capped and uncapped.[22] Normally, telomeres with a few mutant repeats could hold their own. But when the mutant RNA template was combined with a slightly impaired Rap1, the telomeres catastrophically lost all length control and protection. Granted unrestricted access to the telomere, telomerase added mutant repeats without restraint, constructing telomeres as much as fifty times longer than normal. At the same time, the telomeres became prey to enzymes that chewed back the DNA, exposing long overhangs of the strand added by telomerase—probably because the cell's DNA repair machinery attempted to fix what it recognized as a broken end. Not just telomere length but the capping status of the telomere determined whether it functioned to protect the integrity of the chromosome. The study also illuminated the resilience of this homeostatic system in response to a variety of molecular insults—until just one more change acts as a "last straw" and causes a disastrous collapse.

McEachern's and Krauskopf's experiments, along with similar research on cells that normally have plenty of telomerase activity (*Tetrahymena*, yeast, stem cells, and cancer cells), dramatically changed the way researchers understood the homeostatic system of telomeres and telomerase. A system propped up by various, interlocking, and even redundant components will not collapse until several of these components have been disturbed. Classic genetics has a bias toward examining certain kinds of processes—the obvious difference in phenotype that results from a single mutation, a *readable* phenotype—which may overlook the less apparent consequences of mutation in so complex a system. Once again, in biology first principles do not necessarily have strong predictive power; they can only get you so far. A mutation to a gene that seemingly produces no consequences might easily be dismissed as uninteresting, yet the absence of any effect may paradoxically be a clue that some compensatory mechanism exists for a gene that plays a crucial role or a signal that fluidly interconnected factors influence a given process.

When a mutation doesn't result in a phenotype change yet the gene seems to have conserved sequences, a geneticist may well have sniffed out something important enough to be provided with backup.

As researchers put together a picture of an increasingly complex system, the initial iconic notion of the telomere as a fortress fortified against all comers by a complex of proteins that collaborated in its defense gave way to a far less rigid conception of the molecular processes in the cell. The fortress metaphor could not account for the process that enabled telomerase to get on and off the DNA strand sometimes and *only* sometimes (at least during cell replication), though researchers could find no other convincing explanation.

A fortress wall that could be breached posed the question of how such a system could contend with the problem of broken chromosomes, which has been studied extensively in yeast and human cells. The cell possesses exquisitely sensitive machinery for sensing a break. In the best circumstances broken ends are quickly rejoined, but until that happens, the cell postpones dividing, since a broken chromosome without a centromere won't get inherited properly. Though normally broken ends are rejoined where they were severed, a broken end will join to any end of a chromosome, thanks to the work of DNA damage-repair proteins. Telomeres, as a specialized chromosomal end, are rarely mistaken for broken ends or fused to other ends, because such errors can lead to genomic instability and cell death. The only means of reliably preventing this seemed to be a barricade that prohibited access to marauders, the DNA damage-repair proteins.

Strong evidence challenging the fortress model came in 1995, when Tom Petes, a yeast geneticist at the University of Chicago, used genetic screening to find any factor that made telomeres shorter in various collections of mutants and then tracked which gene produced which mutation. He identified one such gene that encoded a protein he called Tel1. The analogous protein in human cells turned out to be ATM, which normally responds to DNA damage—in other words, it repairs broken ends. This was a complete surprise, given the expectation that the protein cap on telomeres would prevent such proteins from acting on telomeres. Over the next few years, other genes previously implicated in various aspects of DNA damage response started to show up as the cause of altered telomere maintenance. Against expectation, researchers found that many DNA damage-response proteins associated with telomeres.

The job of the telomere, it turns out, is not to fend off these proteins but to co-opt them—practicing a form of jujitsu that uses the force and energy of an opponent to defeat that opponent. These DNA damage-response proteins actually help telomerase to act on telomeres in yeast by interacting with telomeres in some way to make them "available" to telomerase, though researchers do not understand exactly how the cell regulates such interaction. In Blackburn's lab, Simon Chan showed that in yeast the DNA damage-response protein Tel1 prompts telomerase to work on telomeres, but how the protein does so remains a mystery.[23] It may make the enzyme more active, alter the DNA strand, or alter the complex of proteins that protects the strand. Corroborating work in Blackburn's and other labs found that in yeast and mammals, the DNA end-joining damage-response protein Ku is located at the telomere and required for telomere maintenance. The DNA damage response elicited by an uncapped telomere or broken end brings enzymes to these ends, and an uncapped telomere has the special property of diverting this response into telomerase action, which can keep even a short telomere capped and therefore stable. In a review, Blackburn suggested that a mutually reinforcing complex of telomeric DNA and proteins, including damage-response enzymes, modulates the cell's recognition of a telomere as capped or uncapped and triggers responses that maintain the telomere, including telomerase access and alternate backup mechanisms such as recombination.[24]

Throughout the 1990s researchers amassed evidence in a variety of organisms, but they could not confirm that the differing complexes of proteins in different organisms might function in universal ways. Scientific conferences provided an invaluable opportunity for comparing notes, and Blackburn regarded a 2000 conference on telomeres at Cold Spring Harbor Laboratories as instrumental to consolidating this knowledge: "Some of the proteins associated with telomeres and telomerase had been found in a ciliate here, a yeast there, and it was unclear whether they were an oddity of their system. What has become clear is that all of these components are found universally."[25]

In an exciting breakthrough first announced at a scientific conference in December 2002, Mark Zijlmans, a researcher in the Netherlands, used sophisticated imaging methods to examine live human cells in which selected proteins were tagged with fluorescence and observed in real time. Zijlmans found that one of the main proteins protecting telomeres (analogous to Rap1) came on and off the DNA strand every nine seconds. In his conference talk, he described this complex of proteins as resembling a cloud.[26] Major protective proteins exchange on and off the telomere, so that the protein complex resembles bees buzzing around a hive rather than a rigidly interlocked defensive wall. The proteins that coat telomeres do not fend off all comers but negotiate with them for access to the telomere. Though researchers had once assumed cell processes would prevent the access of the DNA damage-repair proteins, they visit telomeres all the time, yet *don't* cause the damage it was feared they would.

Understanding of this homeostatic system does not evolve smoothly, and often it is hotly contested as well as continually revised. In 1998, de Lange's continued work on telomeric proteins in human cells led her to an intriguing collaboration with Jack Griffith, an electron microscopist, in order to look at the actual structure of the telomere. In particular, de Lange was curious about the very tip of the telomere, which might differ structurally from its internal regions, but she credited Griffith for suggesting the experiment: "When I was exploring telomere structure, Jack Griffith asked me, is there a little nubbin at the end? Why don't we look for that? I didn't come up with this idea, because I was influenced by the dogmas in the field, the way we were drawing telomeres on paper. Someone from the outside, asking seemingly naive questions—how do you know it looks like that?—helps you to realize that you don't know something. In biology, breakthroughs come when you question what you think you know."[27]

This was "a horrendously difficult experiment," de Lange noted. "A human cell has 6 billion base pairs of DNA and only about 6 million bp of this is telomeric DNA—less than 0.1 percent." To purify the telomere DNA away from all the other DNA, de Lange used restriction enzymes that recognize sequences not present in telomeres—enzymes that could cut everywhere else in the genome but not in the telomere, so that telomeres would constitute by far the largest fragments. A series of steps, including gel filtration chromatography, separated the larger fragments of telomeric DNA from the smaller fragments of DNA and purified them a thousandfold. De Lange reported that when Griffith examined the telomeres using electron microscopy, "he found a very specific structure that makes a lot of sense from a functional viewpoint. The telomere end turned out to be folded back and tucked in to the rest of the telomere."[28] Work in de Lange's lab showed that a telomeric protein, TRF2, could make these types of telomere loops in the test tube.[29]

Armed with a working model that in the cell TRF2 makes the t loop at the chromosome end and so protects it, de Lange exhaustively tested the hypothesis: "We used endless controls to convince ourselves that the most obvious artifacts were not in play. About 95 percent of our experiments were experiments to address artifacts, and only 5 percent were discovery experiments." In her lab, de Lange is still exploring how this loop might protect chromosome ends: "One of the good things about an architectural solution—to tuck in the end and hide it from enzymes that could act on the terminus—is that if you mess with the machinery that makes the loop, things will go wrong and enzymes will start acting on the end of the chromosome. That is what we've found. When TRF2 is deleted from the mouse genome, all the chromosome ends get fused together. Enzymes that repair DNA go berserk and ligate ends of DNA together, and the whole genome is one big chromosome."[30]

This idea initially met with resistance, as de Lange had expected when she proposed a new model. Years before, Blackburn and Gall had hypothesized that the tip of telomeric DNA might end in a tiny folded back loop, held together by a different kind of base pairing. And McEachern, in presenting his telomere homeostasis model, had drawn the whole telomere as a folded back loop, with the looping mediated by proteins.[31] Yet initially Blackburn was among those who greeted de Lange's findings with skepticism. Her lab had done experiments in the test tube in which even double-stranded telomeric DNA could be invaded by complementary oligonucleotides added to the solution in the test tube, giving it the appearance of a loop. Since the results had been published obscurely, Blackburn sent them to de Lange and told her an in vitro artifact might account for these findings: "Titia and Jack's results were purely in the test tube, and our results had mimicked those they had found with telomeric DNA stripped of its proteins. Titia took this as resistance—but then, so would I, if I were in her shoes. People have now done electron microscopy where they didn't strip proteins off and can still see the loop, which makes it less likely an artifact was in play."

Blackburn was careful to explain why it's important to challenge a new model: "You want a decisive experiment in which you attack the hypothesis. You can also look for evidence that is consistent, which has been done for the loop idea. But as yet no one has been able to design an experiment that could knock down the hypothesis if it's untrue. You try to train students to seek not only more positive evidence but evidence that will cut the hypothesis off at the knees. If the hypothesis is right, it will stand. Less than this is *not* confirming proof."

Like de Lange, Blackburn has also experienced resistance to her new ideas. Since the discovery of telomerase by Greider and Blackburn, biologists had assumed that it served only to make telomeres longer, and any change they observed in the cell as a result of altering telomerase resulted from the consequent change in the length of telomeres. Much like Krauskopf's work, Lundblad's experiments with the Est phenotype showed that the effects of mutations might become observable only over several cell generations. Because individual cell lines survived for some time as the telomeres dwindled in length, researchers assumed that cell senescence occurred only when increasingly shorter telomeres dropped below some set minimum.

But by the late 1990s, subsequent studies hit on anomalies that challenged any simple equation between shortened telomere length and cell senescence. Blackburn was the first to propose that telomerase might have other functions in the cell in addition to maintaining the telomeres, an idea that even members of her own lab resisted. Two studies in her lab led her to this view. In the mid-1990s, Prescott worked on yeast mutants from which the normal telomerase RNA had been removed, thus inactivating the enzyme. These yeast mutants ordinarily stop dividing after a period of telomere shortening. But when Prescott added less active (hypomorphic) mutant versions of telomerase to these cells, the telomeres shortened and eventually achieved equilibrium at this new length, shorter than the minimum at which cells lacking telomerase RNA normally ceased dividing.[32] Yet the cells kept on dividing vigorously. Blackburn's lab then collaborated with Bishop's lab to conduct similar experiments on human cells. Working with Bishop's postdoc Jiyue Zhu, Blackburn's postdoc He Wang added a hypomorphic form of telomerase to cultured human cells that normally had a small amount of the enzyme or none at all. As a normal (nonmutant) cell line approaches senescence, the telomeres begin to fuse, but this did not occur in the mutant cells. Instead of following the usual curve (a gradual decline to some point at which the cells cease to replicate), the telomeres hit a low point of length and then the cells *continued* to divide for longer than expected.[33]

The expression of hypomorphic telomerase thus could result in short but stable telomeres rather than the gradual dwindling of telomeres and eventual cell senescence. The cells in both studies were seemingly immortal—the human cells grew for 450 days as opposed to a couple of months, and the yeast replicated forever. These findings uncoupled *bulk* telomere length from the ability of the cell to divide, breaking down a simple causal relationship between telomere length and cell senescence. For Blackburn, these results pointed to one conclusion: the role of telomerase in the cell extended beyond simply keeping the bulk of telomeres long. The hypomorphic telomerase that made much shorter telomeres than were normally required by the cell could still sustain cell life much longer just by being present. In other words, the enzyme functioned somehow to protect the integrity of telomeres, not just to elongate them. But these experimental findings had yet to be explained in molecular terms: what set of interlocking processes could account for them?

Blackburn's scientific thinking has a cyclic nature, something like contraction and dilation. Prescott remarked her skill at "asking a very pointed question in a way that will give you an answer, whether expected or not," and also recalled that she had "very strict standards of proof"; if he brought her unusual results, "it often took a few experiments to convince her."[34] Yet Blackburn's skepticism has qualities of radicalism as well as conservatism. Given her intimate familiarity with different model systems and her alertness to anomalies, she can fluidly transition from a narrowly constricted, descriptive mode to a risk-taking stance—a seemingly prescient leap to a new or revised model. Part of the process of revising or refining an existing model—or constructing a sound alternate—requires fitting together diverse data, and Blackburn felt studies on the protein component of telomerase (TERT) corroborated the need for a new model of how telomerase functions. Researchers had found that because of the way the gene for TERT is expressed, it can make different spliced products. When transcribed into RNA, the same gene often produces two or more types of messenger RNA. While some forms of messenger RNA that encode for the protein make a typical TERT with all its reverse transcriptase intact, other forms of messenger RNA make a TERT in which this reverse transcriptase and other portions have been snipped out. These proteins are continuous, though a portion is missing—evidence of alternative splicing. Blackburn interpreted this fact as support for the notion that telomerase has multiple functions: "Well, what can these differently spliced proteins be doing if they don't encode for a job telomerase does in the cell?" At least three labs reported on this phenomena, publishing in relatively obscure journals, but "because no one had a context for this data," Blackburn explained, "the field was sitting on these discoveries that could be exceedingly revealing—or not. We don't know."

To Blackburn's frustration—"These results tormented me"—all these findings on the unexpected effects of altering telomerase were regarded as anomalies: "The whole field seemed to be ignoring a clue that was staring everybody in the face." In reviews she wrote from the late 1990s on, including a 2000 review in *Nature*, she proposed a model of telomerase functioning that would incorporate this new evidence.[35] When Greider and Blackburn discovered telomerase, they met with similar resistance, since they were positing an enzyme that didn't fit with the customary understanding of how DNA was replicated in the cell. But Blackburn had been on the conservative side of the argument with respect to de Lange's experiments on the t loop. Blackburn sees this battling over the validity of new explanations for biological process as a necessary, if at times harrowing, test: "New ideas are problematic, to be approached cautiously and suspiciously. Have you thought about this straightforward explanation? Yet openness is necessary—we have to consider how it *might* work. These two modes bounce off each other. The tension point for me is when, dammit, I can see there's enough phenomena going on and controls are in place, enough to show the conservative explanation is not working anymore. You get mad at people—you idiots!—for not seeing this, yet you have to respect the reluctance to shift from a straightforward explanation."

Whether telomerase has other functions in the cell in addition to lengthening telomeres still remains an open, much-debated question, but contention has not dimmed Blackburn's interest in the notion: "I like this kind of question—one that hasn't made it into the canon yet. Maybe the explanation will turn out to be unexciting, but we still don't know." Seemingly known terrain had suddenly been translated yet again into unknown territory. A number of provocative questions followed on these findings: Does telomerase sit on telomeres and protect them? And if it is associated with telomeres at times other than when it is adding to the DNA strand, how does the cell distinguish between the enzyme's visits to perform this task and other kinds of visits? Work by Gottschling had already demonstrated that telomerase could add to the telomere only during a short period just before cell division took place, so beginning in 2001 Blackburn and PhD student Chris Smith investigated whether telomerase associated with the telomere at other times. Smith studied *Saccharomyces cerevisiae* (baker's yeast) cells at a stage in the cell cycle known as G1, just after two new cells have been formed and telomerase is incapable of polymerizing DNA. He used a chemical cross-link to detect the presence of the enzyme, adding to the cells a chemical that would make two nearby molecules stick together only if they were proximate to each other. By this biochemical means he confirmed that telomerase was present on the telomeres during this stage, functioning in some way unrelated to adding to the sequence. Like a number of proteins, the enzyme comes off the chromosome during mitosis but then immediately loads back on. This study provided concrete evidence that telomerase could actually be protecting the telomeres in some way.[36]

Further confirmation that telomerase did more than simply add to telomeres came from a study begun in 2001 by Chan in Blackburn's lab. Chan set out to examine what might be happening to telomeres just after mutation had deleted telomerase from yeast cells and long before catastrophe occurred. Normally, only a tiny fraction of telomeres in a normal cell undergo shortening—fewer than one in a hundred thousand, illustrating how rarely the cell's homeostatic system breaks down. Chan used an extremely sensitive quantitative assay to determine whether any individual telomeres had suffered shortening in the absence of telomerase. He found that at this early stage, one in several thousand was affected—a tiny increase over the norm within a single cell but potentially a significant increase in an entire colony or an organism with billions of cells. The shortened telomeres also started to fuse with other telomeres or their DNA strands began to break—a genomic instability that often presages cancer in more complex organisms. In the absence of telomerase, even long telomeres became unstable and incurred a slightly increased risk of becoming uncapped and chewed down by other enzymes.[37]

Current research on Rap1 in Blackburn's lab again illustrates how any notion of homeostasis must be continually revised in light of new evidence. Rap1, shown by McEachern and Krauskopf to protect telomeres by binding to telomeric DNA, has been purified by Tanya Williams and Dan Levy and they are investigating whether it also serves as a vehicle for bringing Rif1 and Rif2 to the telomere. Rap1, literally clamped on to the DNA strands, may provide some first layer of protection by itself, and then offer an additional layer when it binds the Rif proteins, which do the "heavy lifting," protecting telomeres from too much unbridled addition by telomerase. Rap1, when it binds elsewhere on the chromosome, acts as a repressor-activator protein, which can either block or activate transcription of DNA, depending on the interaction of multiple proteins. Similarly, Rap1 may serve different functions on the telomere, depending on its proximity to certain other proteins.

The good news is that so complex a system preserves the integrity of DNA via multiple pathways. The bad news is that this complexity undermines any notion of telomerase as a magic elixir that can simply be juiced up or down to serve therapeutic purposes. Yet the advance in basic scientific knowledge begun during the 1990s exposed tempting targets for therapeutic intervention, particularly with respect to mutations that resulted in the addition of the wrong sequence or uncapped the telomeres, provoking cell suicide. The basic science research conducted in Blackburn's lab naturally led her to explore the parallel strand of medical implications for human health.

9 | Dr. Jekyll or Mr. Hyde?

Since the early 1990s, when Blackburn first moved to UCSF, her lab has worked on telomerase in human cells in culture. In 1995, she began collaborating with clinical researchers to explore the potential therapeutic value of telomerase and derive a clearer understanding of its role in health and aging. The magic elixir West had once dreamed of would prove to have potential for good *and* evil in human health. Research from the late 1990s on confirmed that telomerase plays a protective role in the cell, stabilizing telomeres even when it is not lengthening them, yet the very high levels of telomerase present in cancer cells clearly contribute to the proliferation of the disease. In a 1999 review titled "Telomerase: Dr. Jekyll or Mr. Hyde?" Blackburn and her coauthor, postdoc John C. Prescott, warned that any therapies that manipulated levels of telomerase in the cell might be as likely to cause damage as to cure ills: "In the end, we would do well to keep in mind the words of Robert Louis Stevenson's Dr. Jekyll, 'I hesitated long before I put this theory to the test of practice. I knew well that I risked death; for any drug that so potently controlled and shook the very fortress of identity, might, by the least scruple of an overdose . . . utterly blot out that immaterial tabernacle which I looked to it to change. But the temptation of a discovery so singular and profound at last overcame the suggestions of alarm.' "[1]

Blackburn's comparatively small lab had managed to be enormously productive even when it came to competing with larger teams of researchers, and she liked her autonomy, in particular the freedom to explore a wide range of scientific questions in her work. When it came to scientific competition, she preferred to have elbow room, and as her former postdoc Mike McEachern put it, "If you go into the mammalian field, there's the risk someone is already doing what you are doing."[2] Yet increasingly biomedical research demanded that researchers wed differing areas of expertise. Dave Gilley, who now studies the role of telomerase in human breast cancer tissue, worked on *Tetrahymena* as a postdoc in Blackburn's lab from 1992 to 1997, and he spoke of the 1990s as an era marked by a crucial shift toward translational research:

> The field has changed dramatically since I was a grad student. It's been harder for people, maybe harder for Liz, to go from a culture of science where we could work independently on model organisms within the four walls of our own lab and do everything ourselves. That's not really possible today, and not desirable, because the science doesn't move forward. Research today is all about collaborations. The tools have become more abundant and more complicated, so that every lab cannot be expert on every tool—microscopy, clinical oncology, or mass spectrometry. I can go across the street or cross the world to form a collaboration and accomplish something in two weeks that would take me two years to perfect on my own. There used to be a big divide between PhDs and MDs, and there are still some old-timers reluctant to respect expertise in each field, but today we've learned how to work together more and more. Collaborating with a surgeon, I can get epithelial cells from a hundred different women. An assay I'm developing can in turn be useful to a surgeon, and he could never develop it. Liz has embraced this change after maybe some delay.[3]

Basic science research continues to provide a foundation for exploring questions of human health, and Blackburn's established reputation has helped her to continue to obtain funding to pursue both types of research. Often the basic science research in her lab intersects with her clinically related collaborations. For example, in the mid-1990s, when Prescott found that expression of hypomorphic telomerase in yeast mutants sustained shorter telomeres as cells divided indefinitely, researchers in the Blackburn and Bishop labs unexpectedly found similar effects in human cells. When hypomorphic telomerase was artificially induced in certain types of human cells that don't normally express it, the cells became immortal, without the telomere fusion and genomic instability that characterize cancer cells. The suggestive implications of such research are concentric rather than linear, providing a new understanding of the protective properties of telomerase, qualifying the notion that shorter telomeres can be equated with cellular senescence, and posing provocative new questions about the role of telomerase in cancer.

Questions about the relationship between telomeres and aging had perplexed researchers for some time, and what was understood about cell senescence in yeast and *Tetrahymena* paved the way for considering more complex organisms. Studies of human telomeres in white blood cells and fibroblasts (from connective tissues, usually skin taken from a biopsy, that are easy to study experimentally) showed that telomere length drifts downward over a lifetime. Telomeres are thousands of bases long in humans, and in adults they shorten by roughly thirty to sixty base pairs a year. Yet this varies greatly from one individual to the next—an eighty-year-old person's telomeres might be longer than those of a forty-year-old. Studies on twins led researchers to deduce that 40 percent of the variation in telomeres among people of the same age could be attributed to inherited (genetic) factors—which left 60 percent unaccounted for.

West had championed the theory of telomeres as a biological clock ticking down, and though this model still influences some researchers, Blackburn maintains that research by her own lab and others has already disproved so simple a causal relationship between the downward drift in telomere length and aging. A 1998 clinical study qualified even the notion of a steady, gradual decline. Working with the Blackburn lab, pediatrician Kevin Shannon examined blood samples taken from newborns to determine the length of their telomeres. Blood samples were drawn from these children for at least the first five years of life and also drawn from their siblings, parents, and even grandparents. Shannon found that in children from birth to age five, telomeres shortened at a *faster* annual rate than they did later in life, with the annual loss slowing to a steady rate in adulthood.[4] The notion of telomeres as a clock ticking down doesn't make good sense if their length declines so quickly in healthy, growing children.

Does this downward drift in telomere length matter in terms of health and disease susceptibility? What do we die of when we get old, and are

short telomeres a cause of mortality or not? If shorter telomeres correlate with advanced aging, so does wrinkled skin, but no one would
suggest wrinkles cause aging. Most elderly people die from cardiovascular disease (including heart attacks and strokes) or cancer, but the risk
factors that influence mortality don't seem to be relevant for centenarians, some of whom drink and smoke yet live much longer. Their unusual
longevity—only about one in a thousand of the total population in the
United States—remains an anomaly, while the thicket of risk factors for
age-related diseases entangles everyone else. Only if telomerase and
telomeres prove to affect such risks can they have a significant impact
on therapeutic treatment of age-related disease.

Researchers could ask more focused questions about this possibility
by studying a model system such as mice. Mice live only two years, in
contrast to the human life expectancy of eighty years and a maximum
life span of around one hundred years. Yet mice, like humans, have a
low activity of telomerase in most types of cells, so researchers can artificially create genetic mutations in mice on the qualified assumption that
similar processes occur in humans. After Geron researchers cloned the
human gene for telomerase RNA, by homology Greider was able to clone
the corresponding gene in mice. Greider collaborated with Ron DePinho,
whose expertise in mice systems complemented her expertise in telomerase, and by the late 1990s they and their postdoctoral fellows genetically engineered laboratory mice in which the telomerase RNA gene had
been deliberately knocked out, a process different from selecting for vari-
ants such as *est* mutants, in which the researcher waits on chance.

Just as in Lundblad's studies on ever shorter telomeres in yeast, the
damaging effects of the loss of telomerase in these mice did not become
obvious for several generations. The telomeres shortened only gradually
as the mouse cells divided. With every successive generation, the knockout mice had shorter telomeres, but because the telomeres in these lab
mice are naturally very long (five to ten times longer than in humans),
the shortening continued for several generations until in the sixth generation, the knockout mice lost the ability to breed.[5] The absence of
telomerase has the greatest impact on self-replenishing cells and germ-
line cells, in which it is more active. Germ-line cells multiply many times
before they give rise to an egg or sperm; in the absence of telomerase,

they lose some telomere length each time, so each fertilized egg begins life with dramatically shorter telomeres than its parents. Germ-line cells are by necessity picky about their genetic material and will commit cell suicide (apoptosis) rather than develop a genetically damaged egg or sperm. As soon as some threshold is crossed, a mouse without telomerase loses fertility.

These findings confirmed that telomerase is important for continued generations of a mammal. Greider and DePinho contend that the loss of telomerase has no major effect until the telomeres become really short. Conversely, Greider's former postdoc, Maria Blasco, argues that the absence of telomerase affects a mouse cell even before the telomeres become drastically shortened, because other signs of genomic instability cropped up before the mice became completely infertile. As early as the second and third generations, the mice had prematurely exhibited some of the characteristics of aging—heart problems and weakened immune systems that made them vulnerable to infection. Blasco, working in Greider's lab and later in her own lab in Spain, found that in a different strain of mice, problems cropped up in embryos and mice became infertile in still fewer generations.[6]

If studies on mice suggest that telomeres and telomerase play a role in aging, it took some time before studies on humans produced corroborating results. But by the late 1990s, one study in particular suggested a close relationship between telomere length and disease. When the average telomere length of arterial endothelial cells taken from heart patients was compared with that of healthy people the same age, the heart patients' telomeres were shorter by the equivalent of eight years. Richard Cawthon, at the University of Utah, developed a simple, reliable method for measuring telomere length that didn't require large amounts of DNA to provide an accurate reading. By a stroke of luck, Cawthon got the opportunity to study DNA samples taken from elderly people in Utah in the early to mid-1980s; once frozen, DNA could be preserved without deterioration. These subjects had been followed for seventeen years; many had died of natural causes. Now Cawthon could ask if any correlation existed between their telomere length at the time blood was drawn and their risk of dying from various causes later. Those people who had shorter telomeres went on to have three times the mortality rate from cardiovascular disease and eight times the mortality rate from infectious diseases such as pneumonia. Strikingly, over a seventeen-year period, those who had shorter telomeres when their blood was drawn in the 1980s had a higher rate of mortality overall than those whose telomeres were longer.[7]

Though such studies show a statistically significant correlation—cautiously termed an "association" in clinical studies—they still don't clarify cause and effect. Were the telomeres shorter before these people became susceptible to disease or as a consequence of the body's effort to fight off these diseases, which burdens cells in ways that might reduce telomere length? For example, infection strains the self-replenishing cells of the human immune system, which are among the few types of human cells in which telomerase is active, and under such stress, telomeres might run down because there is not enough telomerase to replenish them.

Opportunism again advanced biologists' understanding when Collins astutely conjectured a possible connection between a rare congenital disease and a small patch of human telomerase RNA. People who inherit dyskeratosis congenita, so named because its visible symptoms include skin disorders, die prematurely in middle age or sometimes earlier. They die of infections—essentially, bone marrow failure when their immune system wears out after trying to cope with challenges that other people would normally conquer. Clinical researchers had traced one form of this disease to a gene variant on the X chromosome (making it an X-linked disease, like color blindness, inherited by males from their mothers) and mapped it all the way down to the gene encoding for the protein dyskerin. Collins had already observed that one of the extra bells and whistles that kept human telomerase RNA stable was a region of the RNA with a structure similar to that of a special class of RNA that binds dyskerin. This special class of RNAs puts the finishing touches on the assembly of ribosomes, which conduct protein synthesis in the cell. The similar RNA structural motifs led Collins to wonder if the disease was caused by a problem with telomerase functioning. She found suggestive evidence that this might be so: patients with this X-linked disorder had shorter telomeres than did their unaffected family members. And when she measured the amount of telomerase RNA in these patients' cells, she found it was lower than normal.[8] But mutated dyskerin in these patients affected two aspects of cell biology: telomere maintenance and the maturation of ribosomes. Thus, shorter telomeres might be a side effect of problems with ribosomes rather than a cause.

This tantalizing mystery remained unsolved until an international team of clinicians published an independent study in 2001. This study had followed patients with a different form of dyskeratosis congenita, conducting gene mapping within families to identify the chromosome region that carried the disease-causing gene variant. A comparable gene-mapping technique, which compared diseased persons with healthy relatives, had been used by Mary-Claire King in the 1990s to identify a breast cancer gene, BRCA1, on chromosome 17. The 2001 study on families with a form of dyskeratosis congenita traced the mutation to a region of chromosome 3—*not* to the X chromosome, where the dyskerin gene identified in the previous study resides. This region of chromosome 3 includes about a hundred genes; one is the telomerase gene. The strong hint supplied by Collins's work enabled the researchers to home in on this candidate without having to winnow out the other genes, and it turned out to be the culprit, mutated in each affected person in three families.

Each of the three families had a different mutation on the telomerase RNA gene, and each rare but serious mutation had happened independently by chance. In one family, two changed bases in the telomerase RNA sequence lay in a conserved structure called a pseudoknot. Blackburn's universal model of the common core structure of telomerase RNA demonstrated that this structural motif, common in many types of RNA, was always present in telomerase RNA. Because telomerase RNAs vary naturally from one individual to the next, the changed bases might be just a harmless variation or truly the cause of the disease. Blackburn and her lab group designed experiments on the changed bases in this pseudoknot to determine which was the case. After they synthesized the telomerase RNA gene with the exact base change that occurred in one dyskeratosis congenita family, graduate student Melissa Rivera assembled the mutated telomerase RNA and TERT in a test tube and then conducted activity assays. In a related experiment, postdocs Lifeng Xu and Hinh Ly inserted a synthetically mutated gene into a cell, let the cell assemble it, and then did activity assays. Both in vivo and in vitro assays showed that the mutated gene produced less than 1/100th of normal enzyme activity; in other words, this mutation made telomerase all but dead.[9] When Collins studied the genetic variation in another of the three families that participated in the clinical study, she traced the mutation to a different essential region of the telomerase RNA gene. Later research on affected members of other families with the congenital disorder identified additional mutations in essential regions of the telomerase RNA gene. Recently, mutations in the gene encoding for TERT, the protein component of the enzyme, have also been associated with the disease in some families.[10]

These "clean as a whistle" conclusions linked the disease to telomerase functioning. Each person carries two working copies of the telomerase RNA gene, inheriting one from each parent, but people with dyskeratosis congenita inherit just one working copy (because the other is mutated) and can produce only 50 percent of the normal level of telomerase, a defect known as haploinsufficiency. Consequently, carriers of the disease, who might have a normal childhood, cannot survive to old age. These people die young because of a deficient immune system, though insufficient amounts of telomerase also result in faulty regeneration of other self-replenishing tissues or organs, from the skin disorder that characterizes victims of the disease to scarring in the liver and pulmonary fibrosis in the lungs. Significantly, the patients' symptoms resemble those of aging—blotchy skin, gray hair, and the inability of the bone marrow to supply the body with enough stem cells and progenitor cells to sustain a healthy immune system.

Telomerase sustains healthy telomeres not just by maintaining their bulk length but also by preserving the protective protein cap, so both these factors may play a role in the disease. In a 2001 longitudinal study of patients with dyskeratosis congenita and aplastic anemia (a similar disorder), clinician Monica Bessler analyzed telomere length in the patients' blood cells, with results that underscore this dual role. Patients always had shorter telomeres than healthy people, but as they aged, the bulk length of their blood-cell telomeres did not follow the typical gradual curve of decline that occurs in a healthy person, but remained at a steady, low level. It took years before this low level of telomerase resulted in blood cells that could no longer divide, though during that time bulk telomere length did not decline markedly, the expected signal of an imminent crisis.[11] A strong clue to the reasons for this apparently puzzling data comes from Simon Chan's 2001–2002 study on yeast, which showed that even if bulk telomere length fell within normal range, lack of telomerase resulted in a low fraction of telomere ends becoming uncapped, and even a low fraction of uncapped telomeres signaled the cell to stop dividing. Over time, perhaps not only shorter telomeres but also the uncapping of chromosomes causes catastrophe.

The results of these studies on dyskeratosis congenita contribute to a growing body of evidence that the damaged functioning of telomeres and telomerase bears some causal relationship to aging or age-related diseases. Shorter telomeres correlate with a shorter life both for patients with dyskeratosis congenita, whose telomeres are much shorter than those of their relatives, and for the elderly individuals studied by Cawthon. Significantly, dyskeratosis congenita can be pinned to a specific genetic change that cuts the body's capacity to produce telomerase by 50 percent. Though this is not final, confirming proof, it sharply increases the likelihood that the function of telomerase in maintaining healthy telomeres bears a causal relationship to age-related diseases such as heart disease and those that stem from the worsening ability of the immune system to fight off infection.

The concept of telomeres as a biological clock ticking down, with nothing to counteract this inevitable "mechanical" failure, has to be replaced by a model that accounts for the key role played by the presence of telomerase in the cell, which has a protective value that may fluctuate and might possibly be modulated. Thus, it may someday be possible to tweak telomerase production up a notch in patients with dyskeratosis congenita. The clear evidence that the haploinsufficiency of telomerase can cause serious illness has also led Blackburn and other researchers to wonder if natural variations from one individual to the next, not so extreme as to cause a clinical disorder, might also affect disease susceptibility. Researchers have already found such variations. While some of these may be harmless, others may bear a crucial, causal relationship to differences in how people age, as Blackburn's study on the pseudoknot in the RNA template of telomerase suggests. Studies on patients with aplastic anemia, while not yet fully conclusive, already support the notion that subtle variations in an individual's telomerase RNA genes may correlate with certain propensities for disease. For this reason, Blackburn ardently encourages clinical researchers, especially those in the field of geriatric medicine, to study telomerase genes when they analyze blood samples. Today, polymerase chain reaction (PCR) techniques provide a quick, efficient way to amplify DNA in a blood sample and sequence it, and the comparatively short sequence for telomerase RNA can easily be mapped by this method. Once enough individuals have been tested to create a large, statistically significant sample, it may be possible to identify a spectrum of variations that influence disease susceptibility. Research has already confirmed that centenarians do have longer telomeres than typical, and their children and siblings often demonstrate this tendency to live long, suggesting that the good fortune of a long life may lie in part in the genes.

The limited number of times human cells can divide has probably evolved to provide sufficient "reserves" for self-replicating tissues in a human who lives from thirty to forty years. In the last century or so, as the human life span has increased, these reserves may prove inadequate, causing diminished tissue maintenance and other physiological problems of aging. Conversely, whenever the number of cell divisions surpasses this limit, it poses a heightened risk of cancer.[12] As Dr. Jekyll, telomerase functions to maintain telomere length and the protective cap on telomeres, but as Mr. Hyde, telomerase run amok contributes to the abnormal cell growth that leads to cancer.

As telomere length gradually declines, cells face a succession of checkpoints at which they might stop dividing. Shortening telomeres may serve as protection against tumor growth by triggering cell senescence, but if this checkpoint is passed, telomere instability, in combination with many transforming events, may contribute to cancer progression. Bishop and Varmus, who shared a 1989 Nobel Prize in Medicine for their work, explored one such transforming event, identifying how viral oncogenes could transform a host cell into a tumor cell. Ordinarily benign genes, called proto-oncogenes, in the host cell can be captured by a virus or damaged in some other way so that their expression goes haywire. When researchers make human cells in culture precancerous by inserting viral oncogenes that prompt the cells to keep multiplying, disaster eventually occurs in most cells, with telomere fusion and broken ends resulting in cell suicide. Those cells that survive this crisis can catapult toward another—cancer—and grow and establish colonies. In the early 1990s, Greider, Harley, Bacchetti, and their colleagues showed that some of these precancerous cells activated their telomerase to avert fusion and broken ends, but others relied on backup mechanisms when telomerase activation failed, similar to a phenomenon first reported by Blackburn, Lundblad, and McEachern in studies on yeast lacking telomerase. Telomerase had been implicated in 90 percent of human cancers, but as yet it remained an open question whether highly active telomerase played a causal or corollary role in cancer progression.

During the 1990s, research on the activity of telomerase in cancer cells was a natural outgrowth of basic science experiments in yeast and *Tetrahymena*, which, like cancer cells in culture, normally have a high degree of telomerase activity and grow forever. Yu's 1990 study and similar studies of mutations that triggered cell senescence by disrupting telomere maintenance pointed the way to exploring how telomerase inhibitors might stop cancer cells from growing. Later experiments with cancer cells in culture and preclinical animal model studies showed that inhibiting telomerase activity in cancer cells reduced their capacity to multiply. But studies by Carolyn Price in 1999 and William Hahn in 2002 clearly demonstrated that telomerase was active, albeit in trace amounts, in human somatic cells, so that cancer therapies targeting telomerase might also affect healthy cells, just as chemotherapy does. Researchers could, however, possibly exploit a key difference: Telomerase can often be ten to one hundred times more active in cancer cells than in normal human somatic cells.

While the principle of inhibiting telomerase activity in order to slow or halt the growth of cancer cells seems straightforward, Blackburn emphasized that "the picture is not nearly so simple." Over time, research has uncovered byzantine complexities in the workings of telomerase in the cell. The fact that the enzyme carries its own RNA template makes it unique among enzymes, and thus its behavior can't simply be predicted by precedent. The enzyme adds to telomeres but only under certain conditions, and complex, interlocking factors, including the protein cap, regulate its access to the telomere. By itself, the bulk length of telomeres is not necessarily a marker of a cell's health, since telomerase has other protective functions, still not fully understood. A seeming discrepancy hinted that the enzyme might play some as yet undivined role in cancer or that unknown factors might be involved: though the telomerase level is high in late-stage cancer cells, the bulk length of their telomeres is often shorter than in normal cells.

Given her reservations about any straightforward means to exploit telomerase in cancer therapy, it's easy to see why Blackburn, as Gilley noted, did not rush to collaborate with clinical researchers on potential treatments. McEachern, another former postdoc, offered a perspective on Blackburn's reluctance that contrasted with Gilley's: "Liz has never really been at the forefront of pushing the connection between human cancers and telomeres but has been cautious about making a statement linking telomerase with cancer cells. She doesn't want to be on the wrong side of the argument, and she's got nothing to gain from pushing this connection too hard."[13]

But having little to gain gave Blackburn a unique vantage point on the hype, and her characteristic skepticism, complete with fulsome qualifications for any declarative statement, had hidden within it a wellspring of creativity. In 1995, Blackburn was asked by members of a UCSF consortium of breast cancer researchers and clinicians to study the effects of telomerase inhibitors on cancer cells. She recognized that any investigation of potential cancer therapies, which requires a wide range of scientific expertise, posed both risks and benefits for her lab: "Intellectually, my lab very much stood on its own two feet by this point. It would be a big departure to be interacting with a range of people. I would be a rookie in this new area, with all the attendant dangers of being a rookie. The work of my lab might become diluted if we branched out into unknown territory. Yet it would be impossible for me and my lab group to understand all the complexities of this disease on our own, and here were all these rich resources at UCSF. I could hook up with people whose expertise would enable me to investigate further than I could on my own."[14] Blackburn also had strong personal motives for her long-term participation in this consortium. Her faculty colleague Christine Guthrie had recently been treated for breast cancer, and in 2003, Anat Krauskopf, whose work had demonstrated the two-state nature of telomere capping, would die from the disease.

Blackburn anticipated that telomerase inhibitors, which impede the enzyme's ability to add the correct sequence to the DNA strand, might not be advantageous as a cancer therapy, primarily because their effects would cause telomeres to shorten only gradually, so that it would take several cell generations before the cancer cells might die. Triggering cell senescence, in which cells cease to divide but may survive for some time, provides a less certain means to combat cancer than any method that provokes apoptosis, or cell suicide. Research already going on in Blackburn's lab suggested a way to get faster results, so she told her new collaborators, "I have a more exciting idea. Let's get the highly up-regulated telomerase in cancer cells to make toxic telomeres." Firmly grounded in Blackburn's basic science research, this proposal was also a highly inventive, daring leap simultaneously spurred by Blackburn's reservations about the efficacy of telomerase inhibitors. Gilley remarked that Blackburn, like McClintock, had "a feeling for the organism," a "feeling for the biology plus an incredible ability to sympathize with all the different angles and approaches simultaneously. She's a real scientist because she's very, very creative, and that's what real science is all about, not collecting technical data but a very creative pursuit of the truth about the physical nature of the world. Liz is in that pursuit, and she has the curiosity and the talent and capacity to do it."[15]

Blackburn's interest in making toxic telomeres stemmed from years of related experiments in her own lab. When Yu found some mutations to the telomerase RNA template that caused *Tetrahymena* cells to sicken very quickly rather than gradually, he had dubbed them "monster cells." In the mid-1990s, Gilley, also working with *Tetrahymena*, had looked at base changes in the RNA template that made the enzyme copy the wrong sequences, and he coined the term "toxic telomeres" for these mutant telomeric sequences. At the same time, McEachern had discovered that in the unrelated *K. lactis*, he too could make mutant-template monster cells. Visibly deformed when looked at under a microscope, these mutants, created simply to prove the function of the template, produced a serendipitous finding: when telomerase adds certain mutant DNA sequences to telomeres, cells die rapidly, in sharp contrast to the gradual decline that occurs when telomerase does not add DNA to telomeres and they shorten over several generations.

Blackburn immediately wanted to explore this possibility in cancer cells—if organisms as distantly related as yeast and ciliates malfunctioned in similar ways, the mutation was likely to affect human cells too. Since cancer cells already exhibited revved-up production of telomerase, when a mutant-template RNA was introduced into the cells, they would avidly assemble this with the protein component of the enzyme and hasten their own end. While her clinical collaborators developed a method for delivery to cancer cells, over the next decade, Blackburn's lab worked on devising a mutant form of telomerase that would create toxic telomeres.

The task of exploring how mutant repeat sequences might crash cancer cells first fell to Rivera, a doctoral candidate in Blackburn's lab. A recent graduate of MIT and the first member of her family ever to get a college degree, Rivera had just arrived in Blackburn's lab, despite her New York–based family's reluctance to see her relocate to the West Coast. Working with human breast cancer cells, Rivera tried to mutate the RNA template of telomerase in half a dozen different ways, each designed to synthesize mutant DNA and consequently disrupt the binding of telomere structural proteins. Her experimental procedure included a mechanism to turn this mutant gene on or off through the addition of a specially tailored molecule that could be added to the cell culture. Before she could even conduct the experiment, she had to build an elaborate system. To introduce the gene into the cells, she created a plasmid vector (as had been done in Blackburn and Szostak's yeast experiments) that could be integrated into the cell's DNA. Next, to select for cells that had integrated the genetic material from the plasmid vector, she inserted a drug-resistance marker. She also had to track cells with five different mutant templates as well as a control that had wild-type telomerase RNA. In addition, for each cell line to which she'd added the mutant telomerase RNA, the wild-type gene had been left intact on the assumption that once turned on, the mutant version would overwhelm the wild-type gene, as had happened in Yu's experiments with *Tetrahymena*. After painstakingly treating the cell cultures, Rivera succeeded in creating a few hundred cell lines—a huge number to manage.

In the days before Blackburn began seeking her own funding for clinical research, her lab shared a tissue culture room with colleagues down the hall, and the resulting busy traffic increased the risk of contaminating a cell culture with mold or fungus. When some of Rivera's cell lines began to grow poorly, she feared that their slow growth had resulted from contamination, though she could not be sure of the source of the difficulty. The frustration and the apparent failure of her desperate rescue efforts convinced Rivera to take a year off after she passed the qualifying exam for a PhD. She found work in a biotech company, troubleshooting for the big machines that do DNA synthesis, a comparative respite from the pressured world of the research lab.

At about the same time that Rivera took a leave of absence, Strahl left the lab to care for her husband, stricken by cancer, and to replace her, Blackburn hired lab technician Inna Botchkina, who had extensive expertise in human tissue culture. Botchkina began attending to Rivera's cell lines to see if they could be salvaged. She soon told Blackburn she'd observed an interesting phenomenon: while the dozen or so wild-type cell lines (the control) were growing well, all the cell lines with mutant RNA templates were not. This gene expression should not have happened; since these mutant genes were supposedly still turned off, they should have become active only when the "trigger" molecule was added to their nutrient broth and absorbed by the cells.

Blackburn knew the genetic system that Rivera had used was not watertight—a low amount of gene expression occurred before the gene was activated. Precisely because telomerase is *so* active in cancer cells, these cells had voraciously incorporated the mutant RNA. Botchkina's observations immediately suggested that even though a cell's wild-type telomerase RNA gene remained active, adding just a tiny bit of mutant telomerase made a significant difference, lowering the growth rate of the cancer cells tenfold.[16] Most likely, the mutation somehow disrupted the protein cap on telomeres, and if even a few telomeres in the affected cell became uncapped, the cell would commit suicide.

When Rivera came back to the lab after a year, the work she thought had been for nothing became the basis for her continued research. This time, she avoided working directly with the tissue culture and instead did biochemical reactions with the mutant telomerase complexes in these cells, eventually teasing out roles for the bases in the RNA template and pseudoknot in telomerase's enyzmatic activity.

Roughly parallel with the work taking place in Blackburn's lab, Greider and DePinho continued their collaboration, working with mouse models to study cancer. By the late 1990s, their knockout mice, which could not produce telomerase, had also been genetically selected for a proclivity to develop tumors. Greider and DePinho measured tumor growth in these mice and discovered it did not differ greatly from tumor growth in mice with telomerase.[17] Since telomerase is highly up-regulated in human cancer cells and frequently in mouse cancers too, the susceptibility of the knockout mice to tumor growth was puzzling.

How could Greider and DePinho's findings be reconciled with the fact that in the later stages of human cancers, telomerase is highly active, transforming into a Mr. Hyde that contributes to accelerated cell growth? A number of studies, including the collaboration between the Blackburn and Bishop labs in 1999, have shown that inducing telomerase activity in certain types of human cells can immortalize them without necessarily triggering cancer, confirming that multiple transforming events must occur before cells become cancerous. In a different experiment on another strain of cancer-prone mice, Greider and DePinho demonstrated that telomerase might act as a Dr. Jekyll by providing some protection against the early stages of cancer.[18] Hints that this might be the case in humans have come from studies on dyskeratosis congenita patients who prove to be more cancer prone than unaffected family members who have a full complement of telomerase.[19] These suggestions reinforced Blackburn's notion that creating toxic telomeres might prove a more effective cancer therapy than devising telomerase inhibitors, which could diminish any protective function of telomerase.

By the time a human tumor is detectable, many genetic mistakes and changes to cell characteristics—epigenetic changes—have already occurred, so investigating any causal relationship is tricky. Either telomerase was up-regulated in advanced cancers because it helped to trigger cancer or sustain abnormal cell growth, or the abundance of telomerase in cancer cells was merely one of many changes consequent on abnormal growth, and the up-regulation didn't play a causal role. Though Greider and DePinho suspected the latter, Robert Weinberg at MIT was unconvinced; to generalize from the mouse model to humans was to compare apples and oranges. Starting in 1999, he began to mutate particular genes in human cells in culture and then to "interrogate" them for their cancerousness. Using viral oncogenes to trigger cancerous growth in human cells in culture, Weinberg traced the genesis from precancerous cells to cancerous ones. He found that to be most cancerous, cells had to have telomerase; if the viral oncogene could trigger rampant growth, only by inserting telomerase into the cells could Weinberg sustain this growth.[20]

Yet even this evidence is not the end of the story. Telomerase may prove necessary to the growth of cancer cells for two possible reasons: because of its ability to add telomeric DNA or because of some other function of the enzyme that affects cancer progression. Assays for telomerase showed that one-third to one-half of the human cell colonies Weinberg studied relied on a backup mechanism instead of up-regulated telomerase. In a more recent study, Sheila Stewart and William Hahn, postdocs in Weinberg's lab, compared these two types of cell colonies. The most cancerous cells were those with telomerase, though in a lab dish both types of colonies had initially grown at the same rate. By itself telomerase is not the cancer-causing property, but its presence makes a stark difference in the ability of cancer cells to proliferate.[21]

The artificial insertion of oncogenes into cells in culture differs from the slow pathway to cancer that might occur in living human beings, who may not fall prey to the disease until long after their exposure to carcinogens. In the later stages of cancer, cancer cells not only multiply rapidly but also migrate throughout the body (metastasis) and blithely continue multiplying in the absence of signals from surrounding cells to do so. Research into the role telomerase plays in cancer cells had focused exclusively on its effect on the cells' ability to proliferate, but not these other properties.

In 2001, experiments in Blackburn's lab unexpectedly implicated telomerase in the metastasis of cancer cells. Postdoc Shang Li, who joined Blackburn's lab in 2000, continued the experiments on mutant-template telomerase RNA begun by Rivera. He conducted a simple experiment that he expected to have a simple result. Working with a variety of cancer cells, including those that Rivera had worked on, Li decided to "knock down" the resident wild-type telomerase RNA, already present at high levels in cancer cells, using a newly developed molecular genetics trick, RNA interference. Since the mutant-template telomerase RNA has to edge out the resident RNA in order to be assembled with telomerase protein into active telomerase, Li was effectively wiping out the competition so that the cells would succumb even more quickly to the mutant-template RNA. Li had anticipated cells would respond gradually to RNA interference, as in previous experiments in which telomerase inhibitors had limited the capacity of cancer cells to multiply. Instead, the cancer cells in which the telomerase level had abruptly been lowered quickly stopped dividing within a few days, *before* mutant RNA was even introduced and *before* their telomeres had time to shorten or become uncapped. To entirely rule out this last possibility, Li conducted a sensitive assay, adding to his cell samples fluorescent antibodies that travel to uncapped telomeres and looking under the microscope to determine if this had happened. It hadn't.[22]

For some unknown reason, then, these cells had stopped dividing almost instantly, leaving Blackburn and Li initially at a loss. But they knew that sometimes cells behave differently than predicted because their gene-expression pattern has changed. Any number of interlocking factors can influence whether a gene in a cell is turned on or off, and now Blackburn and Li wondered if the high level of telomerase in cancer cells might influence the expression of other genes. By this time, the human genome had been mapped and new gene-expression profiling technology was available, so they could cast a wide net to test this hypothesis. They profiled the expression of all the genes in the affected cells, using microscopic robotic systems to lay out DNA from every gene in a cell on the surface of a slide. In this process DNA molecules are prepared in solution, and each gene adheres to the slide in a separate, minute droplet. The technology is so precise that a researcher can choose to lay out genes on a grid according to which chromosome they can be found on and then locate a gene by its row and column on the grid.

Working with UCSF clinical collaborators Julia Crothers and Christopher Haqq, Li next extracted the cell's RNA, a complex mixture of all the transcripts from all the genes—essentially, a representation of how actively each gene is expressed in a given cell. The researchers added a fluorescent tag to all this RNA. When this mixture, in solution, was added to the DNA already immobilized on the slide, each fragment of RNA could find its matching gene sequence by base pairing, bonding with the sequence from which it was originally copied. If a gene was turned off in the cell (and thus not making RNA), no fluorescent RNA would attach to it. If a gene was highly active, it would now have a bright fluorescence as several strands of RNA attached to it, and other genes fell along a corresponding spectrum of brightness. Now the researchers had a complete picture of the gene expression in the cell and could compare cells in which telomerase had been knocked down with a control group of cells.

The question that Li and his collaborators asked was neutral, without any assumptions: Are any genes more or less active in one group of cells than the other? In the cells in which telomerase had been knocked down, about seventy-three out of approximately ten thousand genes interrogated had a lower level of expression than in the control. Dramatically, one of the genes whose expression was dampened by telomerase deprivation had been implicated in metastasis.[23]

In previous experiments, telomerase inhibitors had prevented the enzyme from carrying out its job of adding correctly to the DNA strand but did *not* change the amount of telomerase in the cell. To actually "silence" the gene in cancer cells so that it did not make telomerase had not been tried. And this study also showed that telomerase knockdown abruptly stopped growth in many types of cancer cells, including bladder cancer and breast cancer cells, in which the enzyme was up-regulated. These results opened up new avenues for exploring how telomerase levels might influence the expression of other genes in the cell, particularly those implicated in tumor growth and progression, and also identified an Achilles' heel of cancer cells: a sudden drop in the level of telomerase had dramatic, quick effect.

Li soon teamed up with Mo Kashani-Sabet, a clinical researcher who studied cancer in mice in another lab at UCSF, to study melanomas and hunt for any evidence that the cancer cells' susceptibility to telomerase deprivation held true for living organisms as well. They devised a way to knock down telomerase RNA in melanoma (skin cancer) cells and then injected these cells into the veins of mice. After several weeks, they killed the mice and analyzed their lung tissue to determine if the melanoma cells had metastasized to form tumors, which appear as black spots on the lung. (When metastatic cells travel in the bloodstream, they can zero in on tissue, normally lung cells.) When Li and Kashani-Sabet compared these results to those of a control group, they found that mice injected with melanoma cells in which telomerase had been knocked down developed fewer metastatic tumors.

Though Li and Kashani-Sabet had no idea how this mechanism worked, the swift, dramatic decrease from the normally high level of telomerase in these cells to a very low level rapidly changed gene expression. Lowering the amount of telomerase in cancer cells interfered with their metastatic capacity—an impact that has *nothing* to do with telomere length or uncapping.[24] Blackburn likened this difference between cancer cells and normal ones to the difference between an average person and a heroin addict: "Generally we can do well without heroin, but addicts become so adapted to the drug that when they go cold turkey, they suffer severe physical symptoms of withdrawal. Perhaps the cancer cell is like a heroin junkie that can't function without its drug—telomerase. The homeostasis of its physiology is adapted to a high level of telomerase, and if that's taken away, the cell goes into shock."

Thus, for completely unexpected reasons, reducing the level of telomerase in cancer cells may one day provide effective cancer therapy, especially if it targets the metastatic properties of cancer. Blackburn, working with other members of the UCSF breast cancer consortium, is continuing efforts to develop clinical applications for the findings of Li and his collaborators. Recent research suggests yet another new direction for cancer research. Since the bulk of cancer cells multiply rapidly (aided by the up-regulation of telomerase), most anticancer medications target this proliferation of cells. But the very slow appearance of cancers, which can take decades to metastasize, has led some researchers to believe there are cancer stem cells, few in number, that do not proliferate quickly. Though embryonic stem cells can potentially create any type of human cell, adult stem cells are specialized, capable of creating only a single type of cell and responsible for replenishing tissues.[25] About forty years ago, researchers first began to suspect that specialized stem cells repopulate cancer, and they confirmed that as some cancer cells die out, stem cells resupply the population.[26] These cancer stem cells, which may originate from adult stem cells, can lurk within a tumor, so that even if medical therapy eradicates fast-growing cells, the stem cells remain, capable of dividing forever, if at a slower pace, and eventually produce more cancer cells, including those capable of metastasis and rapid growth. Telomerase has been found to be active in these cancer stem cells or their earliest progeny, which opens up yet another new opportunity for developing effective cancer therapy.

The relationship between basic science and clinical research cannot be mapped as if a clear route leads from one to the other. No one could have imagined that Blackburn's first exploration of the gene sequence of a tiny, quirky protozoan could have led to significant clinical research on human aging and cancer. And who would have drawn a direct line from Yu's monster cells to cancer treatment, or from Blackburn's effort to characterize the core structure of telomerase RNA to a disease that inflicts premature aging on its victims? In biological research, chance plays such a large role—unexpected findings that redirect an experiment, frustration in one line of investigation that forces someone to try another, or felicitous timing in which one study illuminates the potential or rewrites the implications of another. But exceptional scientists have a fortunate and receptive relationship with chance—a facility for recognizing the right time to take a gamble.

Blackburn has contributed enormously to biomedical research not only because she has made novel discoveries but also because a combination of bold inventiveness and meticulous attention to detail has enabled her to contribute substantially to the emerging picture of an enzyme that plays a crucial role in human health. Gall, Blackburn's former mentor, viewed the clarity of Blackburn's thinking as an essential impetus for the field:

I believe firmly that progress in medical matters ultimately rests on information gotten through basic research. At the same time, every advance in research doesn't have an application right around the corner. Clearly, understanding the structure of chromosomes and how they replicate has to be part of the underpinning of any understanding of abnormal growth. The ends of the chromosomes clearly presented a molecular problem in terms of how cells maintain and replicate chromosomes, and Liz has provided a very clear understanding of how that process works. This doesn't lead immediately to a new drug or cure for cancer but provides the basic theoretical understanding needed to think about any advances. Her work is extremely important. It wouldn't at all surprise me if a Nobel Prize came out of it.[27]

Gilley, a leading telomere researcher in his own right, emphasized that subsequent investigations have depended on Blackburn's initial forays: "Because Liz's foundation was so incredibly strong—as opposed to building research like a house of cards, when you find everything collapses when you get to the top, try to pursue larger implications— hundreds of researchers can go forward from her early work to explore really important aspects of human health and understanding biology. She'll very likely win the Nobel Prize in the future. If not, something's wrong—the prize is not always given to the right person." Tellingly, Gilley added, "I don't really think she would care much about that. If she thought about it, she'd probably be very embarrassed by it."[28]

10 Members of a Guild

When Dana Smith joined Blackburn's lab as a staff research associate in 2000, one of her first tasks was to organize twenty-five years of lab notebooks, valuable hard evidence of the lab's research, which had been stored in random order. Given the sheer physical volume of the lab notebooks—produced at a rate of about three shelf feet a year—this assignment proved daunting. As Smith explained, "The stored notebooks are a kind of running timeline. Postdocs might be in the lab for a two- to six-year stretch of time that overlaps with the tenure of other postdocs, making it even harder to organize the notebooks chronologically. Very few people had actually put their names in their notebooks. I would show a notebook to Liz and ask, 'Whose handwriting is this?' And she'd always know."[1] Smith acknowledges that Blackburn may have taken her cue from the data in the notebooks as well as the handwriting but still marvels at how quickly Blackburn could connect any notebook to the person who had done the work.

Carol Gross, Blackburn's colleague at UCSF, affirmed the intimate nature of collaboration within a lab: "Your lab is like your family. You generate ideas with other people, and sometimes you don't know whether the idea came from one person or the other. You know each other so well that what you're saying explicitly is only 15 percent of the conversation."[2] Ironically, the camaraderie in a lab is shaped in part by the stressful nature of the work and the long, erratic hours. Blackburn's lab, which reflects the cultural norm, includes a common room with a refrigerator, microwave, tables, and cushioned benches where people can sleep or read while they're waiting on experimental results. Thrown together for long hours, bonded by the highly specialized nature of their work, which not only doesn't translate to cocktail party chat but might be arcane even to a scientist in another field, lab members dwell in an insular social world. Exciting results only fitfully interrupt tedious work at the bench, and the kind of personality that thrives in such conditions is that of a devotee—a grad student in Blackburn's lab had the double helix tattooed on his ankle. Even the humor consists of insider jokes. When Blackburn's lab hosted a monthly get-together for labs on the same floor at Mission Bay, flyers were posted announcing "Blackburn lab isolates strain of killer yeast that lives forever . . . and they ferment some darn good beer."

But if a lab is like a family, it also functions as an intensely competitive business with the idiosyncratic peculiarity that competitors cannot survive without each other. Blackburn noted an old adage: "In business, you want to kill your competitor. In science, you want to keep your competitor alive."[3] Mentoring is not a sideline but the central mechanism by which science progresses, with the work it produces generating continued funding for the lab. And market demand *does* influence research, via the editorial predilections of high-impact journals and the priorities of the primary source of public funding, the NIH. Though a research lab with a well-established reputation attracts the most ambitious graduate students and postdocs, a successful primary investigator faces as many competing demands on her time as a corporate CEO. In May 2005, for example, Blackburn's calendar included giving a keynote address at the Human Papilloma Virus Conference in Vancouver and another talk at the Cold Spring Harbor Telomere and Telomerase meeting, where she also chaired a session. She traveled to Bethesda, Maryland, to attend a meeting of the National Advisory Council on Aging, which serves the National Institute on Aging. The council, made up of senior researchers in the field, provides oversight on the policy and direction of the institute, one of the National Institutes of Health, and like many of its members, Blackburn also gave a talk at a scientific symposium held in conjunction with the council meeting. When she was not away at scientific meetings, Blackburn conferred with other members of the E. B. Wilson Medal Selection Committee, reviewed grants for the Stanford Cancer Center, served on UCSF thesis committees, and attended a faculty retreat. During the same month Blackburn also had to meet two grant-reviewing deadlines. Her one-on-one meetings with lab members must be scheduled on a calendar so crowded her assistant even pencils in the hours allotted for writing a textbook.

And if a lab is like a business, it also has much in common with a medieval craft guild: the value of the work is collectively determined by guild members, who rise through the ranks in a series of steps designed to demonstrate mastery. Those who become masters are obligated to teach their trade, but graduate students and postdocs, the underpaid apprentices of research science, are vulnerable to exploitation and at the same time eager to claim individual credit for their work. Conducting scientific research requires the same odd combination of repetitive work and inspired creative effort once demanded of an apprentice medieval craftsperson whose assigned rote task might ultimately constitute a component of a work of art. The same graduate student who must dump out a box of toothpicks before sterilization and sort them in a jar with their broader tips down (so that he won't poke holes in the soft plate when streaking yeast cells) must also be capable of making sound inferences based on the data the experiment nets. Like members of a medieval craft guild, lab members are motivated by altruism as well as ego: ambition coexists with a collective investment in discovering answers to questions that will expand scientific knowledge.

The purely intellectual aims of mentoring exist within a matrix of pressures that constrain free-ranging inquiry. Yet breakthrough scientific work so often depends on a hunch or a gamble. Blackburn fittingly described her crucial collaboration with Szostak as a "cockamamy idea," pursued "just for the fun of it." In 1996, when Lundblad identified the gene for the protein component of telomerase, she succeeded because she was willing to sift through a genetic screen of *one million* yeast colonies to find the one colony with the *est2* mutation, and years later she still winced at the odds: "It's scary when you realize how much of your career rests on picking one in a million."[4] Lingner, who as a postdoc combined his biochemical results with Lundblad's genetics to confirm the gene for this protein component, at one point in the preliminary research feared he had identified the wrong candidate for the protein, wasting years of effort and sharply diminishing his chances of finding a job as a researcher.[5] Early in her career as a primary investigator, de Lange gambled her entire NIH budget for a year on purchasing a thousand liters of cultured cells so she could attempt to purify a protein that *might* bind to telomeres, and though she succeeded, she said, "I would never advise anybody to do what I did."[6] How can a primary investigator cultivate risk taking when, in pitching ideas to graduate students and postdocs, she may be pressed to make conservative choices in order to serve both the lab's broader interests and the student's career prospects?

Though she claims to prefer to avoid crowded races, Blackburn's work has remained central to an increasingly competitive field for over thirty years, despite the comparatively small size of her lab. The tension between the competitive toughness required and Blackburn's self-described inward-looking and inner-directed drive is embedded in her approach to mentoring. How Blackburn interacts with members of her lab illuminates not only her own particular qualities as a researcher but also the nature of the mentoring process itself. Because Blackburn has trained a number of leaders in the field—Greider, Shippen, Lundblad, McEachern, Gilley, and Wyman—and each of them has trained others, her influence as a mentor extends beyond the parameters of her own lab and has helped to shape a field in which women have been singularly successful.

Because of the way UCSF structures its graduate programs, students "are king," according to Blackburn—much in demand and free to pick which lab they'll join. In contrast, people seeking a position as a postdoctoral fellow undergo an arduous vetting process before joining a lab. During a daylong lab visit, they give talks and interview with all lab members, which Blackburn described as a mutual "courtship": "We're trying to pick and choose the best match, while the applicant is picking up the sensibility of the lab. I interview potential postdocs, but so do other lab members. How will this person influence the personality of the lab? I would override the views of the other lab members only if I thought someone was really wonderful."

Prescott and Gilley, both members of the Blackburn lab in the mid-1990s, vividly recalled their first visits to interview for a position. Prescott was surprised that a "big name like Liz" made him feel she would be flattered if he joined her lab: "I was coming from a good but small lab, and interviewers with much less stature felt it their place to be hard ass. Liz can pick and choose and doesn't have to be nice to get good people. Halfway through my daylong visit, she took me aside and said, 'I just want you to know, I'd be very excited if you came to the lab.'"[7] When Gilley visited the lab, Blackburn put him up at her own house. Like Prescott, Gilley was impressed by Blackburn's lack of pretension—"There I was just showing up at the doorstep of this really famous person I'd idolized and read about during my previous training"—and he also immediately registered that "she doesn't really separate her personal and professional life":

At the time I stayed with her, her son Ben was fairly small, about four or five. She was so out of touch with popular culture she didn't even know who Michael Jordan was, and in the early 1990s, everybody knew who he was. She's kind of known for this. I worked with John Preer as a graduate student. When Liz was a postdoc in the Gall lab, they had overlapped—Preer did a sabbatical there at the time. John and his wife, Bertie, had visited Liz and told me she and John were so focused on science—the sink in their bathroom had a plugged drain, broken for years, and they didn't want to take time to get it fixed. Instead they put a butter knife beneath the drain plug to keep it working. I noticed the plugged drain on my visit too. At their bedside, instead of what most people have, like newspapers or TV, both Liz and John had stacks several feet high of scientific journals. I think having a child had a balancing effect on their lives. When Liz went off to put Ben to bed at around nine at night, John was watching a baseball game, and I sat down next to him and he said, "I don't know anything about the rules of baseball, but I'm watching to learn the rules so that when Ben gets old enough, I can teach him."[8]

The personal style of a primary investigator profoundly influences the culture of a lab, and Greider, who felt that "Liz's boundless enthusiasm helped a lot" when they collaborated on the discovery of telomerase, observed on a recent visit to Blackburn's lab that "it seemed like that optimism and enthusiasm is still a primary attraction."[9] Gilley regarded Blackburn as "incredibly dedicated to her science and to her life in science," yet this did not translate to a driven, directive mentoring style.

According to Shampay, a graduate student of Blackburn's in the mid-1980s and now a professor at Reed College, "There are many ways you might define a good mentor. Some people are very explicit, others are just themselves, open to talking to you but not advocating in any overt way. Liz belongs in the second category, very casual, doing what she's doing and showing it can be done."[10] Tet Matsuguchi, a graduate student in Blackburn's lab in 2005, gave a similar assessment of Blackburn's style: "One type of mentor is a micromanager, and the other gives you more freedom to decide what you want to do. Liz is the second type, but she is always there to help. She gave me the option of deciding on my own what to pursue or having her suggest ideas. For a month or so we met two to three hours a week to discuss what I wanted to do. She likes two types of experiments: very direct, in which you know exactly what's going on but you may face technical challenges, and fun experiments, when you don't know what's going to happen at all but it's conceptually very interesting. We discuss both."[11]

Blackburn confirmed that she emphasizes individual "ownership" of projects: "You want postdocs to engage with something already going on in the lab and take it somewhere new—become original as they learn from the work itself." Former postdoc Shippen, who moved with Blackburn from Berkeley to UCSF in 1990, underscored the risk a primary investigator takes in living up to this aim: "There's a big temptation in a large lab to put three people on a project if things aren't moving. Graduate theses and postdoctoral projects aren't considered as important as the big discovery. Liz gave people ownership of their project and respected it. In addition to being generous, she was very respectful of people. She treated people in her lab as colleagues, not underlings."[12] Smith, who has worked with Blackburn for over six years, observed that her nondirective approach, while it can be challenging, fosters autonomy: "A lot of primary investigators are capable only of wanting to make scientists in their own image. Liz is really good at encouraging people to become the scientist they can become, the scientist they want to be. One complaint you hear from students starting out is, 'I wish she'd give me more guidance, tell me what to do.' That's not Liz's style—to stage every experiment—though she's eager to absorb any data you want to show her."

Blackburn's optimism has its risks, as does her reluctance to frame her work in terms of career considerations. A primary investigator might see her role as that of steering a student toward a likely prospect—an experimental question neither too peripheral to matter much nor so centrally significant that many researchers are already investigating it. This obligation seems inconsistent with simply giving people their heads, yet Smith suggested Blackburn tends to do so: "At telomere conferences, within any given area, eight to ten people are giving very similar talks. It's more difficult than it used to be to carve out your own niche and not step on the toes of someone else doing similar research. Liz doesn't worry about that but encourages people in the lab to pursue their own interests." Blackburn's own view of how she directs her lab qualifies Smith's assessment along an unexpected axis: "I'm tired of people breathing down each other's necks in our field. I like being off where it's new. I'm always the one who wants the postdocs to do crazy things. If you do something rational, it's likely that others are working on it too." This discrepancy hints that as a mentor, Blackburn struggles with divided impulses and potentially contradictory responsibilities to guide students toward success as well as foster their originality. She has found a structural remedy that enables her to juggle these two imperatives, organizing the lab so that Smith, promoted to the rank of specialist, pursues experiments that might pose a risky or unpromising choice for a postdoc under pressure to publish: "Dana does my off-the-wall experiments for me."

Given the need for collegiality within the lab and the larger scientific community, a primary investigator must maintain a distinction between autonomy and territoriality for ethical as well as practical reasons, mediating potential conflicts when people pursue questions that overlap. Weekly lab meetings, the norm in virtually every research lab, can be either an arena for competitive maneuvering or a central venue for establishing an ethos of community service. Blackburn has a rare absolute rule on attendance: no one in her lab may miss a meeting except in an emergency. She brought to her own lab a cultural memory of the spirited debates in the MRC cafeteria at Cambridge, in which a graduate student could challenge Crick provided he could back up his assertions, and of the egalitarianism of Gall, whose lab meetings incorporated presentations by all lab members, including lab tech Truett. A lab meeting provides a forum for people to learn how to question their own and each other's data, and the presenter's work often touches on the projects of other lab members in ways that provoke useful thinking. By the 1990s, some researchers in Blackburn's lab studied yeast and others studied cancer cells, and frequently these two groups had important advice to impart to one another. Cherry, a graduate student in Blackburn's Berkeley lab in the early 1980s, recalled that "Liz had a strong sense of service. You had a responsibility as a scientist to help others. If you got this technique to work, you had to help others, whether you were a postdoc or a lab tech."[13]

Consistent with her nondirective approach to experimental questions, Blackburn tends to let the critiquing process prove itself in the lab meetings. She noted that a presenter who founders in the details of her particular area of expertise undercuts the quality of feedback she receives and learns directly from the process how to adjust the presentation. Yet the primary investigator, as a repository of a vast amount of information and experience, has a unique contribution to make. According to Lundblad, who now runs a lab at the Salk Institute, "When a result pops up in my lab that triggers something, it's part of my job to remember there was a paper years ago that had data that didn't totally make sense, but given your result, you should go look at that. We might go back five or ten or fifteen years." Blackburn also viewed this responsibility as central—"I bring a broader perspective and a memory, years of familiarity with a variety of systems and questions"—but added, "I don't necessarily bring a more honed critical sense. Someone who is ready to write up results in a paper can solicit feedback as the paper unfolds—on everything from missing information to the clarity of the figures. The best meetings are those in which people pitch in as equals and pool resources to help the presenter see how to refine the work."

Just as she tends not to presuppose a hierarchy of importance for data, Blackburn eschews a hierarchical approach to experimental collaboration. She termed her one-on-one relationships with lab members "synergistic," in which each person contributes intellectual resources:

At the bench you're cooking a complex soufflé that requires your attention. One person can think about only so many things. Once your primary task is to think, you are free from having to be in command of every detail of the lab work. Instead of focusing on when to add buffer to the solution, you can concentrate on the design of experiments and how to ask questions. You really do think better about the significance of experiments and anticipating where they might lead, and you also make sure people in the lab get better at doing this too. You stand back and ask someone questions that help her to refine the experiment, and she's collecting information that feeds your thinking, and in this engagement, you can also challenge each other.

This approach requires giving younger collaborators the chance to make discoveries for themselves. When Shampay collaborated with Blackburn and Szostak, she found evidence that a yeast telomeric sequence was being added to a plasmid introduced into yeast cells. Shampay reported that "Liz came in one day and said, 'Let's try to think of a model for the replication of telomere DNA that would allow for the addition of sequences to the end of the chromosome.' She asked me to think through the problem—gave me room to consider it. I suggested, 'Maybe it's adding to the 3′ end to make that overhang longer.' She got me to say it. That's proper training."

Former members of Blackburn's lab portrayed her as remarkably open-handed both personally and intellectually. According to Shippen, Blackburn had a relaxed notion of her own property rights: "Liz never thought of ideas as proprietary. There were people in her lab who were very closed about their ideas, but she was completely open and generous." Gilley, at first awed by Blackburn's prodigious memory, relished their wide-ranging scientific conversations:

It took about two years in her lab before I could have a really good scientific conversation with Liz, where I could feel I was up on it. She has a phenomenal memory of every technical detail and bit of data for years—essentially a photographic memory. She can go through a mental filing system of some kind and remember the intensity of a signal on a result from ten years ago. It takes time to get up to speed on that. As a scientist, there are certain types of people you just click with. You can just sit and talk, and you're flowing in this kind of beautiful way, free and noncompetitive, just throwing out ideas. My mentor John Preer was the same way, and I think Liz looked up to him. It's an old-time way to be, and there aren't too many of those these days. Many people, they might as well be selling shoes. The science is just a vehicle, more like a business to them.

How one asks a scientific question is so essential to successful research that the phrase recurs like a refrain when scientists talk about their work.

After working with Szostak, Lundblad came to Blackburn's lab in 1988 to complete a second postdoc "because it was *the* place for working on telomere biology. The most helpful thing was watching somebody else's process: how Liz identified a problem that was interesting and decided on what approach to use to tackle that problem and how she interpreted data. These are all not completely linear strategies. You can always design an experiment that gets results, but it may not be interesting. Successful researchers can define a question that they're pretty certain will yield an interesting answer." Lundblad also emphasized that a question-centered approach to science does not respect boundary lines between disciplines or technical approaches but opportunistically seizes the best method at hand for addressing a problem: "I don't think either Liz or I define ourselves as 'I do genetics' or 'I do biochemistry.' You follow the path that helps you look at the next big problem, no matter what it is."

Scientific objectivity depends on inculcating certain attitudes as well as developing qualities of intellect. Cherry, now managing a genetics database in the Department of Genetics at Stanford University, said that Blackburn was particularly good "at helping you consider the ways you were asking a question. In group meetings we discussed our experiments, and Liz would ask questions to get us to think about what we were doing and why, whether the result was really telling us something or we were confusing the issue because we wanted a particular result." Prescott affirmed the rigor of Blackburn's approach: "From her, I learned to be very critical and ethical about how you carry out experiments and interpret data."

Consistently, Blackburn's former lab members remark that she was both scientifically critical and interested in "big ideas" and experiments that took off in an unconventional direction. Shippen suggested that she learned from Blackburn tenacity as well as a mode of thinking about science that depended on both rigor and risk taking:

Sometimes I'd bring her data and be sure she'd be so excited, and she'd look at it a few moments and say, 'No. Dorothy, I don't see that. Here's what I see.' Or I'd go in to her office dejected, and she'd say, 'Oh, look! This is so unexpected. It's so exciting.' The twists and turns were always what excited her, and she'd encourage you to figure them out. Rather than the neat package you wanted, she'd be eager to explore the novel or unexpected result. She'll jump to something that most people wouldn't even think of going to, but her mind works so fast and she has such an open, continuous view of the information she has access to and is completely unafraid to propose something that people wouldn't readily put together. A number of the ideas she pitched to people didn't pan out, maybe because we didn't have enough information at the time, but it was a constant stream of interesting and unconventional ideas. You have to do minutiae to sort out how things work, but she always entertained the idea that this could be a big result.

Shippen's remarks imply that a primary investigator's approach to anomalous or unexpected results can foster risk taking within a framework more expansive than a simple dichotomy between success and failure.

Most established primary investigators don't work at the bench, but Blackburn did so for much of her career. Cherry reported that in her lab at Berkeley, "Liz was quite regularly at the bench. She had her own area in the lab that no one was supposed to mess with—we weren't even sure what she was doing. Only later did I learn that this was unique, not typical for a primary investigator. Liz did this when she was working on some new idea, enough to get an experiment started so a graduate student could come along and continue." A member of Blackburn's lab at UCSF nearly a decade later, Lundblad also recalled Blackburn working at the bench:

Pretty soon after I joined Liz's lab, she had some relief from teaching, perhaps a semester without teaching responsibilities, and she was working in the lab with her technicians a fair bit. To have someone at Liz's level of her career actually at the bench is an opportunity to really learn from someone about how they design experiments. About that time, PCR had been developed, and it was a technology-intensive protocol, requiring a thermostable enzyme and a piece of equipment that would shift your sample to a different temperature every ninety seconds. The equipment was phenomenally expensive at the time, and only the big labs could afford it. Liz decided one day that she was going to do a PCR experiment, and she did it manually, working on the last ten or twelve inches of someone's bench with containers of water and a timer. Every ninety seconds, she shifted the sample manually. I remember thinking, "Boy, she doesn't have much ego." This was what she wanted to do, and if she had to stand there for an hour or two and monitor this timer, that was what she was going to do. She didn't make a lab tech do it, she did it. She really is an experimentalist. She loved being at the bench.

Smith echoed Lundblad's perception of Blackburn's immersion in the work:

Before I came to Liz's lab, I had worked in a number of other labs, where I had good mentors. I also gained enough experience to see how primary investigators differed from each other in the way they ran their labs. One primary investigator gave students exciting projects and worked really hard to connect them with luminaries in the field, to help their future careers. Another primary investigator attracted great people who got along well, and he was driven to ask the big, hard questions, to become famous for his discoveries. Liz *is* famous, and you don't know it when you speak with her in person. It doesn't seem like that kind of ambition is the driving force behind her science. She's really driven by the questions, genuinely excited by telomerase and how it works.

Clearly, Blackburn sustained a pure space for the work and conveyed this by example, but where is the tough competitor, the woman who beat out her collaborator Szostak in identifying both the enzyme that added to telomeres and the process by which it did so? If her toughness was in evidence with her lab members, it seems to have been couched; Shampay recalled Blackburn in lab meetings "gently putting people on the spot," a seeming oxymoron. Early in her career, Blackburn had rejected an authoritarian role: "The old model of mentoring was 'tear them down and build them back up,' only some people don't get built back up. I'm too softhearted to do that." But she can't afford to be too soft to train talented people to compete. It's her job "to see the holes in someone's thinking," but "I try to ask questions that don't insult the person: How will you account for that? Have you thought of this alternative explanation? What could you use as evidence to rule out this alternative explanation?" The personal accommodation Blackburn had worked out for contending in a highly competitive field—an attempt to segregate personal ego from the work itself—turns out to have significant corollary value for her work as a mentor. In the context of Blackburn's idealism, generosity, and receptivity to *any* results, criticism that might be felt as impersonal instead fosters confidence and persistence.

How a mentor imparts the rigor and drive necessary to succeed in research depends on a constellation of personality traits, and Blackburn's approach offers a bracing antidote to the notion that the culture of research science is mercilessly competitive. The warm regard of her lab members pays tribute not just to her intellectual mentoring but to her personal kindness. As a postdoc, Shippen relied on Blackburn's support to continue her research at a time when she suffered from a debilitating illness: "I was diagnosed with Crohn's disease—an autoimmune disorder —as a graduate student. Crohn's disease is characterized by periods of being well and then quite ill. I had a major flare-up while I was at Liz's lab and was out for two months. She was phenomenal during that time. She devoted a technician exclusively to my project to keep it going in my absence. I tried to come back earlier than I should have, and she sent me home." Gilley told a similar story, prefaced by reference to Blackburn's scientific qualities, as if to emphasize that her competitive excellence was not compromised: "The science is one thing: there's nobody better. Liz's intuition of science and intellectual capacity and openness to new ideas—I don't know of anybody I could put up higher than Liz. When I was a postdoc, I got a bad viral infection—a viral meningitis—and was in bed for six weeks and afterward needed time to recuperate. She was incredibly supportive. She never put any pressure on me to return to work. And I had seen her be just as supportive to other members of the lab. She was incredibly supportive to women in the lab who had children or were pregnant. She's really a saint." Like Gilley, Prescott stressed Blackburn's unusual combination of traits: "I have a lot of respect for her as a scientist and a person. There is no shortage of really bright scientists, and Liz is clearly one of the top scientists. But she's also really nice and ethically bound, and that narrows the pool down quite a bit."

Blackburn also satisfies the most competitive measure of a mentor's success—the subsequent achievements of her lab members—though by all accounts, she does not prepare her students to continue on by focusing on the grim realities of a tough profession. Shampay recalled that if anything, Blackburn might have shielded her students a little too well: "I think she protected us from the gory details—like how to write grant proposals—more than she really needed to. As a graduate student, I felt protected. Maybe the postdocs learned more." When Shippen began looking for an academic position, Blackburn offered practical help, but Shippen regarded her optimism as even more important: "She came to my office after my paper came out in *Science*. She had just heard a seminar given by a job candidate who had come to Berkeley, and she said, 'You're just as good as this guy, and you need to get out in the job market now and strike while the iron is hot.' She gave me the chance to talk at a Gordon Conference in her place. I was this little Southern girl, not used to playing in the big leagues, and she said, 'Of course you can do this.' She treated you like you were famous and accomplished, the person you would become." Yet as Gilley noted, Blackburn set scrupulous limits on how much help she provided, a seeming caveat:

Part of the way you're judged or promoted as a scientist is who you train and how successful that person is, so there's a conflict of interest. Nepotism or cronyism hurts the field, but most people try to promote the people who have worked with them. Liz is incredibly sensitive and unbiased in that regard. She is the most fair person that I've ever met in my life. As far as judging and reviewing papers or grants, she doesn't overpromote people who have worked with her. She will recuse herself in many cases if she's had connections with anybody. That's just very unusual. Would I like her personally to go around and give me a shot? Maybe I would, but in the long run it's not required. I'm going to do good work, or I'm not.

This faith that the work will prove itself echoes Blackburn's cherished personal tenet.

The time when a postdoc or graduate student moves on from the lab is rife with potential for conflict, since the ownership of experimental work that may be ongoing has to be decided. According to Shampay, "You hear stories about lab PIs being more dictatorial about what people can work on, what they can take with them when they leave the lab," but Blackburn gave Shampay wide latitude to pursue interesting findings: "My dissertation was focused on cloning and characterizing *Tetrahymena* micronuclear telomeres. But Liz gave me the freedom to follow some observations on the variations in length in yeast telomeres. When the *Tetrahymena* experiments were not going well, I'd turn to yeast, where the blots always worked. This work eventually resulted in a paper coauthored with Liz." When Greider left Blackburn's lab, she and Blackburn amicably divvied up the experimental work that followed on their discovery of telomerase. For all the mythology about cutthroat competition, keeping one's ego in check at this critical juncture is not merely nice but wise: it fosters open communication and mutual respect with one's future colleagues.

The women working in Blackburn's lab as graduate students and postdocs have arrived at a vulnerable moment in their careers—one at which many women drop out or fail to advance to prominence. Blackburn herself almost dropped out when, after completing her postdoc at Yale, she faced a difficult job search and ended up following her husband to San Francisco, secretly relieved that she might be pregnant and so opt out of the game. For decades it has been assumed that the gender gap in the sciences would dwindle as more women earned degrees and received professional training. But many women have now come through this pipeline, and still their numbers in academic ranks, particularly in the ranks of full professors and administrators, have not correspondingly increased.

According to the American Institute of Physics Statistical Research Center, from 1970 to 2001 the percentage of women earning bachelor's degrees in biology climbed from 30 to 70 percent; in chemistry, from 10 to 40 percent; in physics, from 6 to 22 percent; and in engineering, from 1 to 20 percent.[14] Yet the numbers of women in tenure-track positions in the sciences held constant from 1975 to 1995.[15] More recent statistics document a disturbing drop-off in the numbers of women at higher academic ranks in departments at elite institutions (as ranked by the National Science Foundation). Women fare better in biology than in the physical sciences, yet they still do not fare well. While women earn 45.89 percent of PhDs in biology at these institutions, their numbers diminish as they rise in rank, from 30.20 percent of assistant professors to 24.87 percent of associate professors and 14.79 percent of full professors.[16] Women from ethnic and racial minority groups are "virtually absent" from leading scientific and engineering departments.[17] Although women who earned PhDs after 1985 stand a better chance of promotion than older women, they continue to rise at a slower rate than men, and the seemingly healthy increase of women at lower ranks is swelled by older women who have failed to advance.[18]

The particular demands of juggling an academic career and family responsibilities have often been cited as a primary handicap for women, in large part because the tenure system demands the most of women in their peak childbearing years. Martha West, one of the authors of a report on gender discrimination in faculty hiring in the University of California system, notes that female faculty are less likely to have children than their peers in the general population—40 percent as opposed to 85 percent of all women.[19] Women faculty who do become mothers advance more slowly than either men or women who do not have children.[20] A

1998 study of medical school faculty showed that among faculty with children, women reported fewer publications, slower rates of promotion, and less career satisfaction than did men. No similarly significant gender differences were found for faculty who did not have children.[21]

In an effort to recruit and retain the best women faculty, many academic institutions have revised tenure policies to allow for greater flexibility. According to a recent survey, 27 major research universities now permit junior faculty to postpone their applications for tenure after the birth of a baby.[22] The California State University (CSU) system is more successful than its prestigious counterpart, the University of California system, in promoting female faculty, largely because of policies that offer greater flexibility in pursuing tenure; according to Nina Fendel, a regional representative for CSU's California Faculty Associations, CSU's efforts, begun roughly fifteen years ago, have paid the additional dividend of an increase in women in administrative positions.[23]

Yet this disparity between the CSU and the University of California systems is also influenced by the difference in the priorities of the member institutions; the CSU system focuses more on teaching than on research, and the University of California system boasts some of the premier scientific research institutions in the world. Research science imposes more strenuous demands than other academic work, which may induce women with children to choose alternatives, including non-tenure-track faculty positions that emphasize teaching over research. Graduate students and postdoctoral fellows do the yeoman's share of "hands-on" lab science, putting in long and erratic hours, and they face more of the same as junior faculty. The demands of pursuing tenure make it difficult to raise children at the same time unless the other parent takes on a larger share of family responsibilities, but in most cases, the partner who does so is a woman, and because as many as 50 percent of women scientists marry other scientists, they face greater pressure to make an either-or choice.[24] Major-league research science proceeds at a pace that makes it difficult, if not impossible, for researchers to take time out to have children, even if flexible policies are in place. Renowned geneticist Mary-Claire King claimed that "by far the biggest obstacle is how to have enough hours in the day to be the mother of a young child and do the work. . . . You cannot control how experiments will work and how many times you'll have to do them. And there's nothing about raising a young child that you can control. There are no easy solutions to that. . . . I think it is impossible to drop out [of science] and drop back in."[25]

Because even women who do not have children fail to achieve tenure at the same rate as men, other factors influence their dismaying drop-off at the postgraduate level. On the basis of ample exposure to working conditions in their field, women graduate students and postdocs realistically anticipate that they will face tougher career challenges than their male peers. First, women scientists must perform better than men to be ranked equal to them, and their hiring and promotion is subject to unconscious gender bias. A 2006 study by a committee of the National Research Council attributed the disparity in achievement and access to resources to implicit bias in the hiring and promotion of women, concluding that "evaluation criteria contain arbitrary and subjective components that disadvantage women," despite the belief among scientists and engineers that "they are objective and intend to be fair."[26] Second, women often do not receive an equitable share of resources from academic institutions or the NIH, which funds 99 percent of the research sponsored by the Department of Health and Human Services. At MIT, a 1999 study found that female faculty did not receive the same salary, space allocation, awards, and resources as male faculty with comparable accomplishments.[27] A 2005 RAND Corporation report on the distribution of NIH grants found that women applicants garnered only 63 percent of what male researchers received in grant funding and were dramatically underrepresented among the top 1 percent who received the largest grants.[28]

While women's discouragement about their future in science is based on reality, it is a different reality than the one their male colleagues experience, which constitutes an additional burden for those who persist. In faculty surveys conducted at MIT, Princeton University, the University of Michigan, and the University of Wisconsin, women scientists reported more antagonistic behaviors from colleagues than did men.[29] In a 2002 faculty survey conducted at the UCSF medical school, male and female faculty sharply differed in their views on the university as a good work climate; nearly half the women, but less than one in ten men, reported experiencing discrimination, and while many women felt limits were placed on their participation and advancement, men did not perceive that women faced such barriers.[30] Only 45 percent of women faculty at UCSF expressed satisfaction with their prospects for advancement, as compared with 57 percent of men.[31] At the University of Wisconsin, 60 percent of women scientists responding to a 2002 survey did not believe the gender climate at the university would improve in the next few years.[32]

Consistently, women scientists perceive the tension between career and personal life as more problematic than do their male counterparts. In *Why So Slow?*, a study of gender inequity in the workplace, Virginia Valian contends that gender schema greatly influence career choices. Women are expected to place more emphasis on leading a full personal life outside work, so that "a woman who puts her work in second or third place is in harmony with the female gender schema" and likely to receive social reinforcement for this choice. Conversely, men who question the time demanded by their careers are out of sync with the norms of their gender.[33] Ironically, though the gender schema for men is more constricting, conforming often grants men a competitive advantage: "The result, for many men, is a strong commitment to earnings and prestige, great dedication to the job, and an intense desire for achievement."[34] In the 2002 faculty survey at UCSF, more women than men perceived the demands of their job as unreasonable (three-quarters of women as compared to six in ten of men) and believed this took a heavy toll on their family lives.[35] In a talk at Columbia University, Princeton University president and noted biologist Shirley Tilghman cited a similar American Chemical Society survey in which many more women chemists than men reported balancing family and professional life as their greatest career obstacle—21 percent compared to 2.8 percent. For Tilghman, this disparity illuminates the way in which "women and men experience careers in science and engineering differently." She attributes the dropout rate for women in science to the isolation of women in a male-dominated profession rather than an inherently different approach to competition: "Women report a greater level of dissatisfaction and a greater need for mentoring than men. One view is that these are signs of weakness on the part of women, signs that they need more nurturing than their tougher male colleagues. I would argue that young women's dissatisfaction and call for mentoring grow out of their need to have the cultural milieu of science—a culture that was formed when all scientists were men—interpreted for them."[36]

As Valian reports, for a woman to conform to the competitive model of research science requires that she be "out of role"—out of sync with the gender schema—and women who do so often elicit negative responses for behavior that would be rewarded if they were men.[37] Furthermore, lab-rat culture places a high premium on objectively determined merit, and if the male majority of their colleagues do not perceive gender bias as distorting objectivity, women face persistent cultural pressure to blame themselves rather than challenge practices that effectively, if not intentionally, discriminate. Even when flexible tenure policies are available for new parents, women fear they will jeopardize their careers if they exercise this option; according to Joan Girgus of Princeton University, some eligible women thought asking for a tenure extension "would be seen as a sign of weakness."[38] And Martha West also attributed the small percentage of eligible women who took advantage of similar options in the University of California system to the fear that they might hurt their chances for advancement.[39]

In this context, women's success in the field of telomere research is nothing less than astonishing. A 1992 *New York Times* Science Times article on telomere research emphasized that three women—McClintock, Blackburn, and Greider—launched the field.[40] A decade later, a review in *Science* paid special attention to the "unusually large" number of women in telomere research.[41] At scientific conferences devoted to telomere research, half or more of the speakers are women, something Blackburn termed "equity, roughly like the gender breakdown in the general population," but this remains a novelty in the sciences. Lundblad, a leader in the field since its early days, observed that "some people say the field is dominated by women, but that is a skewed frame of reference. It's dominated by lack of gender bias." Whatever makes this field an anomaly might foster conditions that could influence gender equity in other scientific disciplines, and Gilley argued that this could only improve the science: "Certain fields of science are still dominated by men, but they are hurting because of that. They are not making the advances that other fields with more equal percentages of the sexes can make, because they're missing an important component of the scientific community."

Telomere research has an unusually uncluttered genealogy in that so many of the researchers in the field today started out in a core group of labs, including Blackburn's. Lundblad noted that "the field was launched really by a handful of labs. A lot of it has to go back to a single laboratory, Joe Gall's at Yale. He trained many of the people who became prominent leaders—Liz Blackburn, Ginger Zakian, and Mary-Lou Pardue all went through Joe's lab. Tom Cech, who trained with Mary-Lou Pardue, has had a number of successful women come through his lab as well. A particular constellation of people who came through Joe's lab have played major roles in the field." Greider also credited Gall: "Joe was an excellent mentor to women. His was the initial founder effect: prominent women came out of his lab, and then more women went into their labs." The unusual success of women who began their careers in Gall's lab suggests that the gender of a mentor matters less than equitable treatment. The far-reaching effects of such mentorship are illustrated by the active leadership of women in the American Society for Cell Biology, which Gall helped to found. To the extent that Gall's scientific descendants have passed on his traits, they have disseminated an approach to science that supports the success of men and women equally.

In her own lab, Blackburn has sustained Gall's egalitarianism and also extended this family tree so that it branches out into virtually every area of telomere research: Shippen researches plant telomeres; Greider, working with mice as a model system, has focused on cancer-related telomere research; Lundblad has identified proteins that bind to telomeres and play a role in regulating the activity of telomerase in the cell; McEachern has continued work on *K. lactis*, begun in Blackburn's lab; and Gilley studies telomeres in human cancer cells. In turn, each of these researchers is training a next generation in the field. For example, Greider has trained Kathy Collins, Chantel Autexier, and Maria Blasco, and Lundblad has trained Tim Hughes. Blackburn's influence on the field is thus both indirect—via the prolific genealogy of her lab—and direct, via her continued ability to break new ground in her research, an unusual creative longevity. As Harrison Echols points out in his history of molecular biology, "The period of a high batting average for most successful scientists is about 10–15 years, the same as that of a good baseball player."[42] According to Lundblad, Blackburn shows no signs of slowing down: "Liz contributed to the field because of the huge number of people she has trained and because of her staying power. She was very young when she made her two pivotal discoveries, and she has remained in the field and remained as a leader. We're still in the first generation, and Liz is at the peak of her scientific creativity. Her constant presence in the field has been very important."

Felicitous timing may also contribute to the success of women in telomere research. Lundblad noted that the field "has its molecular roots in Liz's postdoc work in Joe Gall's lab," which took place in the early 1970s, just as the number of women earning PhDs in biology began to surge. A relatively young field offers more opportunities to young scientists, as Blackburn explained: "In the beginning, this was a very open field, not crowded. If you flinched at going up against the tough guys, there was room." And the field opened up at a time when research on chromosomes was a hot topic in biology, which, according to de Lange, also drew women to the field: "One possible factor might be that in the late 1980s, if you were a woman and you were interested in chromosome-related issues, you had a choice to work on gene expression, telomeres, or some other chromatin-related work. Chromatin work in those days wasn't very popular, and gene expression was a very competitive, very male-dominated field. So women may have gravitated to a small field in which there were not a large number of very successful men dominating the field. Once you have women in a field, I think it may attract more women, though I have no real proof for that. Certainly I was not conscious of this in making my decisions."

Many women of this pioneer generation echo de Lange's sentiments in insisting that gender had nothing to do with their career choices. Like Blackburn, they were drawn to telomere research out of scientific curiosity and excitement. "It's not been on my radar screen to consider doing *x* rather than *y* because I perceived there was more encouragement for women," said Lundlbad, and she resisted the notion that her career be defined primarily in relation to gender: "There is no way I could say whether a paper got rejected because of gender discrimination or because it was a lousy paper. The only way you can point to gender discrimination is if you collect data and you find patterns of trends. I reject the premise that I've had a miserable time in science because of my gender. I've had a great time in science." Shampay, a colleague of Greider's in Blackburn's lab at Berkeley, once discussed with Greider "whether I had consciously chosen a woman as a thesis adviser. I picked Liz's lab because I wanted to work in lower eukaryotic systems and liked the work going on in the lab. It may have influenced me, but it was not my primary motive. It never occurred to me I couldn't do something because I was a woman." De Lange emphatically countered the notion that women scientists approach competition differently than do their male peers: "I don't see any difference between men and women as scientists. Women are competitive and can play rough." Greider was equally firm on this point: "Women employ the same mechanisms men do to compete." But Blackburn, who came of age scientifically before these younger women, *did* feel constrained by gender. "I disguised myself as a man," she once told journalist Peggy Orenstein. "At the time, I didn't think of it as a sad thing. But it is sad."[43]

Yet all these women share a refusal to see gender as a factor in their own ability to compete. On the one hand, this begs the question of whether the traditional culture of research science may dissuade talented women more often than it does talented men. On the other hand, as participants in a field in which women have true equity, these women no longer have to feel out of role in their peer group. Furthermore, the notion that only a narrowly defined competitive style can succeed in the sciences may be a self-perpetuating myth. Blackburn's reluctance to adapt to an alpha-male model has not proved to be an insurmountable barrier to success. Nor would she define this as a deficiency in her competitive spirit.

In telomere research, women have never been in the minority and thus have had a rare opportunity to perform unencumbered by the burdens that attach to minority status. Tilghman has noted that "stereotype threat," in which targets of stereotypes perform less well in a culture in which they're regarded as less able, undermines women's performance in a male-dominated field; she cited a study in which women scored lower on math tests in the presence of men—a deficit that grew as the number of men in the room increased, though men were unaffected by the number of women present.[44] Other studies show that reaching critical mass increases the participation of minorities in academic and professional groups.[45]

In the early days of telomere research, the network of researchers resembled a closely knit club—a girls' club, for the most part. Until the 1994 Banbury Conference on Telomeres, no scientific conferences were devoted exclusively to telomere research, and individual researchers attended scattered conferences according to how their research intersected with broader questions. Yet perhaps because they constituted an esoteric clique, telomere researchers avidly sought out their peers, and de Lange recalled a strong sense of community: "The telomere field used to be so small that we would love to meet just to find somebody else who was interested in telomeres. Most people weren't, and didn't even know what they were—even into the early 1990s. I'd be completely delighted when I met people who worked on telomeres, and we'd hang out and talk about telomeres all the time." Blackburn remarked that "telomere conferences are fun," more informal than many other conferences she attends, and "there was—and still is—a wonderful camaraderie, even between people who might challenge each other in a session." At the 1996 Banbury conference, Blackburn and Jerry Shay, another researcher, took turns walking the halls with Greider's baby while she talked with other colleagues, and Blackburn suspected "Jerry or I might have felt less comfortable doing that at a different conference—say, in the cancer field with the medical types."

Since the early 1990s, the telomere field has grown exponentially, and a great deal more is at stake in the implications of this research for human health. Telomere researchers hold contrasting views on whether or not the collegiality of the early days has survived. Blackburn's former postdoc McEachern reported that David Sinclair, a highly successful researcher in the field of aging, "had done one paper in the telomere field, and he thought it was too ruthless."[46] In contrast, Lundblad spoke of the field today as "pretty competitive, but by and large highly collegial. I've heard other people talk about their respective fields with some dismay about the darker side of the competitive nature of their field. That hasn't been my experience, though the telomere field has been characterized by very rigorous peer reviews of papers and grants, and as a community the field has kept a pretty high standard of scientific excellence." According to Greider, it has simply become much harder to retain the collegial atmosphere given the size of the field: "When I was in graduate school, the field of transcription factors was a gigantic, tough field. Meetings were contentious, and there was a lot of competition, and you had to think about someone else doing the experiments you were doing. Telomere research has become more like this field just because it's gotten larger." De Lange's remarks suggest that the phenomenal growth of the field has changed its culture: "I used to talk about my data—I actually still do, but I shouldn't—years before it was published, and that is probably not a smart move given the current competitive state of the field."

If the field has grown tougher, women have reached a critical mass that has reshaped the social compact, especially with respect to personal networks that tend to disadvantage women in male-dominated fields. When Tilghman organized the annual Gordon Conference in 1988, 33 percent of the presenters were women, but just two years later, the male organizers of another conference on the same topic included only two women among the presenters.[47] In contrast, de Lange noted with amusement that organizers of telomere conferences face the reverse problem: "It's very nice to say to a co-organizer, 'Let's make sure we give men a sufficient platform time.' There is gender bias that is subconscious: men get to know men better and hang out together more, so if there is a name to fill in a slot at a conference or a name to recommend for an award, they'll think of men first. Similarly, the women in the telomere field will tend to think of other women first. You have to actively counteract this to balance things out."

That changed practices can mitigate against the biases of an "old boy" network was vividly demonstrated when in 2004 the NIH instituted the Pioneer Awards to support scientists who creatively explore hypotheses with a high risk of not panning out. The first awards, accompanied by a hefty half million dollars for research, were granted to nine recipients, all of whom were men. Prompted in part by formal protests—including strong criticism from the American Society for Cell Biology—the NIH revamped the application and review process for the awards, allowing for self-nomination, requesting applications from women and minorities, and recruiting a more diverse group of reviewers. In 2005, six women scientists and one African American scientist were among the thirteen recipients of the award.[48] Changing the review process to minimize bias, particularly by allowing for self-nomination, resulted in changed numbers, and among the award recipients was de Lange.

The link between role models among faculty members and the success of women students also suggests that sheer numbers make a difference. Cathy A. Trower and Richard P. Chait, who reported on faculty diversity for *Harvard Magazine*, emphasize that the absence of role models has a measurable effect on women's decision to pursue academic careers: "In fact, the most accurate predictor of subsequent success for female undergraduates is the percentage of women among faculty members at their college."[49] Greider referred to this as a "jackpot event": "When I was in graduate school, my class was 50 percent women, twenty-some odd years ago, but it was unusual to have women mentors, and there was a period of time when there were only women in Liz's lab. The field was founded by women, and women have tended to go into the labs of other women." Though Shampay did not recall making a conscious decision to work with a woman as a thesis adviser, she suspected the experience had some intangible value: "I didn't care if my professors were all men. But as a teacher, I have had the experience of students thanking me for being their only woman professor. It was important to have Liz be a role model, operating probably at a subconscious level."

For her part, Blackburn recalled Zakian as a crucial role model; she had known Zakian since the time they both worked in Gall's lab and drew encouragement from Zakian's ability to combine highly successful research with motherhood. Successively, each woman represents a link in a chain that encourages younger women to see possibility: Greider later looked to Blackburn as a role model when she decided to have a child. Long before Smith joined Blackburn's lab, she began to have doubts about making a career in research, primarily because she felt it would demand a kind of aggression she didn't have. But after listening to Blackburn give a talk on her research, Smith changed her mind:

The way I met Liz profoundly influenced what I thought about women in science. I was a graduate student at Berkeley, just realizing I wanted a master's degree and not a PhD. I was in a tiny department in which there was a lot of strife, and big science was so much more political than I had thought. It eclipsed doing the science. I decided I didn't want to be a primary investigator. I struggled with judging myself for that. So just about this time, when I was taking a difficult oral exam for my master's, I attended a talk Liz gave on her recent discovery of telomerase. I was blown away by her talk—it was spine-chilling, amazing. After the talk, she took questions. She was questioned aggressively by a faculty member from another department, an established scientist. This woman had prevailed during a male-dominated time but seemed determined to be especially hard on women herself. Liz handled these questions with utter equanimity, readily giving supporting data. I thought, OK, I want to be like her, and not like this other faculty member. It was very inspiring to me.

If Blackburn could not merely survive but flourish without donning heavy armor, so could Smith.

As a leader in her field, Blackburn has continued to provide an alternative role model. As Greider commented, Blackburn "has a reputation among scientists as being very strong in her views," yet she is neither authoritarian nor combative. According to Lundblad, "In many ways, Liz has kind of a light touch. You don't see her standing up at a meeting and pontificating. People who are leaders are so because often they make their presence very dominantly felt, which effectively can sometimes be a nonleadership stance. Liz has a much less overbearing stance in the field, yet has continued to be at the front of the field and producing."

If the sheer critical mass of women in a research field perpetuates the success of others, their presence among its leaders creates further opportunities for change. Greider critiqued prevailing assumptions that a highly competitive culture is simply the nature of the beast: "Science could get done just as well if there weren't so competitive an environment. Women suffer more, if you consider their attrition, but that's not intrinsic to their personal style. It's intrinsic to inequities in our culture as a whole." A policy Greider instituted in her own lab illustrates her point: "A number of lab heads tend to talk to some people more than others, and people sometimes don't get help because their projects are going poorly. In my own lab, to get around this, I established regular individual meetings so that we go over data and people who need attention the most will get it. Before, when I'd come back to the lab from a trip, there'd be this anxiety because a number of people wanted to talk to me and didn't know when they'd get the chance. Scheduling regular meetings removed that level of anxiety."

In positions of authority and freed from the insecurity of a younger investigator, prominent women have key opportunities to counter the win-or-lose notion of competition and the either-or choices many women feel they must make when family needs and career requirements clash. As a young woman, Blackburn was unquestioningly "deep into the whole life" of lab-rat culture. After the birth of her son, she pared back her work schedule so she could spend more time with Ben, yet she and Greider confirmed the existence of telomerase by the time he turned three. In published interviews, Blackburn has repeatedly challenged old assumptions, particularly the notion that research demands excessively long hours: "Because part of the culture of science is that if you're not there until late, you're not really doing it, which is the biggest pile of crap. All these hours and chatting and things like that don't make the science better."[50] More to the point, she runs her lab in a manner consistent with this declaration, to the benefit of both men and women, and her prominence makes it easier to do so. She has instituted flexible hours that make it possible for all lab members to adjust their schedules to the demands of personal life, and a lab member's pregnancy is just another fact of life—the occasion for betting pools on the size, weight, and sex of the baby. Blackburn is not merely being modest when she remarks that the stable funding her lab enjoys grants her greater leeway as a mentor: "An anxious PI might put pressure on postdocs to produce papers in order to get the next grant. I can let a postdoc take time off to care for a sick family member because I can consider her training over the long-term, since my cycle of funding lasts several years. She has ownership of her project—her intellectual investment—and I am not under pressure to take it away unless this is absolutely essential."

Women primary investigators have the power to change lab-rat culture, one pragmatic step at a time, eroding the old ethos that divides professional and personal life absolutely. When Ben was still young, he was anxious about his mother's trips to scientific conferences, and Blackburn's absences constituted an enormous pressure on her family, especially since conferences were often scheduled for weekends, "and we live our family life on weekends." So that her husband and son could accompany her to conferences, Blackburn frequently had to find rooms at a hotel other than the conference hotel, typically not a child-friendly site. With her husband along to care for Ben, Blackburn could attend conference events and spend her free time with her family. Over time, her former grad students and postdoctoral fellows began to join her at an off-site hotel, bringing their own young children with them. If important networking was going on at the conference hotel, then this satellite network could also satisfy the hunger for scientific conversation.

11 | Citizen Scientist

A few weeks after two hijacked airplanes crashed into the World Trade Center on September 11, 2001, Blackburn received an unexpected phone call from Leon R. Kass on behalf of the White House. Kass asked Blackburn if she would agree to be nominated to serve on a newly formed bioethics advisory council, which he would chair. President George W. Bush had announced plans to form the council in an August 9, 2001, speech in which he declared a policy that would limit federally funded research on human embryonic stem cells.

At any other time Blackburn might have refused. She knew Kass had advised President Bush on his recently announced policy, which she viewed as riddled with logical inconsistencies. She could ill afford to add more work to her demanding schedule. Yet as she spoke with Kass, "The horror and turmoil of September 11 were a vivid part of my consciousness. As Kass talked on the phone, the thought went through my mind that normally I would say I was too busy. But at a time like this, how could you refuse a presidential request to serve on a national committee? It is hard to recall now the unmet need, felt by so many people—witness the long lines of blood donors immediately after the attacks—to do anything to serve at that time, to ease the pain through some personal response. So I agreed."[1]

To Blackburn's surprise, Kass seemed eager for her to join the council and minimized any difficulties that might arise in nominating someone who was not a U.S. citizen. Only later would she learn that Kass had approached a number of scientists before he spoke with her, and "knowing his views, they had turned him down." In their initial conversation, Kass assured Blackburn he wanted the panel to represent diverse opinions and would not force consensus. Kass also waved away Blackburn's concerns that she might not be able to attend every meeting, since she'd have to fit into an already jammed schedule frequent flights from the West Coast to Washington. Typical members of President Bill Clinton's bioethics council, he told her, had attended roughly two out of three meetings. Blackburn was further reassured when Rebecca Contreras of the White House Office of Presidential Personnel later told her that the council was expected to meet once every two months.

Blackburn had exceptional qualifications for serving on the council. Her research had implications for human health, yet she was not directly involved in stem-cell research, nor did she have any financial stake in the policy recommendations such a council might issue, having resisted the overtures of biotech companies interested in telomere research. She had a distinguished reputation as a scientist, and her service as president of the ASCB had given her a useful familiarity with legislative debates on biomedical research and awakened her interest in public policy on science. And she felt a sense of duty, according to Carol Greider, who had served on President Clinton's Bioethics Advisory Commission and was thus an important resource for Blackburn: "The commission I was on for several years was a tremendous amount of fun because there were a lot of open-minded people who knew a lot about science policy, which I knew nothing about, and were willing to be enlightened about the science. When Bush chose Kass to head the bioethics council, it was clear that there would be a very specific point of view. What Liz got herself into was a much more difficult political situation than I was in. She knew she was doing that. She said it's really important that someone who knows the science is willing to do this. Many of us in the scientific community are indebted to her for taking that on."[2]

In preparation for serving on the council, Blackburn did her typically exhaustive groundwork, reading bioethics texts that included the works of Kass. Concerned that she would have a special responsibility to help the council access and evaluate scientific information, Blackburn consulted leading experts on biomedical technology and bioethics, including Varmus, former head of the NIH; R. Alta Charo, a lawyer who had served on Clinton's bioethics council; embryologist Bridget Hogan, who could explain the nuances of embryonic stem-cell research; Joe Goldstein, a Nobel laureate; and British embryologist Anne McClaren, who had served on scientific advisory councils in Great Britain. During her term on the council, Blackburn kept in close contact with scientific experts and consulted frequently with Elizabeth Marincola, the executive director of the ASCB, whose legislative expertise had proved so valuable to the lobbying efforts of the scientific community.

Rapid advances in biomedical technology had for some time raised legitimate public concern over potential hazards that accompanied the promise of this research. The mapping of the human genome, completed during the Clinton administration, raised hopes that gene therapy might produce cures for inherited diseases. Yet genome science also raised the possibility of discriminatory practices such as pre- or perinatal screening for certain "desired" genetic traits, not just disease. In 1996 researchers in Scotland cloned Dolly, a sheep created when the nucleus of a sheep egg was removed and replaced with the nucleus of an adult cell from a different sheep (a process known as somatic cell nuclear transfer, or SCNT).

Stem-cell research quickly became a flashpoint for concerns over the ethical dilemmas posed by the potential uses of genetic engineering and cloning, especially as research outpaced common knowledge about science and commonly held beliefs about the very nature of life. By 1998 researchers were working on human embryonic stem-cell lines, created from embryos discarded by medical clinics that perform in vitro fertilization (IVF).[3] Because embryonic stem cells proliferate rapidly and can differentiate into all cell types in the body, they might one day be used to replace injured or diseased tissues in patients with conditions such as diabetes, spinal cord injuries, Alzheimer's disease, and Parkinson's disease. Yet fears about reproductive cloning flow over to cloning for stem-cell research, since the first step in cloning to create stem-cell lines is the same as for cloning to try to make a baby. Both processes begin by replacing the nucleus of an egg with the nucleus of a mature cell (SCNT); after the egg is stimulated to divide, a blastocyst (a hollow ball of a few hundred cells) forms within three to five days. At this stage, these stem cells are harvested to create new embryonic stem-cell lines, resulting in the destruction of the blastocyst. At the time Blackburn joined the council, techniques for using SCNT to develop human embryonic stem-cell lines had not been developed; this goal represented a holy grail for researchers because it would greatly expand the therapeutic value of human stem-cell lines by making it possible to tailor them to individuals (necessary for the treatment of genetic diseases and probably for dodging the immune-system response to foreign tissue).

The strongest opposition to stem-cell research comes from those who claim the blastocyst destroyed by this process is a human being, whether it grows from a fertilized egg or results from SCNT. The opposition of the Christian right to stem-cell research precipitated President Bush's decision to limit federal funding for this research and institute the President's Council on Bioethics. In the new federal policy announced in his August speech, President Bush parodied the wisdom of King Solomon by attempting to propitiate both those who might benefit from the promise of this research and his supporters on the religious right. Although President Bush prohibited federal funding for the creation of new human stem-cell lines, these funds remained available for research on existing lines, and he issued no restrictions on privately funded research. Among the logical inconsistencies that had troubled Blackburn was the decision to allow exceptions to a ban ostensibly made for moral reasons. In addition, this policy allows clinics to continue destroying embryos "left over" from IVF, but does not allow these embryos to be used to produce new stem-cell lines for federal research; to this day, no federal regulations govern the destruction of embryos in fertility clinics.[4]

At the time President Bush formed his new council, both Congress and a number of scientific advisory bodies had addressed the issue of regulating this controversial area of research. In 1994 an NIH panel recommended funding such research with certain provisions, but a subsequent congressional appropriations bill included an amendment, known as the Dickey Amendment, which prohibited the NIH from funding research that created, destroyed, or discarded human embryos, thus banning funding for SCNT. The isolation of the first human stem cells in 1998 soon made the scope of this amendment ambiguous; a potential loophole conceivably allowed publicly funded research to proceed on human embryonic stem-cell lines so long as they were created by privately funded researchers. In 1999 the National Bioethics Advisory Commission appointed by President Clinton recommended a ban on reproductive cloning but supported embryonic stem-cell research, proposing that Congress modify its prohibition on federal funding, and the NIH developed guidelines for research on embryos left over from IVF.[5] But uncertainty surrounding the legality of human embryonic stem-cell research stifled such work in the United States.

Even so brief a history suggests an inconsistency between government policy and the recommendations of scientific advisory bodies or agencies. Scientific policy can be more or less vulnerable to political manipulation, depending on whether a clear distinction is made between information and inferences about its social value (between the purview of scientists and that of policymakers), whether advisory panels or government agencies rely on credible scientific evidence, and whether renewed consideration of controversial issues draws on precedent established by previous advisory bodies or reinvents the wheel. By the time the President's Council on Bioethics was formed, the Bush administration had already developed a reputation for ignoring or altering scientific data that did not serve its political agenda. In April 2001, a coalition of environmental groups cited a series of actions in the first eighty-seven days of the Bush presidency as "political payoffs" to the oil and gas industries at the expense of the environment and human health, including a proposed $2.3 billion cut in spending for federal environment and natural resources agencies and the refusal to ratify an international treaty (the Kyoto Protocol) to reduce greenhouse gas emissions.[6] The Bush administration continued to challenge the adequacy of evidence for global warming despite a 2001 National Academy of Sciences report, which found that global warming "is real" and that greenhouse emissions, a leading cause, resulted from "human activities."[7]

Rather than revising policy in the light of fact, the Bush administration appeared to be revising the facts in light of policy goals, and this included acting in advance of careful, objective reviews of scientific evidence. Even the formation of the President's Council on Bioethics followed a policy decision by President Bush rather than preceding it. The policy the president declared in his speech essentially disregarded a requisitioned report by the NIH on stem-cell research, which made no policy recommendations and noted that more research would be needed before the therapeutic potential of embryonic stem cells could be determined.[8] The Bush policy bypassed further review, restricted research, and also disregarded the detailed recommendations of President Clinton's National Bioethics Advisory Commission.[9] None of the members of this commission would serve on the new council; such a disruption in continuity increases the odds that federal funding for scientific research will fluctuate according to prevailing political winds. In a 2002 letter to President Bush, former president Gerald Ford took issue with the administration's handling of human stem-cell research, terming it a "breakdown of the process" of careful review before regulations were implemented.[10]

The Bush administration's track record aroused fears that the bioethics council would be composed of individuals whose political views, rather than their credentials as legal, ethical, or scientific experts, had been vetted by the administration. Before she was formally appointed to the council, Blackburn received a phone call from the White House Office of Presidential Personnel, and this interview felt like a litmus test of her political loyalties. She was asked whom she had voted for in the last presidential election and what she thought of Bush's stem-cell policy. If she was troubled by these questions, answering the first one was easy: as a resident alien, Blackburn hadn't been able to vote. But to the second question, she gave a far more wary and vague reply.

Within a few years, evidence would come to light that nominees to scientific advisory panels were routinely questioned on whether they had voted for President Bush, and in 2004, Dr. Gerald T. Keusch, director of the Fogarty International Center at the NIH since 1998, would resign in frustration over repeated White House challenges to the institute's nominees for advisory panels. Keusch noted that during the Clinton administration, candidates approved by the NIH director were customarily approved by the White House, but during President Bush's first term in office, the administration rejected nineteen of twenty-six candidates. In one instance, when Keusch questioned the rejection of Dr. Torsten

Wiesel, a Nobel laureate, he was told that Wiesel had signed too many statements critical of President Bush.[11]

Blackburn's appointment to the council met with cautious optimism from administration critics who feared its makeup would be completely partisan. The *Washington Post* reported that "some observers say the president's council is politically stacked," noting that many of its eighteen members were well-known conservative thinkers.[12] Even conservative columnist William Safire conceded that "the majority of members lean conservative."[13] Council chair Kass was then on leave from the faculty of the University of Chicago, where he was the Hertog Fellow in Social Thought at the conservative American Enterprise Institute. In his extensive writings on bioethics, he once declared that "science essentially endangers society by endangering the supremacy of its ruling beliefs."[14] In an interview not long after the council was formed, Kass portrayed scientists as the opponents of those who sought to consider the moral implications of biomedical research, and he cast the debate in terms of a religious crusade: "So, at the moment, we have on one side scientists with prestige, knowledge, and power backed by powerful economic interests. And on the other side there are those of us who are putting hard questions about human values. How many divisions does the pope have? In this discussion, not very many."[15]

Evidence abounded that Kass had recruited council members likely to support his positions. James Q. Wilson had coauthored with Kass *The Ethics of Human Cloning*, and William B. Hurlbut was a friend whose qualification for service consisted of teaching a bioethics course at Stanford. Other members of the new council had publicly expressed opinions that echoed Kass's suspicion of scientists and his assumption that biomedical research was primarily profit-driven. Robert P. George, a theologian, had remarked that with respect to "the integrity of Christian doctrine," "there is, I'm afraid, an 'us' and a 'them.'"[16] Charles Krauthammer, author of a nationally syndicated column, had a clearly partisan pedigree; he named and developed the "Reagan Doctrine" on foreign policy.[17] Although he supported limited federal funding for human stem-cell research, Krauthammer opposed the use of cloned embryos (as opposed to embryos produced by IVF) and characterized proposed regulations for the creation of cloned embryos for "industrial" (that is, biomedical) and research purposes as "a nightmare and abomination," invoking the bogeyman of the "mad scientist" to bolster his argument.[18]

The staff of the council was headed by Executive Director Dean Clancy. Staff members, who wrote working papers for council deliberations (distributed in advance of meetings), functioned as a conduit for information and also wrote drafts of the council's reports. Clancy's background suggested he would provide a strongly biased filter. According to the *Washington Post*, Dean Clancy "is a self-described Christian 'proclaimer' who favors greater religious presence in the schools and who once smashed a roommate's pornographic videocassette with his bare hands." Moreover, Clancy was active in the Separation of School and State Alliance, which "favors home schooling over compulsory public education in order to 'integrate God and education.'"[19]

When the President's Council on Bioethics first met in Washington, DC, on January 17, 2002, it had eighteen members, a mix of legal experts, bioethicists, and scientists, only four of whom were women. (One appointee, Stephen Carter, stopped attending after just a few meetings and was not replaced until 2004.) Most of the council members had academic affiliations. Rebecca S. Dresser, George, Mary Ann Glendon, and Carter were lawyers and university professors. Francis Fukayama, Wilson, and Michael J. Sandel were political scientists, and Krauthammer, a political columnist. Gilbert C. Meilaender, Alfonso Gómez-Lobos, William F. May, and Kass were philosophers or ethicists, and Hurlbut, a trained biologist who taught bioethics. Paul McHugh was a practicing psychiatrist. Three council members—Blackburn, Janet Rowley, and Michael Gazzaniga—were full-time research scientists, and one, Daniel Foster, an active clinician-researcher. Blackburn admired Rowley, a distinguished cancer biologist, pediatrician, and hematologist in her seventies, and was reassured to see her on the council.

The new council had two primary mandates: to consider the moral and ethical meaning of new areas of research and biotechnology and to make recommendations about the role of state and federal government in regulating and funding research. Over a period of roughly two years, the council's deliberations would produce four reports on specific topics:

cloning (including cloning for stem-cell research), biological enhancements, monitoring stem-cell research, and assisted reproductive technologies. Ultimately, policy recommendations were included in only two of the council reports, *Human Cloning and Human Dignity: An Ethical Inquiry* and *Reproduction and Responsibility: The Regulation of New Biotechnologies*. As federal law required, all council meetings were open to the public, and at special sessions members of the public could address the council. Antiabortion activists routinely attended council meetings and even submitted position papers, but the council also heard from representatives of groups that supported stem-cell research, such as the American Society for Reproductive Medicine.

While council deliberations were public—transcripts and reports are available on the council's Web site at <http://www.bioethics.gov>—other communications were not. For each meeting, council members received a binder of readings, mailed in advance, that included working papers prepared by council staff, essays on ethical issues, and research summaries usually compiled by council staff; all staff papers were clearly marked "not for quotation or attribution." Blackburn noted that "when I read the council's first discussion documents, my heart sank. The language was not what I was used to seeing in scientific discourse—it seemed to me to present prejudged views and to use rhetoric to make points."[20] The first binder of readings included a short story by Nathaniel Hawthorne, "The Birthmark," which was discussed at a council session. In this short story, a scientist-philosopher becomes obsessed by a small birthmark on the cheek of his beautiful wife, and he concocts a potion to "correct what Nature left imperfect," but his pursuit of perfection kills his wife. Only by a rather questionable analogy could this attempt to correct a superficial imperfection be equated with the search for cures for devastating illnesses such as Alzheimer's disease. Blackburn found it disquieting that "a Gothic horror story, patently about a mad scientist, was seriously considered as having something to do with sober national policy."

Initially the council deliberated the ethics of therapeutic cloning (SCNT) and reproductive cloning of human beings. All the council members opposed reproductive cloning, but when it came to cloning for stem-cell research, they could not reach a consensus. While all acknowledged that ethical concerns arose in research on human embryos, they ranged widely between two moral poles: absolute opposition and firm support. The blastocyst, consisting of cells as yet undifferentiated, cannot develop into a fetus unless implanted in a womb, and the beginning of the nervous system, known as the primitive streak, does not begin to develop until after fourteen days, long after stem cells are harvested. Science made a distinction between this pre-embryo and an embryo that in a mother's womb would develop into a fetus, but many council members did not. Certain religious beliefs about the beginning of life (by no means universal, since Judaism, for example, determines an embryo is a human being only about forty days after conception) were thus pitted against the increasingly fine distinctions that can be made by modern biology—distinctions that make it possible for biologists to intervene in early life processes that were once a mystery.

The notion of "competing goods" permeated these early council deliberations, with the moral good of respecting the humanity of an embryo pitted against the moral good of research on embryos that might alleviate the suffering of living human beings. Committed proponents of a ban on therapeutic cloning argued for the absolute status of the embryo as a human being, whereas other council members understood this value as relative. In subsequently published comments, May reported that his physician had posed for him a dilemma that illustrated the concept of relative value: "If he happened across a freezer full of pre-implanted embryos and a five-year-old child in a building on fire, neither he nor anyone else would hesitate in choosing to rescue the child."[21] Council member McHugh, a Catholic, was also troubled by arguments that defined the blastocyst as a human being, and later noted in his statement in the council's report that he regarded SCNT as "the engineered culturing of human cells," not the creation of "a new and unique human individual," so that "an overemphasis on potential would lead us to the unreasonable position that since every one of our somatic cells has 'potential' for producing a human, it should receive some reverence."[22]

Even as the council debated whether and how a blastocyst should be protected as a potential human being, a growing body of scientific evidence suggested that a blastocyst created by SCNT could not develop into a normal organism. In late January, Rudolf Jaenisch, a scientist at MIT, had testified on these biological complexities at a congressional hearing—findings that were readily available to the council.[23] (Jaenisch eventually spoke to the council as well, but not until a few months after its cloning report was issued.) While the fertilized egg has special epigenetic programming that allows it to develop, a blastocyst created by the insertion of an adult cell nucleus (which might be taken from skin tissue) and stimulated to grow by artificial means in most cases could not survive beyond the earliest stages of development. If gestational abnormalities did not prevent an embryo from becoming a viable fetus, the rare survivors such as the sheep Dolly were genetically defective and did not live to normal old age.

At one point in the council's deliberation, McHugh suggested that an embryonic stem cell created by cloning be termed a "clonote" rather than an embryo to clarify this distinction, but Blackburn recalled that "Leon Kass rejected that notion out of hand." Much was at stake in terminology, and Blackburn also remembered fierce debates over terms to distinguish reproductive cloning from therapeutic cloning, in which Kass successfully pushed for the term "cloning-for-biomedical-research" to replace "therapeutic cloning." In Blackburn's view, this changed terminology "removed the moral dimension implied by *therapeutic*—that is, cloning with the aim of relieving human suffering as a competing good with concern for the embryo as a potential person."

It soon became clear that some council members, including all the scientists, firmly supported cloning for stem-cell research while another bloc of members firmly opposed it. But a number of members, including Hurlbut, Wilson, Dresser, Fukuyama, Sandel, and May, constituted a "swing vote."[24] According to Blackburn, Kass's worries that allowing cloning for stem-cell research would inevitably lead to reproductive cloning (a slippery-slope line of reasoning), pervaded the council's discussion on this issue. Blackburn also felt the debate was conducted as if no regulations were in place or would be, yet regulations already governed other controversial and risky biomedical advances that, like stem-cell research, posed the potential for abuse, and stem-cell research already fell within certain regulations administered by institutional review boards and the Food and Drug Administration (FDA). Other countries, such as Great Britain and Canada, had allowed stem-cell research to proceed under strict oversight. In Great Britain, the Human Fertilization and Embryology Authority, a review board that draws half its members from outside the medical and scientific fields, oversees research and regulates fertility clinics. British regulations stipulate that the blastocyst used to create stem cells be destroyed after fourteen days, a limit designed to prevent reproductive cloning.

As the council prepared to debate policy recommendations on cloning, Congress also grappled with the issue. In July 2001, the House had passed legislation prohibiting both reproductive and therapeutic cloning, and the Senate was currently debating legislation that might permit cloning for research while banning reproductive cloning. In an April 10, 2002, speech, President Bush preempted his own council's recommendations by calling on Congress to outlaw both types of cloning as a matter of "conscience."[25] Soon after, at the council's April 25 meeting, Kass was reported to have privately told a council member he "didn't want to embarrass the president," and perhaps to test the waters, Kass sought a discussion on four possible positions on therapeutic cloning, which really boiled down to two choices: support (with or without "regret") for a ban or support for continued research (with or without "humility").[26] Though Kass frequently said he wished to hear moral arguments on both sides, he also commented, in laying out these proposals, that those who opposed stem-cell research might also do so with regret, but "I'm sorry, the tears belong on the other side."[27] During the debate, Wilson broke ranks with the conservatives on the council, refuting the slippery-slope contention reiterated by Krauthammer as "an argument that can be used against every advance in medical science that I can think of."[28]

Just as the Senate was debating human cloning and the council seemed to be heading toward a majority approval for permitting therapeutic cloning rather than banning it, the May meeting of the council was canceled. Before the next meeting, scheduled for June 20, council members submitted statements on their positions on the two options discussed at the April meeting. Yet at the June 20 meeting, Kass outlined seven possible public policy recommendations that blurred the either-or choice initially put before the council:

1. Professional self-regulation only;

2. A ban on cloning-to-produce-children, with no position on cloning-for-biomedical-research;

3. A ban on cloning-to-produce-children, plus regulation of cloning-for-biomedical-research;

4. Governmental regulation of both forms of cloning, without legislative prohibitions;

5. A ban on both forms of human cloning;

6. A ban on cloning-to-produce-children and a moratorium on cloning-for-biomedical-research;

7. A moratorium on both forms of human cloning.[29]

Since all council members supported a ban on cloning-to-produce-children, options 1, 4, and 7 were not debated. Instead, council debate focused on two similar proposals: to allow therapeutic cloning to continue with regulations (3) or to impose a moratorium (6). The proposal to ban both types of cloning was never put to an official vote.[30]

Effectively, this meant that many who would have supported a ban voted instead for a moratorium, and those "swing voters" who might have supported therapeutic cloning with reservations could endorse a moratorium as a means to establish clearer regulations before proceeding. The ambiguity of the proposed moratorium thus divided the vote along new and confusing lines. In his account of the deliberations, council member May stated that "the Kass proposal did not seem like a genuine compromise, in which diverse parties agree on a common, albeit imperfect, agenda. The advocates of a ban among those who voted for a moratorium would have no intention of using the time to develop regulations . . . and in conscience, they would walk away from the effort to develop regulation."[31] By May's count, five to seven council members supported option 6 when, if given the chance, they would have voted for option 5, thus diminishing any "majority" in favor of a moratorium.[32] An account in *Science* reported that by the end of the June meeting, at least nine council members (a majority) supported therapeutic cloning with regulations.[33] Blackburn and several other council members estimated that a total of ten council members publicly stated their support for allowing therapeutic cloning to proceed with regulation.

No official vote was taken at this meeting, but on June 27, only eleven days before the report was due to come out, council members received a draft of the policy recommendations (a chapter in the report), along with an official ballot. Council members were asked to choose between two proposals, both of which banned reproductive cloning. The first proposal also imposed a four-year moratorium on therapeutic cloning, and the second permitted therapeutic cloning with regulation. The moratorium would go a step further than President Bush's stem-cell policy by recommending restrictions on both publicly and privately funded research. In his cover letter, Kass projected a changed outcome for the vote, anticipating ten council members supporting the first proposal, six the second proposal, one unknown, and one abstaining. Nothing in council discussions had suggested this tally; if some members had changed their stated positions in the interim, it was not part of the public record. Furthermore, the council had never deliberated on the length of any proposed moratorium, so the four-year term seemed to have been plucked from thin air. Blackburn was not alone in her dismay at this new configuration. One council member decried this process as "backroom stuff"; Rowley publicly stated that she was "really caught by surprise" when she saw the numbers; and Gazzaniga questioned whether the tally reflected the actual positions of council members.[34] What was essentially a split decision—with seven council members favoring permitting stem-cell research, three favoring a moratorium, and seven favoring a ban—became a majority in favor of a moratorium when the votes in these last two categories were combined.

This last-minute draft of the council's report on cloning contained other surprises as well. Additional arguments had been added to a section that outlined the case for permitting therapeutic cloning with regulations. Several members of the minority challenged this new material, since it implied problems with regulation that did not represent their views. An astonished Blackburn even went over the transcripts of the previous meeting to verify her perception that this material had never been discussed. Her problems with the draft of this report presaged persistent difficulties with the accuracy and logic of the council's reports. When seemingly small matters of word choice cumulatively misrepresented the certainty of scientific information or slanted the arguments for and against, Blackburn was more than willing to fuss over adjectives and adverbs. She also suggested cutting entire sentences and paragraphs in which "strong rhetorical language was reserved for the case against therapeutic research." What survived in the report's final draft, after emendations, suggests how loaded the argument actually was. Chapter 6, on the ethics of therapeutic cloning, asks a rhetorical question about hesitation to pursue this line of research: Why should we not put this cup to our lips? The answer, taken from Shakespeare's *Winter's Tale*, is that "there may be in the cup a spider steeped." No such literary references supported the other side of the argument. In fact, the report gives scant attention to the moral case for pursuing therapeutic cloning: exactly five-and-a-half pages, with seventeen pages devoted to the "possible moral dilemmas" and another twenty pages outlining the moral case against it.[35]

Blackburn also submitted an individual statement to be appended to the report, an option made available at the insistence of council members who held what became the minority view. In her statement, she did not allude to value differences on the council but focused on making the case that a moratorium would impede the very research needed to inform any decision in the future. The statement she submitted to Kass pointedly discussed existing regulations that could govern human embryonic stem-cell research and prevent the possibility of using it for reproductive cloning, an issue overlooked in the report. Of the seven members of the minority—Foster, Rowley, Gazzaniga, May, Blackburn, Sandel, and Wilson—six chose to submit individual statements. In his statement Gazzaniga bluntly found fault with the report: "For me it is full of unsubstantiated psychological speculations on the nature of sexual life and theories of moral agency." In his view, this was "particularly troubling, when cast in the large, as a basis for social and even scientific policy."[36]

When Blackburn submitted her editorial changes, she notified the council staff that she could not sign off on the report if it misrepresented the minority argument or contained inaccuracies. Council members usually communicated through e-mails, but Kass soon telephoned Blackburn to convey what he described as the staff's concern about her "threat" not to sign the document. Blackburn and Kass had an otherwise amiable conversation in which they acknowledged a number of shared concerns on biomedical issues, but Kass also challenged the statement she had submitted. He insisted that Blackburn's discussion of existing regulations be left out of her statement and questioned its accuracy. Blackburn customarily and politely gave ground on matters she considered secondary, and in this case, she believed she could avoid being marginalized by keeping the lines of communication open. Knowing that the council was scheduled to revisit the question of regulations for stem-cell research, Blackburn decided "this was not a battle I had to fight now" and yielded to Kass's demands.

Unwilling to stomach any celebration over what seemed a trumped-up victory, Blackburn did not attend the July 11, 2002, meeting at which the council report *Human Cloning and Human Dignity: An Ethical Inquiry* was made public, with members of the press in attendance. After Kass summarized the content of the report, two council members— Sandel and Gazzaniga—referred to a table on page 129 of the preliminary report (page 202 of the final report) that recorded the actual breakdown of the vote when council members chose between a moratorium, a ban, or regulation, reflecting a more divided council.[37]

Although Blackburn had accommodated Kass by changing her statement, she did not hesitate to use her authority as a scientist to articulate in a public forum the minority argument that had not been fully represented in the report. She joined with Foster, Gazzaniga, and Rowley (all the researchers on the council) to coauthor an editorial, published in *Science* in September 2002. The authors noted that the lack of federal funding would not only impede the development of potential medical therapies but also, by limiting research to the private sector, prevent the public sharing of results that drives academic research. Furthermore, the council was "trying to reinvent the wheel" by proposing to delay research until regulations were developed, since a federally commissioned study, incorporating more than fifty thousand comments, had already recommended specific regulations for human embryonic stem-cell research.[38]

In interviews and editorials similar to the one that appeared in *Science*, Blackburn doggedly emphasized accurate representation of the available scientific information. She told a *Washington Post* reporter, "The only way we can get data about what works best is if we're allowed to do the experiments."[39] In a number of interviews, Blackburn was also astonishingly frank. Remarking on the changed results to the vote on a moratorium, she told Stephen S. Hall, reporting for *Science,* "We always feared that the dirty work would happen at the crossroads."[40] In her career as a scientist, Blackburn had always minimized personal slights and refrained from angry challenges. Paradoxically, now that the issue was not personal, she took it personally, because it struck at the core value of her intellectual life: a stance toward the truth that did not allow for tweaking the facts.

Though she continued to prepare carefully, Blackburn had to struggle to attend subsequent meetings, scheduled far more often than she'd been led to expect. In fact, the council was slated to meet ten times in 2002 (though two meetings were canceled). For the crucial June 2002 meeting, when her husband was scheduled to be in Europe, Blackburn had brought her teenage son along; she told Ben to consider this "your civics lesson." He sat in the audience during each session of the meeting, taking notes, and later he told his mother that he had kept a tally of "how many times the head guy [Kass] interrupted people."

For the next several months, council meetings focused primarily on reproductive technologies and biomedical advances that raised questions about the distinction between therapy and enhancement, including choosing the sex of a child, using drugs such as Ritalin to improve children's academic performance, and treating some of the common symptoms of aging, such as memory loss. Blackburn felt that the questions about enhancement had been thoughtfully framed by the council chair, and yet discussions were characterized by an "underlying current of belief that any enhancements were inherently evil or hubristic." During council discussions, Sandel questioned Kass for appearing to equate enhancement with medical intervention: "What I was suggesting was that the reason that Leon [Kass] has offered for objecting to enhancement is problematic because if it were applied to medicine, it might condemn medicine as well as enhancement."[41] In the same session, Blackburn objected to certain language in the working papers under discussion, particularly the use of the terms "magical" and "mysterious" to describe biological processes, which she considered an essentially anti-intellectual stance. Kass quickly fired back: "There is notoriously a disjunction between the way science understands things and the way human beings on the plane of human experience understand things. These are two different languages."[42]

In council discussions of research on aging, a field to which Blackburn's study of telomeres related directly, Kass and several other council members persistently interpreted the aims of such research as an overweening quest for immortality. Blackburn felt that "it was a good thing to discuss in advance of the technology the questions it might raise, but these issues were discussed with prejudice. We didn't really grapple with them." Once again, Blackburn was bothered by the misrepresentation of scientific facts to bolster what amounted to fearmongering: the specter of scientists creating genetically engineered designer babies was raised as an imminent possibility when it is a scientific unlikelihood, because a trait such as intelligence can't be traced to any single gene and any alteration of a single gene could disastrously influence the complex interaction of a number of genes. "We could have been discussing serious bioethics issues," Blackburn said. "Instead we were discussing science fiction, not science."

Confronted by what amounted to science bashing, Blackburn reacted in a manner characteristic of her and probably most scientists—she marshaled evidence. In this respect, her expertise in research on aging constituted an exceptional resource for the council, since as a member of the National Advisory Council on Aging at the National Institute of Aging, she deliberated national policy on such research. If Blackburn hoped that accurate information might persuade some council members to reconsider their judgments, she also understood that the facts alone might not persuade others on the council. Elizabeth Marincola, whom Blackburn consulted throughout her term on the council, said that her advice in their many phone conversations was primarily strategic: "I helped Liz by talking out tactical and emotional issues more than the scientific aspects of the research." According to Marincola, Blackburn also shared her concerns about making an ethical argument:

Liz, if you will, honors her opponent by really trying to understand their argument so she can engage them at the intellectual level. This really came out in her bioethics council work. Instead of invoking specious, vague, or insulting remarks as politicians are so wont to do, she will study the person's point of view so that she can debate them on the merits of the argument. That's one of the reasons people respect her so much even if they don't agree with her. She spent so many hours, even on weekends, discussing stem-cell research with me—"well, they say this, let's talk about the problems with this perspective." She did not attack others as second-rate scientists or politically motivated, which some members of the council were, but put all of that aside to focus on issues and rational argument.[43]

Ultimately, Blackburn came to feel that her efforts to inform the debate and ensure accuracy in the council's reports were rebuffed.

Successive drafts of *Beyond Therapy: Biotechnology and the Pursuit of Happiness* were circulated even as the council moved on to discussions of reproductive technologies and stem-cell research. The council's mandate had been renewed for two years, yet Kass was determined to meet a deadline of October 2003 for releasing the report. Often a fast turnaround was demanded for suggested revisions to drafts, and increasingly Kass pressured Blackburn to sign off on versions of the report before she'd seen a final draft. Rowley, also highly critical of the report's misrepresentation of science, mentioned to Blackburn a similar pressure; the council's executive director, Clancy, had e-mailed her to say everyone else had signed off on a draft, making her a lone holdout, when at the time at least one other council member, Blackburn, had also refused to approve the report without first reviewing editorial changes.

On a year-long sabbatical, Blackburn now had more time to devote to scrutinizing report drafts, and she challenged them in detail. Her exchanges with Kass became testy, but she did not back down: "When the time came to write this report, we began to argue over the science. I wasn't used to contending in this way, so for me, these felt like huge fights." According to Blackburn,

Chapter 4, "Ageless Bodies," misrepresented research on aging, as the title alone suggests. Kass wanted to quote a researcher from outside the field who had said all scientists want to make people immortal—this wasn't evenhanded. I had very good access to aging researchers and to the science, and I tried to convey the correct information to the staff and to Leon Kass. I was informed by e-mail that they had made all my requested changes. When I saw the final draft, I was very angry—substantial inaccuracies had not been corrected. I contacted Kass to complain. With copies of both versions in hand, I wrote Kass describing my suggested changes point by point. Kass tried to rebut, and I wrote again, countering his rebuttal. The staff had made only the little changes I'd suggested.

The report emphasized the claim that an increased life span resulted in decreased fertility, and though Blackburn had passed along current scientific literature that disproved this myth, the claim was not corrected.[44] Bias with respect to the motives of research also survived into the final draft. The chapter on research into aging opens by claiming that "the scientific quest to slow the aging process . . . treats man's mortal condition as a target for medicine . . . it is difficult to distinguish it from a quest for endless life."[45] By this reasoning, the extension of the human life span in the last century, thanks to medical advances such as antibiotics, organ transplants, and improved nutrition, would also constitute a presumptuous quest to conquer mortality. When the report was issued, LeRoy B. Walters, a noted professor of Christian ethics at Georgetown University, commented that its depiction of biomedical advances made it seem "as if most of basic research is dangerous and not directed toward important goals."[46]

Blackburn, Gazzaniga, and Rowley felt that the hasty preparation of this report led to inaccuracies, compounded when many of their comments were not incorporated into the drafts. Concerned that this foreshadowed what would happen as the council prepared its next report on the more pressing issue of monitoring stem-cell research, the three researchers wrote a letter to Kass calling for procedural changes in report preparation and criticizing the politicization of the council's work. They also collaborated on a letter to *Science* in which they expressed similar reservations. "As scientists," Blackburn says, "we didn't want to go on record with misstatements about the research." If there were value gaps between the scientists and several other council members on issues such as an embryo's moral status, another significant value gap existed as well, one that often leads to the charge that scientists are politically naive. From a political perspective, adroit maneuvering is part of a fair fight, and the research scientists, bested in the battles over council reports and recommendations, should have retreated from the field. For scientists, contesting the accuracy of evidence and revising belief in the light of fact is a professional obligation, ideologically neutral, and thus they are unlikely to anticipate the fallout that results when they do not concede defeat.

This helps to explain why Blackburn was unprepared for a political reading of her motive and method when she collaborated on the letter to *Science*. Gazzaniga had been charged with sending Kass a copy of the letter, and Blackburn only discovered that it might have gone astray when, in November 2003, Kass phoned her, furious. He had just learned of the existence of this letter from a *Science* reporter who had asked him to comment on it. Blackburn assured Kass that the letter had been sent to him in advance and promised to forward her correspondence with Gazzaniga on the matter. Kass was not mollified. "He said, 'I thought you were nice,'" Blackburn reported. As Kass accused her of cowardice, uncollegial behavior, and impropriety in making council proceedings public, Blackburn strove to remain calm: "At one point, I held the phone away from my ear because he was yelling. But I wasn't intimidated. Instead I felt a certain distant fascination. He'd lost his usual veneer."

Yet Blackburn was shaken enough by her conversation with Kass to consult Marincola. "She was really upset because she felt she had been berated by Leon Kass," Marincola said. "And of course she was very aware that he was willing to treat her this way when there were no witnesses. He couldn't be held accountable for it. And then he would be all gentlemanly and calm in a public setting."[47] Blackburn wondered whether to resign from the council rather than continue to be engulfed in battle, and Marincola retorted, "Let him fire you," and encouraged her to remain on the council. Marincola gave Blackburn another piece of advice: You cannot document what transpires in a phone call, and you should never accept another phone call from Kass. In counseling Blackburn, Marincola also suggested that she respond with caution to any future antagonism: "I thought Kass was intentionally trying to bait her into saying or doing something that would embarrass her. Not that she would have done that, but she had been so goaded by this person. I really think that was his calculation. She did see it as a power play, as gender politics. What is appropriate? Should I be more of a man than he is, or should I stick to my own values and decorum? She decided on the latter course."[48]

During their tense phone conversation, Kass had ordered Blackburn to withdraw the letter to *Science*, and after consulting with her colleagues, she agreed. Along with Gazzaniga and Rowley, she feared a future, bigger battle over the next report, and rejecting Kass's demand might cast the scientists as obstructionist, diminishing their chances to influence future debate. On the question of the "impropriety" of commenting publicly on council proceedings, she faced a difficult moral quandary: "If the council was a public body, how much of its proceedings should be kept secret? And if attempts to work within the process failed, did I have an obligation to make public information that was suppressed in final reports? If being polite blunted the message, was it time to change the strategy?"

12 | Political Fallout

Blackburn's encounter with Washington politics exemplifies the tension between the scientific ideal of a strictly defined truth and the more ambiguous context in which public policy is generated. Such tension contributes to the vitality of a pluralistic democracy because it fosters debate rather than foreclosing it, but debate founders when evidence is treated as if it were as malleable as opinion. According to a 2004 analysis of scientific advisory policy commissioned by the Federation of American Scientists, the stem-cell debate offers a spectacular example of this: "Both proponents and opponents of stem-cell research can be found using the argument of 'sound science' to mask what are, in the end, basic differences in values. Effective management of these issues doesn't require masking the difference in values, but it does require an effective mechanism for sorting fact from values."[1]

For the sake of this distinction, Blackburn committed herself to persistent conflict with the council staff and chair. At stake was the *survival* of debate on controversial issues that could affect the future of biomedical research; while some value judgments might not change in the face of countering facts, bias in council reports could foreclose the option of a knowledgeable public making informed decisions. Not so much scientific error as the slant on scientific evidence triggered Blackburn's objections to council reports: "They could say, what have we said that's actually wrong? It was about how the evidence was evaluated and presented."[2]

In the arena of public opinion, this matters greatly. According to a report published by the Genetics and Public Policy Center, "By all accounts and every survey of U.S. opinion, most Americans oppose the use of cloning for reproduction. What is lost in that declarative statement—which routinely is echoed in mass media and political rhetoric, and drives much of the policy discussion about cloning—is that Americans actually have much more nuanced opinions regarding the use of cloning for research and therapeutic options, and that these opinions still are anything but immutable."[3] The report, *Cloning: A Policy Analysis,* cites a number of surveys showing that a majority of Americans (from 58 to 75 percent, depending on the survey) support stem-cell research that uses embryos left over from IVF. Opinions on therapeutic cloning fluctuate depending on the wording of questions. In common parlance, the term *cloning* so strongly connotes "cloning to make a baby" that it influences the results of opinion polls. In a Gallup poll, only 38 percent of the respondents favored the use of cloned embryos for research, but when another poll substituted the term "somatic cell nuclear transfer" for cloning, 72 percent of the respondents approved. In addition, support for human embryonic stem-cell research typically increases when surveys identify specific diseases for which it might one day provide a cure.[4] Because how the science is "presented and evaluated" dramatically influences public opinion, it became a bone of contention during council deliberations.

For some time before Kass phoned Blackburn to demand that she and her coauthors withdraw their letter to *Science,* the council had been considering the topic of monitoring stem-cell research, reviving the controversy that had surrounded its debates on cloning. During its July and September 2003 meetings, the council focused intently on the issue, which would be the subject of its next report. Looming over these deliberations was the question of whether the council would impartially assess the facts or tailor them to support the policy that President Bush had already declared. When he banned federal funding for the creation of new embryonic stem-cell lines, Bush claimed that experts had assured him there were seventy-eight viable stem-cell lines already in existence. The scientific community had immediately challenged these numbers.[5] Only nineteen lines were available, and since they had been grown in culture dishes coated with mouse cells (a method designed to nourish stem cells), all carried the risk of viral contamination, making them useless for developing potential therapies.[6] Subsequent studies confirmed that all the approved stem-cell lines had been contaminated with a mouse molecule that could provoke an immune response if the cells were injected into humans in an attempt to treat disease.[7] Yet another study reported that at least a quarter of the approved stem-cell lines were so difficult to keep alive that they had little potential for research, let alone therapy.[8] For some time the Bush administration would persist in referring to these problems with the viability of existing embryonic stem-cell lines as a matter of scientific debate, but in reality they were well-documented facts.

When it came to influencing public opinion and public policy on stem-cell research, how two types of scientific evidence were presented mattered greatly: the possibility of conducting informative research on existing human embryonic stem-cell lines and the viability of adult stem-cell research as an alternative. The July 24 council meeting illustrates how debate on these issues played out. On the question of whether society had a moral obligation to fund research, the council heard from the director of international programs at the Hastings Center, Daniel Callahan, previously recruited by Kass to advise the president on the stem-cell policy announced in August 2001. In a session titled "The 'Research Imperative': Is Research a Moral Obligation?" Callahan remarked that scientists behaved as if they were "entitled" to federal funding for stem-cell research and argued that research be funded depending on the likelihood of medical benefits, though he did not propose how this could be determined in advance or acknowledge the degree to which the promise of any research is speculative. He claimed scientists had yet to show that embryonic stem cells had any advantage over adult stem cells, though research on mouse stem cells suggested otherwise. According to Callahan, the uncertain promise of stem-cell research did not constitute a "duty to carry it out" when the NIH spent millions annually on "other promising research for the very same diseases," and thus to argue for the necessity of stem-cell research implied the NIH was "wasting a whole lot of money."[9]

When council members responded to Callahan's remarks, Blackburn led off. She pointed out that unless the necessary experiments could be done, no one would ever find out the relative advantages of the two types of human stem cells; the promises of stem-cell research could not be determined precisely because of funding restrictions on such research. Kass cut off Blackburn's line of questioning, saying, "At the risk of perhaps deflecting people from where they would like to go, it seems to me the real challenge . . . for us is to think really about the large theme, which is the moral imperative to research." But Rowley once again brought the discussion back to the realities of research science, challenging Callahan's assumption that the efficacy of research could (and should) somehow be determined in advance: "And it seems to me this does not reflect the real world of science or biomedical science as I have lived it for over forty years . . . in the academic world, which is where fundamental research is really flourishing, the scientists are the ones who have the intellectual curiosity and the creativity to say, 'Isn't this an interesting question?' and to follow up on it." To illustrate the point that research could have unanticipated implications, Rowley cited Blackburn's work on the ends of chromosomes in yeast and her discovery of telomerase: "And now this turns out to be a very important enzyme in cancer. But who would have expected research in yeast to then have that kind of applicability?"[10]

Like the other scientists on the council, Blackburn objected to remarks by Callahan that seemed to disparage scientists, especially since they seemed part of an emerging pattern, in keeping with Kass's insistence that scientists spoke a "different language" from others. At several sessions Gazzaniga had emphatically countered the portrayal of scientists as charging recklessly into new areas of biotechnology, and during one session, he emphasized that "the most conservative group of people in this room are scientists. Maybe not the scientists, but science. It moves slow. It checks. It double checks. It's out in the open."[11] The historical record also counters the image of scientists as greedily focused on the quest for knowledge, blind to the ethical implications of research; when recombinant DNA technology made it possible to tranfer genes between organisms, scientists were the first to call attention to the ethical risks, and at a 1975 conference they proposed guidelines later adopted by the NIH.[12]

On the same day that Callahan spoke, the council heard from three stem-cell experts: John Gearhart, Rudolf Jaenisch, and David Prentice. Gearhart, an embryologist affiliated with Johns Hopkins University, and Jaenisch, a researcher affiliated with MIT, had impressive scientific credentials. But Blackburn doubted the qualifications of Prentice, a professor at Indiana University, as soon as she saw his résumé in the briefings for the meetings. He hadn't published research in significant, peer-reviewed journals, and his list of grants did not impress her—"the occasional little thirteen thousand dollar grant from the NIH, another for eleven thousand. Consistent funding at a much higher level is the usual criterion of serious research."

Gearhart stressed the profound medical benefits of human embryonic stem-cell research and the need for a diversity of stem-cell lines, noting that "enormous differences" between the embryonic stem cells of mice and humans necessitated work on human cells. He concluded that "the greatest impetus that I can think of for making advancement in this field is funding through the National Institutes of Health."[13] Jaenisch explained in detail why a cloned embryo derived from SCNT was not biologically like a fertilized embryo but more like a "laboratory artifact," largely because it stood little chance of developing into a healthy individual.[14]

Only the third speaker, Prentice, presented adult stem-cell research as a viable alternative that could supplant the need for embryonic stem-cell research. Adult stem cells can create certain types of cells that need to be replenished throughout life, such as blood-forming cells in bone marrow, skin tissues, and cells lining the intestine; stem cells gone awry are also implicated in cancer. Bone marrow transplants already capitalize on the repair capabilities of blood-forming stem cells to treat leukemia. In his presentation, Prentice referred to a 2001 study at New York Medical College, which concluded that blood-forming stem cells had repaired heart damage in mice, suggesting that adult stem cells might have the potential to differentiate into many types of cells in the body, just like embryonic stem cells. But in fact, the experimental evidence was questionable: the results of this study proved impossible to replicate in other labs, and further experiments showed that because the stem cells had not first been purified, it was impossible to tell which cells had traveled to the heart muscles of the mice and fused with them. A later study found that blood-forming stem cells only rarely fused with other cell tissues and were thus unlikely to repair damaged tissue.[15]

When Blackburn questioned Prentice on the evidence he'd presented, he admitted that so far only one really good source of adult stem cells existed—a cell line developed by Catherine Verfaille, who had spoken to the council at an earlier session—and that adult stem cells had proved difficult to work with in the lab.[16] In reality, adult stem cells are rare, do not live as long as embryonic stem cells, and are hard to culture in amounts adequate for experimentation. To illustrate how distant their medical benefits may be, Blackburn explained, "If you wanted to harvest your own stem cells in the hope of transplanting them to regenerate damaged tissue—say, from a spinal cord injury—you'd have to wait years before you could accumulate enough cells."

At the September 2003 meeting, the council heard from a researcher in the private sector who, like Prentice, touted adult stem-cell research yet could not provide adequate evidence for its efficacy. William Pursley, president and CEO of the biotech company Osiris Therapeutics, Inc., had supplied the council in advance with a review by researchers at Osiris claiming wonderful results in adult stem-cell research—none of which had been published in peer-reviewed journals. Although Pursley cited a number of studies, Blackburn regarded this as "zero evidence," since no hard data supported the claims: "The innocent nonscientists wouldn't know that. It's important for scientists to participate in policy debates because they can evaluate the data. It's what I thought would be my contribution when I joined the council." Putting a spin on scientific findings bothered Blackburn as much in this instance as it did when Geron had hyped the promise of telomerase.

At this same meeting the council also heard from Tom Okarma, president and CEO of Geron, which had been at the leading edge of embryonic stem-cell research since 1995. When mailing briefing materials, the council staff had gone to the trouble of forwarding Pursley's review separately because of his concern about keeping this proprietary information from the public record, but none of the many published research papers by Geron scientists had been included in the briefing materials because, the cover letter noted, "these were too voluminous to include." Okarma, whose company had long championed public funding for stem-cell research, reported that the FDA had just qualified two of Geron's stem-cell lines for human use and outlined their clinical potential. Asked what obstacles his company faced, he answered that while many technical barriers remained, "Our major problem is funding. . . . The political uncertainty of this field not only turns off investors but also turns off the other source of funding for biotech, which are pharmaceutical partners." Okarma also compared his company's stem-cell research to its cancer program based on telomerase. Though at one time Geron had a testy relationship with academic researchers investigating telomerase, eventually the company collaborated with researchers around the world. As a consequence, Geron's cancer program rested on a "very deep" scientific understanding, in contrast with its program in embryonic stem cells, "where we have a small number of collaborators. . . . So the narrower science base in embryonic stem cell research increases risk of technical failure and exposes patients to greater risk from the experiment." As Geron derived any viable new embryonic stem-cell line, it would be forbidden to share it with academic researchers: "Because it was derived after 2001 in August, the NIH will be prohibited from studying it."[17]

Conditioned by her lab work to qualify assertions and to meet a high standard of scientific proof, Blackburn was troubled that "iffy science" supported claims for the viability of adult stem-cell therapies. She was even more disturbed that justifications for a ban on federal funding for stem-cell research slanted or overlooked evidence in the service of ideology. "The real argument," she said, "was often simply an antiabortion argument, based on the assumption that a blastocyst deserved protection as a full human being." The bioethics council had a mandate to consider values, but increasingly the chair and staff seemed to promote a particular set of values at the expense of accuracy and objectivity, potentially undermining the council's findings both ethically and factually.

Council debates were often framed as though the only moral high ground belonged to opponents of embryonic stem-cell research. During the September council meeting, Kass posed the issue not as a debate over competing goods but as a "contest between right and good" in which "one side at least claims to uphold some kind of absolute moral principle . . . whereas, while others might defend it in absolute terms, it's very hard to say that the case for doing medical research has the same kind of moral absolute status."[18] In a later session of the same meeting, May worried that the council's purpose might be to offer an "apologia" for already declared policy rather than to deliberate in good faith what advice to offer the president. If, as Kass said, an absolute right was in contention with a relative good, no compromise could be ethically acceptable. But May argued that Kass's comment as well as staff briefing papers predicated a dangerously simplistic dichotomy between "on the one hand . . . those who are duty oriented in their thinking, principle oriented—and those duties do not admit of any exception—as opposed to those who are merely awash in the sea of the relative, and all the precariousness of cost-benefit analysis."[19] Blackburn immediately seconded May's concern for contending with ethical complexity, pointing out that a large number of pediatric vaccines were derived from fetal cells, "and so we have made that decision in this case where I think the issue that you raised of relative potential goods, you know, has been weighed, and the decision has been made very clearly."[20]

As Blackburn reviewed the drafts for *Monitoring Stem Cell Research*, the report on these discussions, she found that evidence countering sweeping claims for adult stem-cell research was not incorporated in the report drafts. Instead, the drafts presented the viability of using adult stem cells for medical therapies as a matter of scientific debate when, once again, by scientific standards no conclusive evidence as yet confirmed their efficacy. Verfaille, a leading adult stem-cell researcher, has acknowledged that "it's too early to tell what will and will not be possible."[21] Other leading researchers have reiterated that only further research on embryonic stem cells can illuminate the complexities of adult stem-cell function.[22] For Blackburn, this uncertainty demanded support for further research rather than restrictions. Researchers had not, for example, been satisfied to call into question the results of the 2001 adult stem-cell study cited by Prentice but had pursued the curious phenomenon of cell fusion. Such open-ended exploration would be far less likely if federal funding were prohibited for research on embryonic stem cells, and by leaving private industry unregulated, President Bush's policy fostered exactly what it was intended to prevent: a scenario in which embryos might be treated merely as commercial products and with far less oversight than that demanded by the NIH.

By now it had become routine for staff to forward drafts to council members with only a brief turnaround time; council members never saw a final draft and never quite knew how it would appear. So many substantial errors of scientific fact marred drafts of *Monitoring Stem Cell Research* that several council members insisted the deadline be postponed. A paper by Prentice, commissioned for the report, was criticized as inaccurate by the stem-cell experts who read it. Worse, the report offered no clear statement on why scientists thought embryonic stem cells were advantageous over adult stem cells. When Blackburn wrangled with Kass over editorial changes, he characterized her point-by-point criticisms as "being nasty," and she responded—in writing—that disagreeing with him was not the same as "being nasty." In the back and forth over what constituted corroborated fact, the research scientists on the council repeatedly suggested that the reports be reviewed by outside stem-cell experts, as was done in the National Academy of Sciences. Blackburn even provided Clancy with a list of five reputable experts; to her knowledge, he consulted only one of them, whose comments may or may not have been incorporated in the final draft.

Once again, the report made a loaded argument. The draft report's introduction, for instance, suggested that donating tissue from an aborted fetus for research purposes might encourage abortion—a startling proposition complete with a hidden, unquestioned assumption that abortion was morally wrong. Federal funding for biomedical research was described as "this extraordinary show of public largesse," which Blackburn amended to "this public expenditure." In giving reasons for her change, she pulled no punches; having served on NIH study sections, which imposed stringent criteria for funding, she tartly observed that "such support is not offered indiscriminately."

Policy debates on assisted reproductive technologies (ARTs), which had been ongoing, intensified as the council prepared to issue its report on this topic. ARTs intersected with controversies surrounding cloning, stem-cell research, and enhancement, so discussion revived the thorny issue of the moral status of the embryo, which Kass had earlier acknowledged as "the elephant in the room." Though Kass emphasized that he didn't want this debate to "obscure or prevent an inquiry into a range of other important questions," bias against ARTs cropped up repeatedly in papers prepared by council staff. Reviewing a document purported to provide a neutral depiction of current practices, regulations, and ethical considerations, Wilson criticized it for ignoring ethical arguments for the benefits of assisted reproduction, and he singled out certain statements as sounding like "slogans." Rowley protested that the document offered "a very negative point of view" of ARTs, failing to acknowledge how genetic diagnosis, for example, had reduced the number of children born with terrible diseases such as Tay-Sachs.[23] At the October 16 meeting, Gazzaniga complained that documents presented ART in terms of a "freak show" of horrendous, remote possibilities (such as the creation of chimeras) rather than addressing the serious, immediate issue of regulating medical practices already in effect at fertility clinics.[24] While *Beyond Therapy* and *Monitoring Stem Cell Research* had contained no policy recommendations—thereby obviating the need for a potentially divided vote on the issues—this report would contain recommendations. At the insistence of several council members, individual statements would be reinstated in this report.

While Blackburn concurred with the recommendations of the council, aimed at improving the care and rights of patients attempting to have children using ARTs, she once again was astounded by the report's bias. Drafts of the council's report expressed concerns about the lack of research documentation on ART procedures but did not mention a primary cause: the lack of adequate funding for such medical research. This was an inconvenient fact, since prohibitions against federal funding for any research that resulted in the destruction of an embryo meant that many ARTs, which posed such risks, could not be studied, nor could researchers investigate how to determine which preimplantation embryos were more viable. Early drafts of the report used the phrase *child to be* or *future child* in place of the word *embryo*, consistent with a pattern in which biased language in the report drafts imposed a moral viewpoint as if it were a matter of fact.[25] (Even the title of the report, *Reproduction and Responsibility*, was eerily reminiscent of a slogan for an abstinence campaign.) Blackburn prepared for the report a statement taking issue with scientific inaccuracies that might influence policy or leave scientific research "vulnerable to control by a vocal minority." If the unanswerable insistence on the status of the embryo as a full human being determined how scientific research was regulated and funded, it would come at great social cost.[26]

Once again, Blackburn faced the quandary of whether to go public with her misgivings about council process and reports or to maintain a politic silence. In her line of work, one did not dismiss inconvenient facts but contended with them. Bad science meant ethical disgrace, and her own credibility as a scientist was on the line if she acquiesced to inaccuracies in council reports. Respect for the confidentiality Kass had insisted on conflicted directly with the council's mandate to carry out its decision making in public and fully inform public policy. Blackburn felt she was "answerable not just to the scientific community but to a broader constituency."

With Rowley, Blackburn wrote a critique, "Reason as Our Guide," which she and Rowley planned to submit to the journal *PloS Biology*.[27] The critique incorporated comments on both *Beyond Therapy* (the focus of the withdrawn letter to *Science*) and *Monitoring Stem Cell Research*. Blackburn and Rowley charged that *Beyond Therapy* misrepresented research into prolonging healthy life as being dominated by scientists striving for immortality and that *Monitoring Stem Cell Research* had overstated the current research promise of adult stem cells and played down the potential of embryonic stem cells. The authors also listed specific instances in which their requests for accurate representation of crucial scientific data in both reports were denied. Raising a persistent concern among scientists on the council, Blackburn and Rowley argued again for the importance of federal funding for stem-cell research: only federal resources could support a sustained effort, ensure reliable and unbiased experimental design, and guarantee that findings were publicly shared.[28]

Blackburn sent Kass a copy of the critique before submitting it for publication. In correspondence with Blackburn over the withdrawn letter to *Science,* Kass had offered to comment on any future editorials, and she invited his comments, prepared to listen if any obligation to confidentiality were at issue. Kass did not counter a single one of the authors' charges, complaining only that they had referred to "three scientists" on the council when by his count, the correct number was five, including Foster and McHugh. The authors incorporated an additional minor suggestion of his before submitting the critique.

Not long after this, *Monitoring Stem Cell Research* was released at the January 2004 meeting of the council. Far more space in the report was devoted to justifying the Bush administration's stem-cell policy against all counterarguments than to discussing any actual or proposed regulation of this research. Almost immediately, administration critics declared that the report was an apologia for President Bush's policy, echoing the term floated by May at a council session. Less than a month later, in mid-February, a team of researchers in South Korea, led by Hwang Woo Suk, announced that it had cloned the first human embryo (using SCNT) and extracted stem cells. Therapeutic cloning, which Blackburn had vigorously championed in several editorials after the council imposed a moratorium, was back in the headlines, and President Bush and Kass promptly condemned the research. Although in late 2005 the findings of the South Korean team were called into question, the results initially fueled a renewed public debate over U.S. policy.

On the afternoon of Friday, February 27, 2004, at the request of a White House aide, a mystified Blackburn telephoned the White House Office of Presidential Personnel at a prearranged time. The director informed her that the White House had decided to "make changes" to the bioethics council; she had to press him to clarify that she was being dismissed from the council. Blackburn felt shocked but relieved: "I hung up the phone thinking, Yes! I'm off the council! I no longer have to do battle." She also felt an adrenaline rush: "I sensed that the rash decision to dismiss a scientist from the council would have political import."

"I thought I might have to handle phone calls for two or three days," she said, "and then I'd get back to work." On the following Monday, she was already "back at work," entirely submerged in science. As a member of the scientific advisory board of the Gladstone Institutes, a freestanding, privately funded research institute with close ties to UCSF, she took part in a daylong site visit at which researchers presented their work to be evaluated by the advisory board and the institute's trustees in a closed session. Throughout the day, Blackburn discreetly departed at breaks to take one phone call after another. Though she rarely carried a cell phone, she was forced to rely on one to return calls from journalists, California representative Henry A. Waxman (who wanted to know if Blackburn would testify before Congress if necessary), and a staff member at the California Stem Cell Initiative, anxious for Blackburn to do a videotaped interview that same day. While she served on the council, Blackburn had turned down overtures from this organization, which was campaigning for state funding for human embryonic stem-cell research (an effort that would be successful). Now she felt free to agree to this request.

At the end of an exhausting day, Blackburn rushed from the executive session with Gladstone Institutes trustees to her lab, where producer Jerry Zucker waited to tape her interview. No one else in the lab had any idea their visitor was a Hollywood powerhouse; the producer of *Rat Race*, Zucker ended up in the lab that day because he had a child with juvenile diabetes, which might one day be cured thanks to stem-cell research. The next day was the Super Tuesday primary, and Blackburn went to vote for the first time in the garage that served as her neighborhood polling place. She had only just become a citizen, thanks to a change in Australian law that now permitted Australians to hold dual citizenship.

Despite the efforts of the White House to minimize media attention to its decision—the prearranged phone call with Blackburn took place on a Friday afternoon, a traditional "burying ground" for bad news—newspapers around the globe reported Blackburn's dismissal. When Blackburn learned from the newspapers that the White House had simultaneously announced the departure of council member May, who also opposed a moratorium on therapeutic cloning, her initial relief turned to outrage and disbelief. She did not feel May's departure was merely coincidental, a sentiment she shared with bioethicists and political leaders throughout the country. Even Senator John Kerry, then campaigning for the Democratic presidential nomination, weighed in on the issue through a spokesperson quoted in the *Boston Globe*: "We have diseases that can be cured, and we have a president who has kicked two people off the commission because they happen to think we ought to be doing stem-cell research and other kinds of research, and he doesn't want that outcome. It is clear that the administration has no respect for science."[29] In a letter to the president, over 170 bioethicists protested the change in council membership.[30] Many scientists and their national professional organizations, including the American Society for Cell Biology, also decried the firing of Blackburn. Her fellow council member Gazzaniga remarked that he was "very disheartened" by Blackburn's dismissal and said of the critique she coauthored with Rowley, "I 100 percent support what they are trying to do."[31] Rowley declared that "Liz is an important example of the absolutely destructive practices of the Bush administration," and Nobel laureate Cech protested that "this is not just a decent scientist, not just someone who has made some contribution that was abruptly dismissed. . . . This is a very smart and successful scientist working at the very highest level. She's one of the top biomedical researchers in the world."[32]

Blackburn's dismissal suggests that her qualifications as a top biomedical researcher were viewed as a liability rather than an asset. In initial votes on a ban on therapeutic cloning, the research scientists on the council had almost won the day; minority dissension strengthened the appearance of unbiased deliberation, but if it became a majority opinion, the council might embarrass the president. Why fire Blackburn and not one of the two other research scientists who had firmly supported stem-cell research and publicly challenged the council's recommendations? In council sessions, Gazzaniga had had a number of testy exchanges with Kass in which he countered and even ridiculed statements about the wonder of life, disparaging the notion that a blastocyst was a human being. Rowley was Blackburn's coauthor for the forthcoming editorial in *PloS Biology*. But Gazzaniga had shown himself to be a formidable opponent, and Rowley was a colleague of Kass at the University of Chicago. Blackburn's repeated protests over procedure and scientific misinformation so far were coupled with public civility and eschewed ideological arguments, and her conflicts with Kass had taken place in private.

In addition, Blackburn conveyed absorption in her very active research and a reluctant attitude toward publicity. Yet these qualities burnished the respect she commanded from the scientific community, not only for expertise and integrity but also for impartiality, Elizabeth Marincola emphasized: "Liz's independent scientific reputation is exemplary."[33] Scientists who knew Blackburn well trusted her, which explains why Joe Gall found it "easy to believe" her dismissal was politically motivated: "I can't imagine Liz not providing the most accurate, most considered advice. I would have total confidence in her ability to see all sides of a question and to present them in a reasonable and rational fashion."[34]

In announcing Blackburn's dismissal, a White House spokesperson said only that her two-year term had expired.[35] Kass, pressed for reasons for the departure of Blackburn and May, claimed that May had intended to serve on the council for only two years and never provided a clear reason for Blackburn's dismissal.[36] In an editorial for the *Washington Post*, "We Don't Play Politics with Science," Kass, who had initially assured Blackburn there would be no more than six meetings a year, implied that her absences from council meetings were relevant without actually citing them as cause for dismissal: "Although her important work kept her from attending many council meetings, Dr. Blackburn contributed a great deal of expertise and insight, and charges that her replacement is in any way connected to opinions she expressed are simply false." (Blackburn had missed about half the council meetings, more than any other member except Wilson, who had been ill, but this tally is deceptive, since other council members who attended only some sessions of a two-day meeting were counted as present for its entirety.) The only explicit reason Kass offered for Blackburn's departure was at best specious: "Most fundamentally, this change reflects the changing focus of the council's work, as we move away from issues of reproduction and genetics to focus on issues of neuroscience, brain and behavior."[37] It is difficult to imagine that Blackburn's scientific acumen would have failed to prove useful in this context, or that lawyers and philosophers on the council would have been better prepared to grapple with these new topics.

It is still more difficult to imagine that the three people Kass chose to replace Blackburn, May, and Carter (who had attended only the first few meetings) were chosen for their specialized expertise. The new appointees were Peter Lawler, a professor of government whose writings praised the views of Kass; Diana Schaub, a political scientist who had already condemned embryonic stem-cell research; and Benjamin Carson, a pediatric neurosurgeon who advocated for a larger place for religion in civic life.[38] Lawler's writings include the comment that "abortion is wrong"; Schaub has depicted stem-cell research as "evil"; and Carson, an opponent of therapeutic cloning who gave motivational speeches, complained that "we live in a nation where we can't talk about God in public."[39] Noted bioethicist Arthur C. Caplan characterized the reconstituted council as "a council of clones" and suggested that the administration "got rid of people who did not echo the neoconservative views of Dr. Leon Kass, the council's chair, and the majority of council members, and replaced them with three people who are much more closely aligned with the conservative majority."[40] As a footnote to the question of whether the new council would fully air debate, Blackburn wryly noted that a statement she had worked on for several weeks for the council report on reproductive technologies was not included in the report, published only a month after her dismissal. (She was later invited to publish this statement in *Perspectives in Biology and Medicine.*)

Blackburn steadfastly and publicly maintained that her dismissal was politically motivated, and she singled out Kass for criticism, claiming that he had been unwilling to accept competing views. It is a measure of her commitment and of her revision of her own notion of being "nice" that Blackburn willingly remained in the fight when she might have quietly retreated from battle and devoted her full attention to her research. Commenting on her replacements on the council, Blackburn did not equivocate: "I think this is Bush stacking the council with the compliant," she told a *Washington Post* reporter.[41] Deluged with media inquiries, supportive letters, and speaking invitations, Blackburn tried to respond to every request—a tall order; a log kept by her assistant, Maura Clancy, records over forty interview requests within the first few weeks of her firing. Blackburn's instinct in the past had been to dodge the limelight: "You don't give interviews because you're a pure scientist and shouldn't pander to the press or promote yourself." But she felt the issues were no longer personal, and she remembered science journalist Natalie Angier, in a talk at UCSF, admonishing women scientists to give interviews in order to be heard. Blackburn was amused to receive a hasty education in the wiles of journalists: "Whenever I was asked whether I felt my dismissal was politically motivated, I'd begin by saying, 'Let's look at the evidence.' But my efforts to appear reasonable often wouldn't be in the article. The reporters bided their time and asked leading questions until I said something pungent. A reporter might ask whether I thought Leon Kass had an appetite for diversity or nausea, and given those choices, I'd answer, 'nausea.' And that would be what I'd see in the paper, as if the term had been entirely my idea."

Ironically, Blackburn's firing gave her a public platform from which to criticize the Bush administration's science policy. As Marincola pointed out, "That she was fired from the council ultimately was good for our cause because it angered many people and exposed the political motivation of Kass and other council members."[42] Blackburn kept talking to reporters because she hoped to address "questions that mattered, on substantive issues, not muckraking." She readily agreed to write editorials for the *Washington Post* and for the *New England Journal of Medicine*, which took the unusual step of publishing her editorial online so that it would reach a wider audience. In this editorial, Blackburn noted that leading scientists routinely volunteered their expertise to the government and argued that the Bush administration had done severe damage to this long-standing civic tradition: "It has been the unspoken attitude of the scientific community that it is our duty to serve our government in this manner, independent of our political affiliations and those of the current administration. But something has changed. The healthy skepticism of scientists has turned to cynicism. There is a growing sense that scientific research—which, after all, is defined by the quest for truth—is being manipulated for political ends."[43] Because of this, Blackburn declared, leading scientists would be increasingly reluctant to participate in government, further eroding the quality of scientific advice. Ominously confirming Blackburn's prediction, a leading adult stem-cell biologist slated to speak before the bioethics council at its April 2004 meeting—an invitation inconsistent with Kass's announcement that the council was shifting to new topics—called Blackburn to tell her he was considering declining because he feared his findings would be distorted to make a case against research on embryonic stem cells.[44]

The counterargument to Blackburn's claims about her dismissal is, of course, that she is the one who has a political agenda in this matter. Kass has been characterized as fair-minded by a number of council members, including May, who expressed "great respect for the intelligence and humane sensibility of our chair," despite differing with him on a number of issues.[45] One of the few reporters to whom Blackburn did not grant an interview published an article in the online journal *Salon.com* challenging Blackburn's assertions that she had been dismissed for political reasons and noting that no other council members supported her claims that the council was run unfairly. (A surprising conclusion, since Rowley, Foster, Gazzaniga, and May had all publicly questioned certain practices of the council and its chair.) By this time, "Reason as Our Guide" had been published online by *PloS Biology* on March 5, 2004. When the reporter persuaded Kass to comment on Blackburn's charges in the critique, Kass denied them, though he had offered no factual corrections when Blackburn sent him the advance copy. Accusing Blackburn of substituting politics for science, Kass said that only one of her suggestions for changes to a report had not been accepted, and only because Blackburn "wanted the scientific chapter to issue in a political conclusion, namely, that we now know that embryonic stem cells are more valuable than adult stem cells."[46]

Perhaps one of the strongest arguments that can be made for characterizing Blackburn's dismissal as politically motivated is that it fits a larger pattern within the Bush administration, which has consistently waved aside the views of scientists who challenge administration policy and altered or excised inconvenient scientific data from government reports. Just a week before Blackburn was fired, the Union of Concerned Scientists, in a statement signed by over sixty influential scientists (including twenty Nobel laureates), accused the Bush administration of manipulating scientific evidence for ideological reasons. (By July 2004, an updated version of this statement, citing Blackburn's dismissal as an example of the politicization of science, had been signed by four thousand scientists.[47] By 2007 the Web site of the Union of Concerned Scientists listed 11,000 signatories.) The statement accused the Bush administration of systematically distorting scientific information to serve political goals, and a lengthy accompanying report documented instances in which administration officials revised or censored scientific reports and imposed political litmus tests on prospective members of scientific panels and advisory committees.[48]

Many of these instances made the national news, as did other attempts by the Bush administration to suppress scientific evidence. The administration's willingness to doctor scientific reports by government agencies is particularly alarming. Government employees provided the *New York Times* with copies of original drafts of global warming reports by the National Oceanic and Atmospheric Administration (NOAA) that had been altered to minimize the scientific findings. An NOAA press release originally headed "Cool Antarctica May Warm Rapidly This Century, Study Finds," was retitled "Study Shows Potential for Antarctic Climate Change."[49] In another case, the Environmental Protection Agency was forced to accept changes to a 2001 report so that its conclusions on climate change were rendered equivocal.[50] More than two hundred scientists in the U.S. Fish and Wildlife Service, defying an agency directive not to reply to a survey, reported that they had been ordered to alter scientific reports when the conclusions pitted species protection against commercial interests.[51] In a particularly egregious example of political interference with scientific integrity, in 2004 the administration blocked FDA approval for over-the-counter sales of the "morning after" contraceptive pill, which conservative groups regard as a form of abortion, despite the recommendation of the agency's professional staff.[52] Ominously, a *New Yorker* article documenting the Bush administration's "war on the laboratory" concluded that "many types of scientific analysis and research are proscribed almost wholly on religious grounds."[53]

The Bush administration typically responds to criticism of its science policy by discrediting the source without providing any countering evidence. John H. Marburger, director of the Office of Science and Technology Policy, challenged the criticism in the Union of Concerned Scientists report, describing it as "sweeping generalizations based on a patchwork of disjointed facts and accusations."[54] Defending the administration's decision to alter a report on global warming, Marburger said,

"This administration also tries to be consistent in its messages. It's an inevitable consequence that you're going to get this kind of tuning up of language."[55] Just this kind of "tuning up" had outraged Blackburn as she read drafts of council reports.

To distort or suppress scientific advice reflects a fundamental distrust of the democratic process, as it essentially attempts to impose policy for which consensus might not be obtained if all the facts were known. Scientific integrity anchors debate on public policy *especially* when emotionally charged values are at stake, since, as Jacob Bronowski observed in *Science and Human Values*, "in a world in which state and dogma seem always either to threaten or cajole, the body of scientists is trained to avoid and organized to resist every form of persuasion but the fact."[56] The Bush administration's record does not constitute politics as usual but marks a dangerously unprecedented shift in the relationship between scientists and government, as the Union of Concerned Scientists has noted.[57] In its 2004 report, the nonpartisan Federation of American Scientists concurred: "The election of George W. Bush in 2000 changed the landscape fundamentally."[58] Many other national organizations, including the American Association for the Advancement of Science, the Center for Science in the Public Interest, and the American Civil Liberties Union, have voiced similar challenges to the Bush administration's politicization of science.[59]

None of the recommendations or findings of the President's Council on Bioethics was binding, but this in itself helped to fuel debate over whether and how the findings of similar advisory panels should be used to formulate government policy. In his remarks before the bioethics council, Gearhart had voiced the dismay of many scientists: "The number of federal Blue Ribbon panels that have been set up, at the NIH level, to discuss issues that seem to be ten years away, recommendations were made by our leading scientists, our leading scholars, only to be ignored. And then we find ourselves in a pot once these things happen."[60] Concern over how scientific policy was derived prompted the Federation of American Scientists to recommend in its 2004 report specific systemic changes to provide Congress and the president with competent, independent scientific advice and ensure that "individual citizens and nongovernment organizations have the information they need to conduct analysis and participate effectively in the public debate."[61] Blackburn voiced a similar warning in her *Washington Post* editorial: "Enlightened societies can only make good policy when that policy is based on the broadest possible information and on reasoned, open discussion. Narrowness of views on a federal commission is not conducive to the nation getting the best possible advice."[62]

Blackburn returned to Washington, DC, in early June 2004, to attend a scientific symposium on cancer research, sponsored by General Motors and held on the NIH campus, where a large staff of scientists made for a ready audience. The symposium culminated in a gala black-tie dinner and awards ceremony honoring recipients of three prestigious General Motors Cancer Research Foundation awards: the Alfred P. Sloan Prize for contributions in basic science related to cancer research; the Charles S. Mott Prize for research related to the cause or prevention of cancer; and the Charles F. Kettering Prize for contributions to the diagnosis or treatment of cancer. Only three years earlier, Blackburn had been awarded the Sloan Prize, which is accompanied by a gold medal and a cash award of $250,000.

As a former medal recipient, Blackburn now served on the assembly that selected winners of the annual awards. The nomination and evaluation process for these awards illuminates the kind of meritocracy that operates in research science and also underscores how stringent evaluation procedures help to ensure objective judgment. The corporate sponsorship of such prizes—Monsanto and Heineken also sponsor prestigious awards—is limited to financial underwriting, and qualified scientists independently conduct the evaluation process. Confidential nominations for the awards mitigate against politicking, and a preliminary committee of experts exhaustively evaluates nominees for a given medal, with the prime, demanding criterion being the degree to which the research has changed scientific knowledge. These committees present a short list of finalists in each category to an assembly made up of past medal winners, whose differing areas of expertise broaden the context in which the research is judged.

By this time, Blackburn had come to be as much at home at cancer symposia as she was at telomere conferences. Her work had been recognized in both of these now overlapping scientific communities; since 1998, she has collected major scientific honors every year—so many that the overflow of plaques, medals, and statuary has been relegated to a closet. At the symposium she could be immersed in science, freed of the entanglements of politics, and yet politics would once again intrude. During the week of the symposium, former president Ronald Reagan died after many years of living with Alzheimer's disease. On the day of the awards ceremony, Reagan's body was brought to Washington, DC, to lie in state. The city's streets were jammed, with traffic almost at a standstill as heightened security measures went into effect.

When Blackburn attended the awards ceremony at the Four Seasons Hotel, the hotel was crowded with an influx of guests. Over cocktails, many people approached Blackburn to speak to her about her dismissal from the bioethics council, while outside the hotel, foreign dignitaries in national costume arrived in limousines, here to pay their respects to a former president whose wife and son would use the occasion of his death to plead passionately for funding for stem-cell research that might one day cure Alzheimer's disease.

Even the master of ceremonies for the awards, Sam Donaldson, brought with him the aura of Washington politics. Washington political luminaries often mixed with scientists at this ceremony; the year Blackburn received her award, Katherine Graham, under whose courageous stewardship the *Washington Post* had defied the Nixon White House, had introduced herself to Blackburn. At the Four Seasons banquet, Blackburn made her way toward Tom Kelly, cowinner of the Sloan Prize, which he shared with Bruce Stillman for their independently pursued work on DNA replication and how it is controlled in cancer. Blackburn's personal connection to Kelly was yet another in a string of hidden symmetries this evening: he had recruited her former collaborator and long-time friend Greider to Johns Hopkins University when he directed a department there.

After Blackburn congratulated Kelly on his award, he too turned the subject to politics, in a way that underscored how far Blackburn had come from her modest assessment of herself as an "inward-looking researcher." Kelly simply asked her, "What does it feel like to be a symbol?"

13 "You Have to Think It's Fun"

In the polarized public debate over stem-cell research—defended as a miracle or denounced as a moral crime—Blackburn held to scientific objectivity as a bedrock principle. But a little over a year after she left the President's Council on Bioethics, the imperviousness of scientific objectivity was dramatically called into question. In May 2005, in a paper published in *Science*, South Korean researcher Hwang Woo Suk and his collaborators announced another breakthrough: the creation of eleven human embryonic stem-cell lines tailored to individuals, a major technical advance in the methods for SCNT. But by November 2005, it came to light that members of Hwang's research team had been coerced to donate eggs, and other women, financially destitute, had been paid to donate eggs. Though these practices did not break South Korean laws, both violated accepted medical ethics.[1] The sole U.S. author on the paper Hwang submitted to *Science*, stem-cell researcher Dr. Gerald Schatten, withdrew from the collaboration and asked to have his name removed from the paper, announcing he had lost confidence in the scientific validity of the research. Soon after, a South Korean colleague accused Hwang of faking scientific data; investigations by *Science* confirmed this accusation, and a panel at Hwang's university, Seoul National University, tested the stem-cell lines and found that all came from fertilized eggs rather than cloned embryos. As a consequence, Hwang's other claims to success—therapeutic cloning of the first human embryo using SCNT in February 2004 and the cloning of a dog in August 2005—came under a cloud of suspicion.[2]

As a serious technical setback and a public relations disaster, the scandal delivered a blow to stem-cell research on two counts. Suddenly, fears that a rogue scientist might violate accepted ethics, including cloning human beings, seemed plausible, not outlandish. Furthermore, opponents of stem-cell research cited Hwang's failure as evidence that the therapeutic promise of human embryonic stem-cell research was illusory. Placed on the defensive, supporters of stem-cell research emphasized that a system with stringent oversight might have prevented Hwang's ethical violations. Sean Tipton, president-elect of the Coalition for the Advancement of Medical Research, argued that "probably the strongest research oversight system in the world is at the National Institutes of Health, but they are pretty much on the sidelines. If you don't allow the best American scientists to do the best—and best overseen—research, you force it overseas and into the private sector, and this is the result."[3]

Hwang's disgrace demonstrates that scientific objectivity is vulnerable to powerful political and cultural forces as well as personal ambition. His early successes earned him the partisan and well-funded support of his government, which hoped his breakthroughs would make the nation an international center of stem-cell research and bring foreign investment.[4] Lionized by the media in South Korea, where he faced mobs of autograph seekers, Hwang was briefly an international celebrity.[5] If *Science* and Schatten had initially shared in this glory, both now came under fire. Schatten drew criticism for his willingness to attach his name to a landmark paper despite having no direct participation in the research.[6] Ultimately an investigative panel at his institution, the University of Pittsburgh, found he had committed "research misconduct" out of the desire to "enhance his own reputation," and in placing his name on a paper without adequately verifying the data, Schatten "did not exercise a sufficiently critical perspective as a scientist."[7] The editor in chief of *Science*, Donald Kennedy, had to deny that the paper had been rushed into print and defend the thoroughness of his journal's review process.[8]

This helps to explain, of course, why the defense of scrupulous scientific integrity mattered so much to Blackburn that she spoke out publicly during her tenure on the President's Council on Bioethics, even when Kass insisted she would undermine the council's work by doing so, and continued to criticize the Bush administration's scientific policy after her dismissal. It also helps to explain why she feels a strong imperative to continue participating in decision making on scientific policy. When she left the council, she was already serving on a number of institutional advisory boards, which typically consult with institutional directors on how their resources will be used and review ethical guidelines for clinical research, including stem-cell research. The wide-ranging character of Blackburn's research is reflected in the diversity of boards on which she serves, from the Stanford University School of Medicine National Advisory Council to the Salk Institute, where as a nonresident fellow, Blackburn functions as part of a scientific advisory board. She continues to serve on the National Advisory Council on Aging as well.

While still on the bioethics council, Blackburn testified before the California State Legislature on stem-cell research, describing the deliberations of the council and enumerating existing federal regulations that would be relevant if the legislature passed a bill making embryonic stem-cell research legal under California state law. Sponsored by State Senator Debra Ortiz, this bill was signed into law, but no funding mechanism was provided until 2004, when California voters passed Proposition 71, the California Stem Cell Initiative. The initiative, which Blackburn publicly endorsed only after she left the President's Council on Bioethics, provides three billion dollars in state funding for human embryonic stem-cell research.

It is a truism that an original scientist spends the early part of her career making great discoveries and the latter part collecting honors for it and serving as a "senior statesperson" in the field, somewhat removed from the fray in which new ideas are hotly contested. Blackburn might seem to have arrived at this point in her career. She has authored over 252 scientific papers and articles, making it something of a feat to publish a paper in the field without citing her work or the research derived from it. She collects several significant prizes and honors every year, often accompanied by the obligation to deliver a lecture, yet another distraction from focusing on the lab.

These awards, chosen by committees of peers from different scientific constituencies, reflect the far-ranging implications of Blackburn's work, but they also suggest a continuum between strictly scientific integrity and more broadly ethical behavior. The ASCB awarded Blackburn its Public Service Award in 2004, the same year she was fired from the bioethics council. In 2005 she received the society's Senior Award, which honors "outstanding scientific achievements . . . coupled with a long-standing record of support for women in science and by mentorship of both men and women in scientific careers."[9] In 2004, Blackburn also received the Dr. A. H. Heineken Prize for Medicine. Winners of this international prize, selected by the Royal Netherlands Academy of Arts and Sciences, receive a cash gift of $150,000. In presenting her with the award, Professor Peter C. van der Vliet noted that "the study of telomeres has become a central component of biology, with major new findings coming from the Blackburn lab," and he also took the unusual step of commenting on Blackburn's dismissal from the President's Council on Bioethics: "It was an event that caused an uproar among scientists in the U.S. . . . Controversial issues need open scientific debate in which all viewpoints are considered, even unwelcome ones. For science, aboveboard information, a broad spectrum of views and a critical attitude are essential. . . . When a friendly person and prominent scientist such as our prizewinner has to stand up in public to defend these principles, something is deeply wrong."[10]

The scientific honors Blackburn has received acknowledge not just her seminal discoveries but the vitality of her ongoing work, in particular the connections she has made between basic science research and clinical research. Like the Heineken Prize, the Kirk A. Landon Prize for Basic Cancer Research, which Blackburn received in 2005, honors contributions to medicine. Sponsored by the Kirk A. and Dorothy P. Landon Foundation, the prize is awarded by the American Association of Cancer Researchers, a large national professional association that encompasses both basic science and clinical cancer research. The Kirk A. Landon Prize honors a researcher whose work constitutes "a landmark achievement" and is characterized by "new thinking and novel concepts," and eligibility requirements stipulate that the recipient be an active researcher with a record of recent publication.[11] The same year that Blackburn received this award, Rowley received a companion award, the Dorothy P. Landon Prize for Translational Cancer Research.

In 2005 and 2006, Blackburn received two major scientific awards, both regarded as strong predictors of a Nobel Prize. In April 2005, Blackburn received the Benjamin Franklin Medal in Life Science in recognition of her "breakthroughs in understanding the protective role of telomeres" and the implications of her research for understanding aging and cancer. The award is sponsored by the Franklin Institute, which has been honoring achievement in science and technology since 1874; 101 Franklin laureates have garnered 103 Nobel Prizes.[12] In September 2006, Blackburn was honored by the Albert and Mary Lasker Foundation, which since 1946 has sponsored awards for research in basic medical science and clinical medicine, judged by a jury of distinguished international scientists. Previous winners of the Lasker Awards, widely regarded as "the nation's most prestigious medical prizes," include Watson, Sanger, McClintock, Bishop, Varmus, Cech, Rowley, and Christiane Nusslein-Volhard.[13] Since 1962, 71 recipients of the research awards have received the Nobel Prize for Physiology, Medicine, or Chemistry, most within two years of receiving the Lasker Award.[14] Blackburn shared the 2006 Lasker Award for basic medical research with Greider and Szostak in honor of their contributions to the discovery of telomerase. The 2006 Lasker Award for special achievement in medical science went to a scientist intimately connected with this discovery, Gall, in honor of his contributions as a founder of modern cell biology and his early advocacy on behalf of women in science.[15] In their acceptance remarks, Blackburn, Szostak, and Greider all emphasized the unpredictable route from their early investigations to clinical applications. Blackburn declared, "Science is as creative an endeavor as the humanities . . . doing science is also letting the imagination be open to new ideas and lateral leaps that might at first seem outlandish," and she added that a healthy science policy depends on "an environment of openness to available scientific evidence and freely shared and expounded ideas."[16] Referring to their collaboration, Szostak remarked: "To me this experiment illustrates perfectly the value of talking to people who work in very different fields, the value of collaboration, and the value of, every now and then, putting a little money and effort into high risk, high payoff experiments."[17] Greider noted that "telomerase beautifully illustrates that you never know where medically relevant discoveries might come from" and spoke of science as "inherently a community activity."[18]

The prestige that accompanies accolades has not slowed the pace of research in Blackburn's lab, though it has significantly affected the angle of her approach. She does not have to lie awake at night worrying about getting out the next paper in order to establish credibility and garner future research funding. She has the luxury to go slow in a very different sense—to look around, to explore experimental questions that may not promise the quick, dramatic payoff that a less established researcher needs. Like many other molecular biologists today, she also collaborates more frequently with clinical researchers, an increasingly necessary route to funding but also a way to amplify the expertise of each partner. In this arena also she is prepared to take unusual risks.

When Blackburn left the President's Council on Bioethics in February 2004, she was still completing a sabbatical for the academic year 2003–4, for the intended purpose of returning to an intense focus on science. Her sabbatical year marked a signal moment in a gradual shift in the scope of her research. In the late 1990s, Blackburn had attended a lecture by virologist Don Ganem, her colleague at UCSF. When Ganem, who also works as a clinician, gave this talk at UCSF, he said that his research had led him to delve into minutiae, such as how a particular nucleotide in a virus interacted with a tiny protein. As his work became increasingly myopic, he rebelled and decided to consider the big clinical questions that had originally drawn him to virology, which led him to study Kaposi's sarcoma at a time when its relation to the HIV virus was as yet unknown. This lecture marked the first time Blackburn had heard someone criticize the career trajectory of pinpointing cellular mechanisms in increasingly fine detail, and it was "eye-opening" for her.

So far, her work had always been a step removed from its ultimate practical goal: providing a greater understanding that might lead to cures for cancer or age-related disease. Even Blackburn's work on human tissue was remote from clinical research that might determine how the mechanics of telomeres and telomerase worked in a living person. In her own research, she would continue studying biology "in tiny detail" because it still compelled her curiosity, but she would also integrate a new emphasis on human health: "In the end, in medical research it's the patient that counts. It's quite a jump from basic science to clinical research, in which you might study actual people to see what's happening at the cellular level."[19]

Prompted in part by Ganem's talk, Blackburn had begun a collaboration in 2000 that would explore the connection between cellular processes and an entire, complex organism, and in 2004 this work produced remarkable results. The collaboration began when Elissa Epel, a senior postdoctoral fellow in clinical psychology at UCSF, e-mailed Blackburn with a question. She was conducting a study on mothers, between the ages of twenty and fifty, who were primary caregivers for their own chronically ill children, and Epel wanted to measure the physiological effects of such chronic stress. Curious to know whether the prolonged stress of caring for a child with autism or cerebral palsy could influence the aging process, Epel asked Blackburn if it might be useful to look at telomeres to assess this at a cellular level. Would she be willing to collaborate? Blackburn leaped at the opportunity: "A mother myself, I was immediately interested."

In addition to her felt connection to the mothers who participated in the study, Blackburn appreciated Epel's experimental design because it recognized motherhood as a potent source of stress when any number of other stresses might have been chosen. Furthermore, this study focused solely on women subjects, though its results might be generalized, when more often, the reverse was true in clinical trials. Treatments for heart attacks, for example, once depended on the results of clinical trials conducted on men, with the result that for years physicians often failed to recognize the symptoms of a heart attack in women, which differ from those for men. For Epel to incorporate into her study data on telomeres and telomerase was speculative, but this too appealed to Blackburn. Temperamentally inclined to explore territory at the fringe rather than pursue already established ideas amid a crowd, Blackburn was not averse to taking a gamble.

During the 1980s and 1990s, only surmise, not hard scientific data, suggested how telomerase and shorter telomeres might relate to aging in humans. Not until 2005 did reliable studies directly demonstrate that telomere length drifts downward in a given individual as one ages.[20] No researchers had looked at telomeres to see if they grew shorter more rapidly when a person was under stress, and to offer even tentative proof for this, an experiment would have to provide careful controls for comparison. Furthermore, between the ages of twenty and fifty, individuals experience, on average, such a minute downward drift in telomere length that any assessment would have to make fine distinctions.

Fully registering the difficulties of such an experiment, Blackburn met with Epel and sought a reference from Nancy Adler, the senior investigator in the clinical department in which Epel worked. Blackburn was impressed with Epel and her expertise. A clinical trial that does not study a large, representative sample can be meaningful only if it is meticulously designed to control against as many potentially confounding variables as possible, so that careful statistical analysis might reveal significant differences among subjects. Over a four-year period, Epel would painstakingly winnow subjects for as long as necessary to design the experiment so it had a rigorously selected control group. Of a total of sixty mothers in the study, thirty-nine were primary caregivers of chronically ill children; mothers in the control group cared for healthy children. Women in the control group were scrupulously matched with caregivers for age range and generally equivalent good health, right down to their exercise and drinking habits.

Each voluntary participant in the study took a battery of tests to measure the impact of stress. All subjects in the study took Cohn's Perceived Stress Test, a widely accepted psychological self-assessment chosen because it had consistently proved to have predictive power—that is, whether a person became physically ill within a month of taking the test correlated with perceived stress. On this test, individuals ranked their stress level on a numerical scale, with a higher score indicating a greater degree of perceived stress. Both the control group and the mothers of chronically ill children spread across the range of perceived stress, but mothers of chronically ill children were more often distributed at the higher end of the scale. In addition, the number of years these mothers had been primary caregivers for chronically ill children was recorded.

Study participants also took a number of physiological tests in the hope that the data might provide clear evidence of any correlation. Clinical measurements were taken to quantify obesity and assess chemicals in blood and urine indicative of oxidative load, metabolism, and stress. These tests included blood lipid profiles to measure cholesterol levels, fasting glucose and insulin levels (predictive of diabetes), white blood cell counts that afforded a snapshot of the general state of the immune system, and a measure of stress hormones, cortisol and the catecholamines, present in urine. Yet another test measured "oxidative stress"—how well the body mobilized its antioxidative response to repair the oxidation that damages cells. (Studies on animal models such as flies and worms had shown that manipulation of the enzymes that repair oxidative damage alters life span.) Another test compared resting rates of cardiovascular functioning with these same rates under psychological stress over the span of thirty minutes. To provoke a stress response, study participants were given five minutes to prepare a speech to deliver before an audience trained to respond only with stony looks, and participants were then asked to solve complex math problems on a test administered by unresponsive evaluators. Cardiovascular reactivity, measured as the subjects completed all the assigned tasks, typically follows a curve from a resting state to performance under stress, and the shape of the curve accurately predicts a person's risk for cardiovascular disease.

Because telomerase and telomere length were truly open questions that would become central to the study, Blackburn's lab came to take a prominent role among the many collaborators in this study. Another collaborator, Richard Cawthon, a geneticist at the University of Utah who for some time had been collecting data on telomere length in individuals, had derived a method for measuring the average length of telomeres based on the tiny amount of DNA that could be obtained from a small cell sample. While Cawthon's lab focused on telomere length, Blackburn's postdoctoral fellow Jue Lin painstakingly assayed for telomerase activity in the white blood cells (peripheral blood mononucleosites) in blood samples drawn from study participants. Once released into the bloodstream, these white blood cells are "resting"—not actively dividing—and are known to have little telomerase activity; some researchers assumed there was none. Blackburn suspected low levels of telomerase activity, which meant that a test would have to detect infinitesimal amounts of the enzyme accurately enough to compare subjects with each other and with the control group.

These measurements of telomerase activity were taken on the premise that the researchers were just "taking a look," without any particular hypothesis that a correlation would be found or demonstrably proved. In fact, given the variability in telomere lengths in the general population, even among individuals of similar ages, Blackburn thought of this as a pilot experiment, an opportunity to perfect techniques and to practice on a small sample unlikely to produce statistically significant results: "You could assume individual average telomere length would be scattered on a graph fairly randomly, and maybe you'd guess there'd be *more* telomerase activity in stressed people on the assumption that high levels of stress activate the immune system, making people more prone to illnesses like colds. You could guess any which way you wanted to, and my guess was that telomerase activity levels would be scattered among individuals in a way that would make it hard to draw any conclusions from only sixty people."

Lin faced a daunting task: a test that could reliably detect and accurately quantify such minute amounts of telomerase had never been done before, and Lin's eventual success constituted a significant technical breakthrough. As Blackburn observed, "Often only technical limitations hold up the exploration of scientific questions—once PCR was developed, it made new inquiries possible. The discovery made possible by a technical advance is not the finish line. It gives you the ability to ask new questions you couldn't ask before." The blood tests conducted by Blackburn's lab, like those of other collaborators, were done on blind samples collected and frozen over the course of four years by Epel's team. Once Lin developed a reliable test, she simply collected the numbers and sent them to Epel for statistical analysis.

In March 2004, shortly after Blackburn was dismissed from the President's Council on Bioethics, Epel telephoned Blackburn, very excited. The first numbers had begun to fall into a clear pattern. Subjects with higher scores on Cohn's Perceived Stress Test or a greater number of years as primary caregiver for a chronically ill child had lower levels of telomerase activity in their white blood cells than was average among the control group. Cawthon's measure of telomere lengths also correlated with either of these factors—individuals with greater levels of stress by either measure (perceived stress test or years of caregiving) had shorter telomeres than average. In addition, for these women with higher stress levels, the oxidative load was also higher. That a lower level of telomerase activity could result in vulnerability to illness was already suggested by studies on dyskeratosis congenita, the genetic disorder in which mutations in the telomerase gene halve an individual's capacity to produce telomerase and result in drastically premature death.

Epel did further analysis to determine if these results could be accounted for by mere chance. Consistently, her analysis continued to show a clear correlation between high stress levels and reduced levels of telomerase—one that could not be accounted for by random occurrence. What did these numbers mean? As a person ages, telomeres gradually shrink. Do they reach a threshold level at which cells die? If so, then anything that accelerates this process essentially speeds up aging, with all the attendant health risks, such as a weakened immune system and increased risk for cardiovascular disease. Because of the carefully established controls in this study, the researchers could show that those women with the highest perceived stress level had telomeres equivalent to those of women ten years older. Though it's widely assumed that stress ages people, this study documented what might happen at the cellular level to cause this, providing the first direct evidence that chronic, severe psychological stress ages the body.

These unexpectedly decisive results prompted the investigators to publish their first paper on the study in the November 2004 *Proceedings of the National Academy of Sciences*, which appeared online before it appeared in print.[21] Their findings made the news around the world—something that Blackburn was not used to in relation to her research. Not only every major U.S. newspaper but also newspapers in China, New Zealand, Great Britain, and Canada reported the results, as did radio and television stations. Ironically, the *San Francisco Chronicle's* front-page report on this research appeared just below an article headlined "Anti-Evolution Teachings Gain Foothold in U.S. Schools."[22] In

December 2004, producers for the CBS program *60 Minutes* approached Epel and Blackburn, interested in reporting on their study. Blackburn was interviewed by Scott Pelley, just back from the ravaged region of Darfur, and the interview was incorporated in a show on the subject of aging and health, aired on August 26, 2005, three days before Hurricane Katrina struck.

Scientific findings rarely make the news unless they promise dramatic medical implications or warn of potential harm, as studies on global warming do. Science is also deemed newsworthy when a study produces readily grasped results that confirm or contradict popular notions. News accounts of the stress study emphasized, to the point of whimsicality, how the results provided the first actual evidence for the cellular mechanism that might underlie folk wisdom—"Stress Speeds Up Aging," "Too Much Stress May Give Genes Gray Hair." Even *New York Times* columnist Maureen Dowd got into the act. In a humorous diatribe on the stress of the Christmas holiday season, Dowd bemoaned her fate: "So now, on top of all the stress related to having a president and vice president who scared us to death about terrorists to get re-elected, I have to be stressed about the fact that my holiday stress might cause me to turn into an old bat—instantly, just like it happened in Grimm's fairy tales, when a girl would be cursed and suddenly become a crone."[23] Members of the scientific community, who generally express their opinions in more cautious terms than news reporters and editorial writers, lauded the study as "a new and significant finding" and its results as groundbreaking. Robert M. Sapolsky of Stanford University, who wrote a commentary accompanying the paper in the *Proceedings of the National Academy of Sciences*, commended it as a "landmark observation": "This is a huge interdisciplinary leap . . . a great study."[24]

Although the numbers were in, Blackburn and Epel had yet to publish a statistical analysis of the other parameters measured in study participants—those parameters that related the risk for cardiovascular disease to lower levels of telomerase in blood cells. Blackburn reported that they deliberately separated their results into two papers because including all their astonishing findings in one paper "would have been like drinking from a fire hose." At the time of the *60 Minutes* interview, Blackburn kept quiet about the results on cardiovascular risk because peer-reviewed journals did not publish studies that had been leaked to the media, and only peer review could authenticate scientific accuracy. In late 2004, a large international study of nearly thirty thousand people in fifty-two countries on the six inhabited continents had identified the major risk factors for cardiovascular disease: smoking, a poor blood lipid profile, increased resting pulse pressure, diabetes, adiposity, and chronic stress. Amazingly, in Epel's study group, *all* the six major known categories of risk factors for cardiovascular disease correlated with low telomerase levels, even when they bore no relation to telomere length.[25] The larger study provided important corroboration for what the numbers told Blackburn and Epel: lower levels of telomerase bear a significant relationship to risks for disease. While in some instances telomere length also correlated with this risk, it proved to be only one readout of the activity of telomerase in the cell and did not tell the full story of its impact on cellular processes.

Blackburn and Epel's study raises a host of intriguing questions for clinical and other interventions. The first is whether counseling and stress-management strategies such as yoga might reduce the health risks for individuals under stress. Not only does stress exact a clear physiological price but also *how* a person copes with stress can affect physical health. Those women who perceived that they were under a higher level of stress fared worse on all the physiological factors, including lower levels of telomerase activity. Even compared to other caregivers of chronically ill children, those women who perceived themselves as under greater stress had significantly shorter telomeres. Blackburn and Epel plan to study whether intervention that treats stress has any effect on telomerase activity in a person's cells.

The study also suggests key experimental questions for those investigating the cellular mechanisms affected by stress. Do the prematurely aged white blood cells impair the immune system, making people more vulnerable to illness, and thus act as the primary agent of cellular damage? Or does stress take a toll on other types of cells in the body as well? For example, since oxidative stress corresponded to lower levels of telomerase activity and shorter telomeres, it is possible that elevated levels of stress hormones such as cortisol damage telomeres or telomerase functioning—a question that Lin is now exploring. As Blackburn pointed out, "The numbers don't tell you about the mechanism at work or its directionality. That's the next thing we have to look at." Blackburn has already been approached by a number of clinical researchers interested in collaborating with her to explore other aspects of the relationship between telomerase levels and health risks.

The collaboration with Epel intersected with a long-established line of inquiry in Blackburn's lab. The correlation between lower levels of telomerase and significant health risks might simply be a harbinger of damage to telomeres, since there's a lag time before enzyme deficiencies affect telomere length. Or it might suggest the enzyme has other functions in the cell besides adding to telomeres. Blackburn had first argued for the need to explore the latter possibility in the late 1990s, and though "it was not a popular idea," she has continued to pursue this line of inquiry. Work in her lab had thrown up the first indications that telomerase may have multiple functions in the cell. In studies later corroborated by work on human cells, John Prescott showed that yeast cells continue to divide indefinitely even though hypomorphic forms of telomerase can maintain only very short telomeres, suggesting that maintaining an average telomere length is not the only contribution the enzyme makes to the survival of the cell. Shang Li's studies, begun in 2001, found that when telomerase is abruptly taken away from cancer cells, cell division slows down immediately, not many generations later, as would be the case if telomere length were the only cause of the cell's decline. His further experiments to identify which genes are more or less actively transcribed showed that the profile shifts dramatically with changes in telomerase activity, even though telomere length has yet to change at all. In 2002, Chris Smith and Blackburn found that during a part of the yeast cell cycle in which telomerase cannot add to the telomeres, it nevertheless remains on the telomere, indicating it may function in other ways to protect the integrity of the telomere. At about the same time, researchers in Zakian's lab also observed the same phenomenon.

Blackburn continues to pursue these suggestions that telomerase influences the physiology of the cell. In seeming contrast to Li's findings for human cancer cells, experiments by Lundblad, McEachern, Krauskopf, and others in Blackburn's lab had shown that in yeast, it takes many cell divisions before the absence of telomerase results in declining telomere length in the bulk of cells. Blackburn has assigned research associate Dana Smith to look at yeast cells to see if she can identify any response when telomerase is abruptly deleted from the cell, even before telomere length has begun to dwindle. Though typically science works in the other direction, evolving from studies on simpler organisms to more complex ones, when an interesting but unexplained phenomenon pops up in a complex organism, questions can often be answered more quickly by research on a simpler organism. As Blackburn commented, "We don't know if yeast cells respond, though we know this is true for cancer cells with high levels of telomerase. If we can use yeast to identify a cell's response in molecular terms, this in turn would give us a clue about the nature of the signals cells are getting. Is it the ability of telomerase to interact with the telomere that's the problem? Or just its physical presence in cells?"

In 2005, Steven Artandi, Kavita Sarin, and their colleagues published the results of a study showing that an excess of TERT (the telomerase protein component) can cause particular stem cells or their early descendants to proliferate more than usual. They used a strain of laboratory mice genetically engineered to produce much more TERT than normal and lacking the gene for telomerase RNA, which makes them incapable of synthesizing telomeric DNA. In these mice, the specialized stem cells residing in mouse hair follicles were stimulated to keep actively proliferating, whereas they normally cycle between an active and quiescent state as they regulate hair growth. Because their hair follicles were continuously active, these experimentally engineered mice grew thicker, shaggier hair.[26] Artandi and Sarin emphasized that although the implications of this study are still highly speculative, the potential effect of TERT in stem cells that renew tissues might lead to medical therapies.[27]

In a review that accompanied the findings, published in *Nature* on August 18, 2005, Blackburn, mindful of the speculative hype that sometimes accompanied previous discoveries about telomerase, had fun dismissing speculation about possible cosmetic applications and underscored the study's more substantive scientific implications: "In ancient Egypt, men smeared their pates with hippopotamus fat in a desperate bid to stave off baldness. Is telomerase the new hippopotamus fat?

Probably not. But this enzyme is already known to be vital in sustaining tissues in health and disease, and we should look beyond its eponymous function to understand the full spectrum of its roles."[28] Yet in an interview Blackburn cautioned that any interpretation of such intriguing data must be carefully hedged: "What these researchers did was very artificial—they forced the action of a lot of telomerase, more than would ever be seen in the organism's cells, which means that the results they observed might bear no relation to the normal mechanism in the cell. So there's always a caveat. But because in the work in my lab we were looking at underactive telomerase, more within the typical physiological range, these results showing the effects of overactive telomerase lend credence to the idea that telomerase may have other functions in the cell."

Asked if the accumulating evidence confirmed her belief that this was so, Blackburn brusquely dismissed the idea that belief had anything to do with it: "I would never say, 'I believe.' I would say the body of evidence built up from a number of observations adds up to a model that we have to take into account." Her stringent notion of how scientific proof evolves highlights the capacity to shift perspective continually and flexibly:

One could still argue that all this evidence has to do with DNA polymerization, and we're just not seeing it the right way. Building a model—refining it or changing it or confirming it—is a slow, slow process. When you see unexpected evidence, you have to ask, What could account for this? What can I rule out? What's still possible? You bring in all you know about the system and what you know about biology. Some of this is easy, some not. It's a real no-brainer to note that parts of the telomerase protein component, besides the reverse transcriptase domain, which does the work of active polymerization, are highly conserved across species, yet we have no idea what they do. Would this model explain it? What other explanations might account for this? You work through all the explanations via what you know about telomerase or polymerase. Why does telomerase sit on the telomeres at a time when it can't polymerize DNA? Chris Smith and Ginger Zakian's lab both got this result, so it's not somebody's little mistake.

If long familiarity grants her a particular facility for interpreting data within a richly detailed context, Blackburn resisted the notion of intuition as firmly as she did the notion of belief in relation to scientific inquiry: "You don't work by intuition, unless that's a shorthand term for something much more involved. There's a lot of processing your mind does after a lifetime spent on the same problems, so when you see something not consonant with your mental construct, you're in a good position to see the questions that must be asked. Biologists make working models all the time, in the effort to approximate better and better to the reality. You have to be prepared to play—you have to think it's *fun* that the model might change."

For Blackburn, a strict stance toward evidence and a sense of play are not in tension with each other but necessarily symbiotic. Paradoxically, it's the second criterion—"you have to think it's *fun* that the model might change"—that proves to be the toughest standard to live by. Like everyone else, scientists, even if unconsciously, invest in the outcome of their work, which can influence objectivity out of purely human frustration with puzzling results or purely human ambition and ego. Blackburn viewed this as a persistent problem in science. Citing the example of South Korean scientist Hwang, she expressed moral indignation at investigators determined to "get these results come hell or high water": "Hwang Woo Suk was pushing a technological advance. That has nothing to do with science. Science finds out how it works, it doesn't push for a result. A vested interest in a certain answer disqualifies you from honestly looking at the question."

Blackburn's willingness to pursue experiments that don't produce immediate, dramatic results is, like her capacity to shift perspective, linked to a sense of play. In Blackburn's lab in the 1990s, Prescott had conducted experiments in which he found that a particular mutation to the RNA template annihilated its ability to synthesize DNA, and in characterizing the effects of the mutation he came across another unexpected finding. He discovered that telomerase in yeast is dimeric—in other words, each enzyme particle contains two copies of its RNA and protein components, all bound together. As it turns out, the enzyme is also dimeric in humans. It is not that unusual for enzymes to be dimers, but nature usually provides such doubling up for a reason, such as switching off between the functioning of one complete unit and another. When Prescott and Blackburn submitted these initial findings to *Nature*, "because we thought it was so cool," the journal promptly rejected the paper. Though Prescott and Blackburn eventually published their findings in another good journal, there was no context for these results and they didn't generate particular interest. But Blackburn remains curious about this property of the enzyme, and members of her lab have since done a variety of experiments on yeast and human cells to explore how this dimeric structure might affect the way the enzyme does its job. Though the lab has accumulated pages of descriptions of its complex properties, testing one idea after another, no answer has emerged. "Very unsatisfactory," Blackburn said. "But an open, fun question."

Working with yeast, researchers in the Blackburn lab are also studying how the telomeres are bound with telomeric proteins, conducting painstaking quantitative biochemical analysis rather than studying living organisms. There is still very little basic knowledge about telomere architecture—how the telomere is built up from its component DNA and proteins—and according to Blackburn, at best, existing models resemble "cartoonlike blobs," with precise knowledge of a few areas but huge gaps in understanding the whole system. The progress made in the field so far is analogous to the state of molecular biology at the time that Watson and Crick's model of the double helix precisely predicted its chemical structure yet did not account for how enzymes would perform the complicated functions necessary for DNA replication and gene expression. Understanding how a telomere is built promises to illuminate how telomerase is recruited to work on the telomere.

One difficulty with refining this model, as with any model of cellular mechanisms, is that researchers can pursue only indirect routes for identifying what occurs in the living cell. In the 1990s, David Shore studied the proteins that bind to telomeres in *Saccharomyces cerevisiae* (baker's yeast) and found evidence that Rif1 and Rif2 interacted with Rap1. In the simplest model, Rif proteins might interact directly with a portion of Rap1, but this is only indirectly inferred—still a "blob" on paper that generally models how the proteins bind to each other or the telomere to shield it from too much attention by DNA repair enzymes, which include even telomerase.

As in yeast, in human cells RIF1 can be found in the vicinity of telomeres that have become uncapped, and researchers have inferred that the cell recognizes such telomeres as broken ends, signaling telomerase to go to work on the telomere. In yeast, Rif1 is always present on the telomeres, as if the cell recognizes it continually as a broken end, but in humans RIF1 visits the telomeres more transiently, so one line of inquiry might now explore whether this is simply a matter of degree or of fundamentally different mechanisms in the two different organisms. X-ray crystallography, which has been used to depict the DNA-binding domain of various proteins, has also been brought to bear on this question of how the proteins are bound to the telomere. But X-ray crystallography also constitutes an indirect method: X-rays scattered by diffraction from the crystallized protein form a pattern on photographic film that is used to generate the equivalent of a topographical map of the protein's structure, which can then lead to a three-dimensional model. Daniela Rhodes at the MRC has used X-ray crystallography to get the structure of the Rap1 DNA-binding domain, a small portion of the entire molecule. These images corroborate findings by Shore and others, but only in a general way, since they show a single binding site and not an entire molecule or series of molecules.

Blackburn suspects that the exact way in which telomeric proteins bind to protect the telomere and regulate the activity of telomerase can only be definitively answered by quantitative biochemical analysis or a slew of genetic screens. Though researchers already know how an individual Rap1 protein binds to the DNA site, they do not know how it works on a telomere or how the multiple binding sites of Rap1 sculpt the DNA in different shapes. Blackburn wants to explore "really basic questions" about this process. In her lab, Dan Levy has been looking at *Saccharomyces cerevisiae,* collaborating with lab member Tanya Williams, whose PhD from MIT trained her for the quantitative analysis of proteins. They are carefully quantifying exactly the DNA environment in which Rap1 binds, which no one has done before. Rap1 molecules bind roughly every eighteen base pairs in yeast. The spacing is species specific; in *K. lactis,* the protein binds roughly every twenty-five base pairs. In the test tube and a living cell, changes to the spacing of the binding sites, within a range of from seventeen to thirty-one base pairs, do not affect the protein binding or the ability of Rap1 to protect the telomere from telomerase action. Levy has already shown that if one molecule of Rap1 binds to the telomere, then the next one binds more easily to the first molecule, and the third more easily still, and so on, with the process cascading so that once begun, each addition after the third is exponentially easier.[29] Next, Levy will try to determine how Rif1 and Rif2 bind to telomeric DNA to which Rap1 has already bound; by adding first one protein and then the other to the test tube, he can determine if either binds directly to the DNA strand or requires a "bridge" protein to accomplish this binding.

Blackburn termed this arduous, painstaking work, a comparatively slower route than other approaches, "almost old-fashioned." Quantitative analysis of the kind she had done as a graduate student at the University of Melbourne, attempting to get the molecular weight of an enzyme, had long taken a backseat to the more glamorous work of molecular biology, sequencing genes and proteins and then using biochemical assays to confirm their behavior in the cell. Not surprisingly, she doesn't find quantitative work mere drudgery, especially given its potential to inform a more accurate model: "This is intellectually very satisfying. When you have hard numbers you can deduce underlying principles in ways you can't without quantitative data. You know the degree of accuracy, the level of variation and error. People in the field have accepted a vague picture, and at some point you have to bite the bullet if you want to find an answer that is truly accurate."

At the moment, Williams and Levy's quantitative analysis suggests two possible answers for exactly how multiple Rap1 proteins bind to the telomere, which means that they must redesign the experiment, homing in on an experimental design that will allow them to distinguish between these two possibilities. They are now using spectroscopy to view the protein in a solution similar to that in which it's bathed in the cell, which should help them to deduce its shape and thus conceptualize how multiple molecules bond together. If Williams and Levy can get the protein to crystallize out of a solution in a form that will enable them to use X-ray diffraction, they will also attempt X-ray crystallography to get the structure of the entire protein.

Other work on the telomeric proteins is being conducted in the Blackburn lab by Lifeng Xu, who is looking at human cells to observe how RIF1 becomes associated in the living cell with uncapped telomeres. These studies show that human RIF1 comes off the chromosomes at the moment they pull apart in mitosis and sits on the microtubial (mitotic) spindle.[30] Such behavior has never been seen for a protein, especially one implicated in telomere functioning, so these observations pose more new questions. Using microscopy, Xu is conducting quantitative analysis, and here again advances in technology have made greater precision possible. Like other researchers in the lab, Xu uses fluorescent tags to mark the substances she is tracking, which requires a microscope with excellent resolution. For some time, lab members had been trekking to a neighboring lab to "borrow" the DeltaVision microscopy system, which allows for the detection of molecules labeled with fluorescent tags and can generate three-dimensional images for time-lapse studies. Normally when ultraviolet light is shone steadily on these molecules to make them fluorescent, the dye fades fairly quickly, but the DeltaVision microscopy system exploits weak bursts of light to achieve the same effect, thereby preserving fluorescence longer in living cells while still yielding sharp images. Now a researcher can observe the cells for a "long time"—a few hours instead of a few moments. This microscope was built by John Sedat and others, who licensed it to Applied Precision, Inc., and stipulated that all earnings from the license be channeled directly back into their research. (Sedat emphasized that absolutely "100 percent" of the returned income goes to research, "No lunches, no dinners.")[31] When Blackburn recently purchased a DeltaVision microscopy system for her lab, Sedat installed and fine-tuned the microscope. He is currently working on a still more refined microscope in the hope of achieving a lifelong aim to picture chromosomes in living cells, not flattened as they must be for existing microscopy but intact and in an undisturbed natural state.

As a member of the UCSF breast cancer consortium since the mid-1990s, nearly every week Blackburn attends meetings that foster networking among researchers. She has continued collaborations based on the experiments begun by Melissa Rivera, which suggested that the quick effects of certain mutant templates might be exploited, and on Li's study of cancer cells (published in 2004), which indicated that depriving cells of telomerase enzyme particles might be a useful adjunct in cancer therapy.

The Blackburn lab continues to pursue work begun in the 1990s with the labs of clinician John Park and basic scientists Chris Benz and Jim

Marx, fellow members of the consortium. For some time, Park, Benz, and Marx had been attempting to target chemotherapy to cancer cells only, in the hope that this would not only reduce side effects but also prove more effective in destroying a tumor. They devised a liposome (a fat globule created on a nanoscale) coated with Marx's simplified form of Herceptin, a commercially manufactured antibody directed to a protein highly concentrated on the surface of cells in some breast cancer tumors. Thanks to these simplified antibodies, the liposome locks on to proteins, the HER1 receptors, on the cancer cell, thus forcing the liposome to fuse with the cell membrane, which is also partly lipid, and empty its contents directly into the cell, something like a smaller soap bubble fusing with a larger one. When the liposome is filled with an anticancer drug, it essentially acts as a "smart bomb" that delivers its toxin only to cancer cells.

Park and Marx tested this method for delivering doxyrubicin, an anticancer drug that causes side effects. In laboratory mice, they implanted established cell lines grown from a human tumor, and when a tumor grew, the mice were treated with an intravenous solution in which the liposomes, loaded up with doxyrubicin, were suspended. Within a few weeks, tumors in these mice shrank more rapidly than tumors in a control group of mice treated with the same drug dissolved in an intravenous solution. Furthermore, when the doxyrubicin was delivered via liposomes, less of the drug could be administered to greater effect. This research constituted a significant first step in the process for obtaining approval from the FDA for cancer therapeutics, and now that Park and Marx have completed the next stage of testing (on toxicology), they are ready to seek approval to conduct clinical trials on human.

The networking fostered by the consortium led Blackburn, Park, and Marx to speculate on a new collaboration that would marry their expertise: What if they could put mutant-template telomerase RNA genes in the liposome to be delivered to the cancer cells? Some would make their way to the cell's nucleus and be expressed, with the potential to stop the rampant division of the cancer cell. Members of the Blackburn lab had already delivered these genes in liposomes; several generations of experimental work conducted by Li, Rivera, and others had enabled the lab to develop this method successfully in tissue cultures grown from breast cancer and prostate cancer cells. What if they now tried this strategy with the smart-bomb liposomes coated with antibodies?

Advances in the technology for monitoring tumor cells enabled the researchers in Blackburn's lab to add a refinement to this experiment. The human cancer cells injected into mice in their collaborators' lab had typically been treated so they would show up as fluorescent under ultraviolet light—but only on tissue samples taken from dead mice. Now it was possible to insert, along with the mutant telomerase RNA genes, a "cassette" of reporter genes, which don't change a cell's behavior but are inserted in its genome when the liposome successfully fuses with the cancer cell. These reporter genes produce firefly luciferase, so that cells emit light in ways that can be observed using sensitive light-detecting cameras. Photographing living mice, researchers can track the migration of cancer cells and detect a larger, brighter area when cancer cells proliferate in the form of a tumor. The growth or shrinkage of a tumor can then be tracked as an increasing or diminishing brightness. The mice in this experiment are monitored daily, both before and after treatment with the liposome. Encouraging results on a small sample of mice have led the researchers to increase the scale of this experiment.

Collaboration on cancer research has changed the Blackburn lab. Although no one in the lab directly handles the mice, participation in preclinical trials on animals requires that everyone in the lab, including Blackburn, receive training on animal-use protocols for safe and humane treatment. The early work on cancer-cell tissue in Blackburn's lab attracted clinical research fellows whose expertise was increasingly necessary to the continuation of the research. Today, the lab routinely has a few clinical research fellows on its staff, but while working in the lab, they wear beepers so they can be contacted for patient care, and they must answer any summons immediately. As a consequence, their work in the lab has to be managed differently, so they are paired with a technician or a postdoc who can provide backup if they have to leave for an emergency. This partnership also provides mentoring, so that clinical researchers, some of whom have an MD but not a PhD, can learn the intricacies of molecular biology from peers with more specialized training.

Clinical collaboration poses intellectual challenges as well. Drug development per se is a technical rather than a scientific pursuit. By definition, it's a plus for this research to be pragmatic—to arrive at a therapy—but Blackburn not only clearly differentiated this from purely scientific aims but posited it as a potential threat: "Developing a drug is anti-intellectual." Her lifelong preference for working in an academic environment rather than a biotech company stems from a similar resistance to pursuing work for the sake of a predefined goal: "I don't want anybody to be telling me what to do. In clinical research, there's a goal and only one answer: it works. To me, that's technology, and I don't find that kind of goal-oriented problem solving as interesting to do, though I want it to be done. It would be great if that's the outcome of this research, but it's not my excitement. In the process of doing work that may turn out to provide useful therapy, we're learning how cancer cells are responding, and that's what really interests me. Logically speaking, if this was all about the goal, you'd focus on that and nothing else. But you'd be crazy not to look around to try to understand what cells reveal." Lateral thinking—"looking around" rather than focusing on a narrowly defined goal—is essential to Blackburn's definition of intellectual scientific inquiry.

Blackburn has mentored many younger women in her lab, and her concern for their fate contributes to her continued determination to challenge inequities in her profession. Even when interviewed on her research or about her awards, she raises this issue, and she also addresses it in public lectures. On January 14, 2005, Harvard University president Lawrence H. Summers issued a painful reminder that advocacy remains a necessity.

Summers delivered a speech to a small audience of about fifty at a conference on diversity in the science and engineering workforce, sponsored by the National Bureau of Economic Research. In giving his talk, Summers carefully referred several times to his intention to be "provocative." He argued that women did not advance to the highest academic ranks, particularly in science and engineering, because of their reluctance to commit to demanding jobs and their "choice" to devote more time to family responsibilities. He also attributed women's lack of success in the sciences to innate "variances" in aptitude between men and women. Women's career progress in academia, he said, owed far more to these factors than to the effects of discrimination, and he also characterized concerted efforts to recruit women and minorities as "fetishizing the search process."[32]

Nancy Hopkins, a member of the audience, lodged the first protest to Summers's remarks by walking out before he had even finished his speech. A biologist at MIT, Hopkins had been instrumental in spearheading the 1999 faculty survey that identified specific evidence of gender discrimination and provoked other elite institutions to make similar assessments and redress any inequities. Hopkins explained her hasty departure in dramatic terms: "I felt I was going to be sick. My heart was pounding and my breath was shallow. I was extremely upset."[33] She soon had company: a number of leading scientists and university administrators immediately challenged the speech. Summers endured a testy meeting with Harvard faculty, and at the insistence of many professors, he released a transcript of his remarks and apologized.[34]

Summers lent the authority of his office to an argument many thought had finally been discredited. He cited gender differences in the science and math test scores of high school students as evidence of differences in innate ability; however, research shows that not only are average test scores the same but also differences in scores have diminished over the past few decades, a shift that cannot be accounted for by genetic traits.[35] To make his point, Summers used the example of his own twin daughters, "who were not given dolls and who were given trucks, and found themselves saying to each other, look, daddy truck is carrying the baby truck," as if the use of familial terms demonstrated an innate inability to master science and math.[36] In what seems to be poetic justice, the furor his speech provoked contributed to his decision to resign on February 21, 2006.

His remarks touched a nerve especially among women scientists because they continue to feel complicated and varying kinds of ambivalence about the role gender plays in their professional lives. For her "emotional" response, Hopkins drew criticism from a number of

professional women who judged her according to a male norm of appropriate behavior, with one computer scientist commenting that "instead
of getting angry and leaving, we should get angry and fight," and a fellow
scientist noting that "this does not help to debunk stereotypes of women
as emotional and incapable of cool logic."[37] Summers's speech provoked
furious e-mail exchanges among Blackburn and other members of the
Women in Cell Biology Committee of the ASCB, particularly in response
to his ill-chosen example of his daughters' "feminine" thinking. This
flurry of correspondence evolved into a sharing of childhood stories in
which one committee member after another acknowledged that she had
played with dolls as a girl, as if these women collectively had to accumulate enough evidence to dispel the notion that this somehow diminished their gifts as scientists.

Only a few months after Summers gave his speech, Blackburn was
invited by Barbara J. Grosz, dean of science at the Radcliffe Institute for
Advanced Study, to give a lecture on her work at Harvard University.
Blackburn was determined to find a way to voice her own protest. In
speaking of her work to a lay audience, Blackburn often used slides that
referenced familiar topics in order to convey the conceptual underpinnings of the work. For her Harvard talk on March 2, 2005, at which
she discussed her collaboration with Epel, Blackburn wanted to use a
well-known quotation to portray the difficulty of moving from identifying a correlation—such as that between telomere length and the risk for
cardiovascular disease—to determining causality. She had researched
carefully to find just the right illustration for the circumstances. The first
slide she projected on the screen presented a quotation from Aristotle,
"One swallow does not a summer make," with arrows pointing from
each word to the next to suggest one possible reading of cause and effect.
The next slide, in which the same words were rearranged to show a different possible causal relationship, read: "One does not have to swallow
Summer(s)." The audience cheered wildly.

Blackburn's personal story is interwoven with so many others: how
molecular biology evolved over the last thirty-five years; how a new scientific field is born; how basic science research contributes to clinical
studies on human health by a circuitous route; and how women have
made advances in a male-dominated profession despite enormous

obstacles. Her service on the President's Council on Bioethics illuminates the dangerous terrain in which science intersects with politics and offers a timely reminder of how scientific integrity contributes to our society's welfare. Blackburn's story also counters the popular myth of the solitary scientific genius and the common notion that only ruthless competition can produce the finest research. She is no lone hero but a highly collegial and collaborative citizen of an interactive community, someone whose idealism does not contradict her competitive spirit.

Notes

Chapter 1

1. All quotations attributed to Elizabeth H. Blackburn are taken from interviews with the author on June 12, June 26, and July 2, 2004.

2. Katherine Marsden, interview with the author, October 3, 2004.

3. Ibid.

4. Ibid.

5. Ibid.

Chapter 2

1. All quotations attributed to Elizabeth H. Blackburn are taken from interviews with the author on June 22 and July 12, 2004.

2. Barrie Davidson, Elizabeth H. Blackburn, and Theo Dopheide, "Chorismate Mutase-Prephenate Dehydratase from *Escherichia coli* K–12," *Journal of Biological Chemistry* 247, no. 14 (July 25, 1972): 4441–4446.

3. Frank Hird, letter to Mrs. M. Blackburn, March 10, 1983.

4. Harrison Echols, *Operators and Promoters: The Story of Molecular Biology and Its Creators*, ed. Carol A. Gross (Berkeley: University of California Press, 2001), 13.

5. Nobel Foundation, "Nobel Chemistry Laureates 1980: Fred Sanger, Autobiography," *Nobel e Museum*.

6. John Sedat, interview with the author, March 5, 2006.

7. Spyros Artavanis-Tsakonas, telephone interview with the author, September 9, 2004.

8. Soraya de Chadarevian, *Designs for Life: Molecular Biology after World War II* (Cambridge: Cambridge University Press, 2002), 267.

9. Quoted in ibid., 267.

10. Artavanis-Tsakonas, telephone interview.

11. Committee on Women in Science and Engineering, Panel for the Study of Gender Differences in the Career Outcomes of Science and Engineering PhDs, National Research Council, *From Scarcity to Visibility: Gender Differences in the Careers of Doctoral Scientists and Engineers*, ed. J. Scott Long (Washington, DC: National Academies Press, 2001), 39, 128 (figure 6–4).

12. de Chadarevian, *Designs for Life*, 249–250.

13. Medical Research Council, "Ratio of Women to Men, 1970s and 1980s" (unnumbered table) in *Medical Research Council Handbook* (London: Medical Research Council, 1990).

14. Quoted in de Chadarevian, *Designs for Life*, 249–250.

15. James Watson, *The Double Helix: A Personal Account of the Discovery of the Structure of DNA* (New York: Atheneum, 1968), 45.

16. Harrison Echols, *Operators and Promoters*, 191.

17. Artavanis-Tsakonas, telephone interview.

18. Sedat, interview.

Chapter 3

1. All quotations attributed to Elizabeth H. Blackburn are taken from interviews with the author on June 26, July 19, and August 6, 2004.

2. Joseph Gall, *A Pictorial History: Views of a Cell* (Bethesda, MD: American Society for Cell Biology, 1996), book jacket.

3. Joseph (Joe) Gall, telephone interview with the author, September 3, 2004.

4. Martha Truett, telephone interview with the author, September 8, 2004.

5. Ibid.

6. Gall, *A Pictorial History*, book jacket.

7. George W. Pierson, ed., *A Yale Book of Numbers: Historical Statistics of the College and University, 1701–1976* (New Haven, CT: Yale University Office of Institutional Research, 1977), table B–9.5.

8. Ibid., 178.

9. Beverly Waters, ed., *A Yale Book of Numbers, 1976–2000* (New Haven, CT: Yale University Office of Institutional Research, 2001), table F–2.

10. Committee on Women in Science and Engineering, Panel for the Study of Gender Differences in the Career Outcomes of Science and Engineering PhDs, National Research Council, *From Scarcity to Visibility: Gender Differences in the Careers of Doctoral Scientists and Engineers*, ed. J. Scott Long (Washington, DC: National Academies Press, 2001), 10.

11. Ibid., 145.

12. Harrison Echols, *Operators and Promoters: The Story of Molecular Biology and Its Creators,* ed. Carol A. Gross (Berkeley: University of California Press, 2001), 192.

13. Diane K. Lavett, telephone interview with the author, January 30, 2005.

14. Gall, telephone interview.

15. Barbara McClintock, "Cytological Observations of Deficiencies Involving Known Genes, Translocations, and an Inversion in *Zea mays*," *Missouri Agricultural Experiment Station Research Bulletin* 163 (1931): 1–30.

16. Barbara McClintock, "The Fusion of Broken Ends of Sister Half-Chromatids Following Chromatid Breakage at Meiotic Anaphases," *Missouri Agricultural Experiment Station Research Bulletin* 290 (1938): 1–48.

17. Truett, telephone interview.

18. Ibid.

19. Lavett, telephone interview.

20. Truett, telephone interview.

21. Gall, telephone interview.

22. Elizabeth H. Blackburn and Joseph Gall, "A Tandemly Repeated Sequence at the Termini of the Extrachromosomal Ribosomal RNA Genes in *Tetrahymena*," *Journal of Molecular Biology* 120 (1978): 33–53.

23. Gall, telephone interview.

24. Lavett, telephone interview.

Chapter 4

1. All quotations attributed to Elizabeth H. Blackburn, unless otherwise noted, are taken from interviews with the author on September 3, 10, and 17, 2004.

2. J. Michael (Mike) Cherry, telephone interview with the author, October 20, 2005.

3. Elizabeth H. Blackburn and San-San Chiou, "Non-nucleosomal Packaging of a Tandemly Repeated DNA Sequence at Termini of Extrachromosomal DNA Coding for rRNA in *Tetrahymena*," *Proceedings of the National Academy of Sciences* 78 (1981): 2263–2267.

4. Elizabeth H. Blackburn, Marsha L. Budarf, Peter B. Challoner, J. Michael Cherry, Elizabeth A. Howard, A. L. Katzen, Wei-Chun Pan, and T. Ryan, "DNA

Termini in Ciliate Macronuclei," in *Cold Spring Harbor Symposia on Quantitative Biology* (Cold Spring Harbor, NY: Cold Spring Harbor Laboratory Press, 1983), vol. 47 (part 2), 1195–1207.

5. Dan E. Gottschling and Thomas R. Cech, "Chromatin Structure of the Molecular Ends of *Oxytricha* Macronuclear DNA: Phased Nucleosomes and a Telomeric Complex," *Cell* 38 (1984): 501–510.

6. Dan E. Gottschling and Virginia A. Zakian, "Telomere Proteins: Specific Recognition and Protection of the Natural Termini of *Oxytricha* Macronuclear DNA," *Cell* 47 (1986): 195–205.

7. Cherry, telephone interview.

8. Office of the President, "University of California, Universitywide New Appointments of Ladder Rank Faculty: 1984–85 through 2002–03," University of California at Berkeley, Academic Advancement (unnumbered table), n.p.

9. Ellen Daniell, "Facing Disaster: Ellen's Story," in *Every Other Thursday: Stories and Strategies from Successful Women Scientists* (New Haven, CT: Yale University Press, 2006), 41.

10. Ibid., 47.

11. Ibid., 51.

12. Ibid., 58.

13. Ibid., 67.

14. Quoted in Diane Ainsworth, "Women in Science: A Scarcity of Females Suggests That Science, Engineering Is Still Difficult Terrain for Women," *Berkeleyan*, June 6, 2001, n.p.

15. Jack W. Szostak and Elizabeth H. Blackburn, "Cloning Yeast Telomeres on Linear Plasmid Vectors," *Cell* 29 (1982): 245–255.

16. Richard W. Walmsley, Jack W. Szostak, and Thomas D. Petes, "Is There Left-handed DNA at the Ends of Yeast Chromosomes?" *Nature* 302 (1983): 84–86.

17. Richard W. Walmsley, Clarence S. Chan, Bik-Kwoon Tye, and Thomas D. Petes, "Unusual DNA Sequences Associated with the Ends of Yeast Chromosomes," *Nature* 310 (1984): 157–160.

18. Harrison Echols, *Operators and Promoters: The Story of Molecular Biology and Its Creators*, ed. Carol Gross (Berkeley: University of California Press, 2001), 153–154.

19. Janis Shampay, Jack W. Szostak, and Elizabeth H. Blackburn, "DNA Sequences of Telomeres Maintained in Yeast," *Nature* 310 (1984): 154–157.

20. Carol W. Greider and Elizabeth H. Blackburn, "Tracking Telomerase," *Cell* S116 (2004): S83–S86.

21. Titia de Lange, telephone interview with the author, October 21, 2005.

22. Carol W. Greider, telephone interview with the author, January 26, 2006.

23. Elizabeth H. Blackburn, "Telomeres: Do the Ends Justify the Means?" *Cell* 37 (1984): 7–8.

24. Elizabeth H. Blackburn, "A History of Telomere Biology," in *Telomeres*, 2d. ed., ed. Titia de Lange, Vicki Lundblad, and Elizabeth Blackburn (Cold Spring Harbor, NY: Cold Spring Harbor Laboratory Press, 2006), 6.

25. Elizabeth H. Blackburn, "Appendix: A Personal Account of the Discovery of Telomerase," in *Telomeres*, 2d ed., 551–553. Blackburn gives a similar account of her thinking in these pages.

26. Andre Bernards, Paul A. Michels, Carsten R. Lincke, and Piet Borst, "Growth of Chromosome Ends in Multiplying Trypanosomes," *Nature* 303 (1983): 592–597.

27. Barbara McClintock, letter to Elizabeth Blackburn, March 11, 1983. Private papers of Elizabeth Blackburn. Available online in the Barbara McClintock Papers, "Profiles in Science," National Library of Medicine.

28. Blackburn, "Appendix: A Personal Account of the Discovery of Telomerase," 553–555.

29. Ibid., 556.

30. Greider, telephone interview.

31. Cherry, telephone interview.

32. Greider, telephone interview.

33. Quoted in Keith Haglund and Jennifer Steinberg, "The Landmarks Interviews: Means That Justified the Ends," *Journal of NIH Research* 8 (September 1996): 58.

34. Cherry, telephone interview.

35. Quoted in Haqlund and Steinberg, "The Landmarks Interviews," 58.

36. Greider, telephone interview.

37. Carol W. Greider and Elizabeth H. Blackburn, "Identification of a Specific Telomere Terminal Transferase Activity in *Tetrahymena* Extracts," *Cell* 43 (1985): 405–413.

38. Ibid., 405.

39. Ibid., figure 1, "Model for Telomere Elongation," 406.

40. Greider and Blackburn, "Tracking Telomerase," S85.

41. Jack W. Szostak, "The Beginning of the Ends," *Nature* 337 (1989): 303–304. Szostak cited the results of a paper by Ann F. Pluta and Virginia A. Zakian, published in *Nature* 337 (1989): 429–433.

42. Greider, telephone interview.

43. Ibid.

44. de Lange, telephone interview.

Chapter 5

1. Elizabeth H. Blackburn, "A History of Telomere Biology," in *Telomeres,* 2d ed., ed. Titia de Lange, Vicki Lundblad, and Elizabeth Blackburn (Cold Spring Harbor, NY: Cold Spring Harbor Laboratory Press, 2006), 11.

2. Elizabeth H. Blackburn, "Appendix: A Personal Account of the Discovery of Telomerase," in *Telomeres,* 2d ed., 556.

3. Kelly Kruger, Paula J. Grabowski, Arthur J. Zaug, Julie Sands, Daniel E. Gottschling, and Thomas R. Cech, "Self-splicing RNA: Autoexcision and Autocyclization of the Ribosomal RNA Intervening Sequence of *Tetrahymena,*" *Cell* 31 (1982): 147–157.

4. For a detailed discussion of RNA splicing and catalytic properties, see Harrison Echols, "The RNA World: New Proteins and Revised RNAs," in *Operators and Promoters: The Story of Molecular Biology and Its Creators,* ed. Carol Gross (Berkeley: University of California Press, 2001).

5. Carol W. Greider and Elizabeth H. Blackburn, "Tracking Telomerase," *Cell* S116 (2004): S85.

6. Arthur J. Zaug and Thomas R. Cech, "The Intervening Sequence RNA of *Tetrahymena* Is an Enzyme," *Science* 231 (1986): 470–475.

7. Janis Shampay and Elizabeth H. Blackburn, "Generation of Telomere-Length Heterogeneity in *Saccharomyces cerevisiae,*" *Proceedings of the National Academy of Sciences* 85 (1988): 534–538.

8. All quotations attributed to Elizabeth H. Blackburn are taken from interviews with the author on November 12, 2004, and December 15, 17, 21, 22, and 24, 2004.

9. Committee on Women in Science and Engineering, Panel for the Study of Gender Differences in the Career Outcomes of Science and Engineering PhDs, National Research Council, *From Scarcity to Visibility: Gender Differences in the Careers of Doctoral Scientists and Engineers,* ed. J. Scott Long (Washington, DC: National Academies Press, 2001), 91, 94.

10. Gillian Gehring, "Mixing Motherhood and Science," *Physics World,* March 2002, n.p.

11. Carol W. Greider and Elizabeth H. Blackburn, "The Telomere Terminal Transferase of *Tetrahymena* Is a Ribonucleoprotein Enzyme with Two Kinds of Primer Specificity," *Cell* 51 (1987): 887–898.

12. Blackburn, "Appendix: A Personal Account of the Discovery of Telomerase," 560–561.

13. Dorothy Shippen (formerly Shippen-Lentz), telephone interview with the author, December 15, 2004.

14. Ibid.

15. Ibid.

16. American Society for Cell Biology, "ASCB Profile: Carol W. Greider."

17. Carol W. Greider and Elizabeth H. Blackburn, "A Telomeric Sequence in the RNA of *Tetrahymena* Telomerase Required for Telomere Repeat Synthesis," *Nature* 337 (1989): 331–337.

18. Ibid., 336.

19. Dorothy Shippen-Lentz and Elizabeth H. Blackburn, "Telomere Terminal Transferase Activity from *Euplotes crassus* Adds Large Numbers of TTTTGGGG Repeats onto Telomeric Primers," *Molecular and Cellular Biology* 9 (1989): 2761–2764.

20. Dorothy Shippen-Lentz and Elizabeth H. Blackburn, "Functional Evidence for an RNA Template in Telomerase," *Science* 247 (1990): 546–552.

21. Shippen, telephone interview.

22. Peter C. Yaeger, Eduardo Orias, Wen-Ling Shaiu, Drena D. Larson, and Elizabeth H. Blackburn, "The Replication Advantage of a Free Linear rRNA Gene Is Restored by Somatic Recombination in *Tetrahymena thermophila*," *Molecular and Cellular Biology* 9 (1989): 452–460.

23. Guo-Liang Yu, John D. Bradley, Laura D. Attardi, and Elizabeth H. Blackburn, "In Vivo Alteration of Telomere Sequences and Senescence Caused by Mutated *Tetrahymena* Telomerase RNAs," *Nature* 344 (1990): 126–132.

Chapter 6

1. All quotations attributed to Elizabeth H. Blackburn are taken from interviews with the author on January 4, 5, 7, 11, 12, 14, and 17, 2005.

2. David (Dave) Gilley, telephone interview with the author, January 11, 2006.

3. Jeffrey Brainard, "NIH Program Seeks to Speed Grant Process for New Applicants," *Chronicle of Higher Education*, December 5, 2005, n.p.

4. Thomas R. Cech and Enriqueta Bond, "Managing Your Own Lab," *Science* 304 (2004): 1717.

5. Sam Kean, "Scientists Spend Nearly Half Their Time on Administrative Tasks, Survey Finds," *Chronicle of Higher Education*, July 7, 2006, n.p.

6. Brainard, "NIH Program Seeks to Speed Grant Process for New Applicants."

7. Ibid.

8. Jeffrey Brainard, "NIH Would Receive No Increase for 2007," *Chronicle of Higher Education*, February 7, 2006, n.p.

9. Elizabeth H. Blackburn, private communication to the author, July 7, 2004.

10. Richard Monastersky, "The Number That's Devouring Science," *Chronicle of Higher Education*, October 14, 2005, n.p.

11. Vicki Lundblad and Jack W. Szostak, "A Mutant with a Defect in Telomere Elongation Leads to Senescence in Yeast," *Cell* 57 (1989): 633–643.

12. Elizabeth H. Blackburn, "Structure and Function of Telomeres," *Nature* 350 (1991): 571.

13. Howard J. Cooke and B. A. Smith, "Variability at the Telomeres of the Human X/Y Pseudoautosomal Region," *Cold Spring Harbor Symposium on Quantitative Biology* (Cold Spring Harbor, NY: Cold Spring Harbor Laboratory Press, 1986), vol. 51 (part 1): 213–219.

14. Carol W. Greider and Elizabeth H. Blackburn, "Telomeres, Telomerase, and Cancer," *Scientific American* 274 (February 1996): 95.

15. Calvin B. Harley, A. Bruce Futcher, and Carol W. Greider, "Telomeres Shorten during Aging of Human Fibroblasts," *Nature* 345 (1990): 458–460.

16. These studies are cited in Greider and Blackburn, "Telomeres, Telomerase, and Cancer," 96.

17. Christopher M. Counter, Ariel A. Avilion, Catherine E. LeFeuvre, Nancy G. Stewart, Carol W. Greider, Calvin B. Harley, and Sylvia Bacchetti, "Telomere Shortening Associated with Chromosome Instability Is Arrested in Immortal Cells Which Express Telomerase Activity," *EMBO Journal* 11 (1992): 1921–1929.

18. Greider and Blackburn, "Telomeres, Telomerase, and Cancer," 95.

19. Ibid., 95–96.

20. Carol W. Greider, telephone interview with the author, January 26, 2006. The numerical estimates are based on a version of this chart, incorporated as figure 3, in Elizabeth H. Blackburn, Carol W. Greider, and Jack W. Szostak, "Telomeres and Telomerase: The Path from Maize, *Tetrahymena*, and Yeast to Human Cancer and Aging," *Nature Medicine* 12, no. 10 (2006): xi.

21. Blackburn, "Structure and Function of Telomeres," 572.

22. David Stipp, "The Hunt for the Youth Pill: From Cell-immortalizing Drugs to Cloned Organs, Biotech Finds New Ways to Fight against Time's Toll," *Fortune*, October 11, 1999, 199.

23. Blackburn and Greider, "Telomeres, Telomerase, and Cancer," 95. Blackburn cited the paper that disproved a simple link between aging and cell proliferation in vitro in "Telomere States and Cell Fates," *Nature* 408 (2000): 53.

24. For an example of Geron's persuasive new arguments that telomerase was "absent" or "stringently repressed in normal human somatic tissues," see the Geron press release, "'Immortalizing Enzyme' Detected in Many Types of Cancer, Reported in *Science*," December 23, 1994, n.p.

25. Vicki Lundblad and Elizabeth H. Blackburn, "An Alternate Pathway for Yeast Telomere Maintenance Rescues est1-Senescence," *Cell* 73 (1993): 347–360.

26. Kyla Dunn, "A Look at . . . Patents and Biotech," *Washington Post*, October 1, 2000, B3.

27. Quoted in Stephen S. Hall, *Merchants of Immortality: Chasing the Dream of Human Life Extension* (New York: Houghton Mifflin Co., 2003), 143–144.

28. Ibid., 86.

29. Ibid., 88.

30. Ibid., 139–141.

31. Quoted in ibid., 141.

32. Stephen Hansen, Amanda Brewster, Jana Asher, and Michael Kisielewski, *The Effects of Patenting in the AAAS Community*, 2d ed. (Washington, DC: American Association for the Advancement of Science, 2006), 7.

33. Ibid., 9, 25.

34. Hall, *Merchants of Immortality*, 58, 84.

35. Ibid., 59.

36. Michael (Mike) J. McEachern, telephone interview with the author, January 24, 2006.

37. Blackburn related this story in an interview with the author, and she gave a similar version in an interview with Hall, reported in *Merchants of Immortality*, 83–84.

Chapter 7

1. All quotations attributed to Elizabeth H. Blackburn are taken from interviews with the author on March 22 and 29, 2005, and June 8 and 21, 2005.

2. Robin Wilson, "Women in the National Academy," *Chronicle of Higher Education*, June 10, 2005, A8.

3. Royal Society of London, "List of New Fellows and Foreign Members for 2005."

4. Heineken International, "Elizabeth H. Blackburn," *Laureates*.

5. Carol Gross, interview with the author, September 26, 2005.

6. Belden, Russonello, and Stewart, *The Climate for Women on the Faculty at UCSF: Report of Findings from a Survey of Faculty Members*, Office of the Chancellor, University of California at San Francisco, January 2002, 3, 64.

7. Gross, interview.

8. Association of American Medical Colleges, *Enhancing the Environment for Women in Academic Medicine* (Washington, DC: Association of American Medical Colleges, 1996), 9 (table 1B).

9. Ibid., 5; see also 8–9 (tables 1A and 1B).

10. Ibid., 4.

11. Ibid., 79.

12. University of California at San Francisco, "Making History: Mission Bay Milestones."

13. Gross, interview.

14. Ibid.

15. David (Dave) Gilley, telephone interview with the author, January 11, 2006.

16. Christine Wenneras and Agnes Wold, "Nepotism and Sexism in Peer-Review," *Nature* 387 (1997): 341–342.

17. Ibid., 343.

18. Peggy Orenstein, "Why Science Must Adapt to Women: An Elite Survivor Assesses the Hidden Costs of Exclusion," *Discover* 23 (November 2002): 60.

19. Marcia Barinaga, "Academic Employment: UCSF Researchers Leave, Charging Bias," *Science* 288 (2000): 26–27.

20. Ulysses Torassa, "Gender Bias Dispute Flares at UCSF," *San Francisco Examiner*, March 26, 2000, n.p.

21. Ibid.

22. Quoted in Ibid.

23. University of California at San Francisco, "Making History: Facts."

24. University of California at San Francisco, "Making History: Mission Bay Milestones."

25. American Society for Cell Biology, "About Us."

26. American Society for Cell Biology, "Past Officers."

27. Elizabeth Marincola, telephone interview with the author, January 6, 2006.

28. Joint Steering Committee for Public Policy, "About JSC."

29. Marincola, telephone interview.

30. Joint Steering Committee for Public Policy, "About JSC."

31. Marincola, telephone interview.

32. Ibid.

33. Ibid.

34. Ibid.

35. Ibid.

36. Public Library of Science, "Board of Directors."

37. Quoted in Orenstein, "Why Science Must Adapt to Women," 60.

Chapter 8

1. All quotations attributed to Elizabeth H. Blackburn are taken from interviews with the author on January 11, 2005, and February 15 and 18, 2005.

2. Vicki Lundblad, telephone interview with the author, October 18, 2005.

3. Ibid.

4. Joachim Lingner, Timothy R. Hughes, Andrej Shevchenko, Matthias Mann, Victoria Lundblad, and Thomas R. Cech, "Reverse Transcriptase Motifs in the Catalytic Subunit of Telomerase," *Science* 276 (1997): 561–567.

5. Thomas R. Cech, "Beginning to Understand the End of the Chromosome," *Cell* 116 (2004): 273–279.

6. Elizabeth H. Blackburn, "Telomerase," in *The RNA World II*, ed. Raymond F. Gesteland and John F. Atkins (Cold Spring Harbor, NY: Cold Spring Harbor Laboratory Press, 1999), 609–635.

7. Daniel P. Romero and Elizabeth H. Blackburn, "A Conserved Secondary Structure for Telomerase RNA," *Cell* 67 (1991): 343–353.

8. Blackburn, "Telomerase," 619.

9. Ibid.

10. Jue Lin, Hinh Ly, Arif Hussain, Mira Abraham, Siran Pearl, Tristram G. Parslow, Yehuda Tzfati, and Elizabeth H. Blackburn, "A Universal Telomerase RNA Core Structure Includes Structure Motifs Required for Binding the Telomerase Reverse Transcriptase Protein," *Proceedings of the National Academy of Sciences* 101 (2004): 14713–14718.

11. Blackburn, "Telomerase," 619–620.

12. Cech, "Beginning to Understand the End of the Chromosome," 275.

13. Anita G. Seto, April J. Livengood, Yehuda Tzfati, Elizabeth H. Blackburn, and Thomas R. Cech, "A Bulged Stem Tethers Est1p to Telomerase RNA in Budding Yeast," *Genes and Development* 16 (2002): 2800–2812.

14. Blackburn referred to this function of the RNA component in David Gilley, M. S. Lee, and Elizabeth H. Blackburn, "Altering Specific Telomerase RNA Template Residues Affects Active Site Function," *Genes and Development* 9 (1995): 2214–2226; Elizabeth H. Blackburn, "Telomerase RNA Structure and Function," in *RNA Structure and Function*, ed. R. W. Simons and Marian Grunberg-Manago (Cold Spring Harbor, NY: Cold Spring Harbor Laboratory Press, 1998).

15. Gilley, Lee, and Blackburn, "Altering Specific Telomerase RNA Template Residues Affects Active Site Function."

16. Titia de Lange, telephone interview with the author, October 21, 2005.

17. Ibid.

18. Michael J. McEachern, telephone interview with the author, January 24, 2006.

19. Michael J. McEachern and Elizabeth H. Blackburn, "Runaway Telomere Elongation Caused by Telomerase RNA Gene Mutations," *Nature* 376 (1995): 403–409.

20. McEachern, telephone interview.

21. Stephane Marcand, Eric Gilson, and David Shore, "A Protein-counting Mechanism for Telomere Length Regulation in Yeast," *Science* 275 (1997): 986–990.

22. Anat Krauskopf and Elizabeth H. Blackburn, "Rap1 Protein Regulates Telomere Turnover in Yeast," *Proceedings of the National Academy of Sciences* 95 (1998): 12486–12491.

23. Simon Chan and Elizabeth H. Blackburn, "Telomerase and ATM/Tel1p Protect Telomeres from Nonhomologous End Joining," *Molecular Biology of the Cell* 11 (2003): 1379–1387.

24. Elizabeth H. Blackburn, "Telomere States and Cell Fates," *Nature* 408 (2000): 54.

25. Quoted in Karen Birmingham, "Elizabeth Blackburn," *Nature Medicine* 7 (2001): 520.

26. Mark Zijlmans, "The Role of Telomeres and Telomerase in Cancer" (paper presented at the conference of the American Association for Cancer Research, December 2000, San Francisco). These results were later reported in Karin A. Mattern, Susan J. J. Swiggers, Alex L. Nigg, Bob Lowenberg, Adriaan B. Houtsmuller, and J. Mark J. M. Zijlmans, "Dynamics of Protein Binding to Telomeres in Living Cells: Implications for Telomere Structure and Function," *Molecular and Cellular Biology* 24 (2004): 5587–5594.

27. de Lange, telephone interview.

28. Ibid.

29. Jack D. Griffith, Laurey Corneau, Soraya Rosenfield, Rachel M. Stansel, Alessandra Bianchi, Heidi Moss, and Titia de Lange, "Mammalian Telomeres End in Large Duplex Loop," *Cell* 97 (1999): 503–514.

30. de Lange, telephone interview.

31. McEachern and Blackburn, "Runaway Telomere Elongation Caused by Telomerase RNA Gene Mutations."

32. John C. Prescott and Elizabeth H. Blackburn, "Telomerase RNA Mutations in *Saccharomyces cerevisiae* Alter Telomerase Action and Reveal Nonprocessivity *in vivo* and *in vitro*," *Genes and Development* 11 (1997): 528–540.

33. Jiyue Zhu, He Wang, J. Michael Bishop, and Elizabeth H. Blackburn, "Telomerase Extends the Lifespan of Virus-Transformed Human Cells without Net Telomere Lengthening," *Proceedings of the National Academy of Science* 96 (1996): 3723–3728.

34. John C. Prescott, telephone interview with the author, January 10, 2006.

35. Blackburn, "Telomere States and Cell Fates"; specific reference to the findings is on page 53.

36. Christopher D. Smith, Dana L. Smith, Joe L. DeRisi, and Elizabeth H. Blackburn, "Telomeric Protein Distributions and Remodeling through the Cell Cycle in *S. cerevisiae*," *Molecular Biology of the Cell* 14 (2003): 556–570. (First published online in *MBC in Press*, December 7, 2002.)

37. Chan and Blackburn, "Telomerase and ATM/Tel1p Protect Telomeres from Nonhomologous End Joining."

Chapter 9

1. John C. Prescott and Elizabeth H. Blackburn, "Telomerase: Dr. Jekyll or Mr. Hyde?" *Current Opinion in Genetics and Development* 9 (1999): 372.

2. Michael J. McEachern, telephone interview with the author, January 24, 2006.

3. David Gilley, telephone interview with the author, January 11, 2006.

4. Robert W. Frenck Jr., Elizabeth H. Blackburn, and Kevin M. Shannon, "The Rate of Telomere Sequence Loss in Human Leukocytes Varies with Age," *Proceedings of the National Academy of Sciences* 95 (1998): 5607–5610.

5. Han-Woong Lee, Maria A. Blasco, Geoffrey J. Gottlieb, James W. Horner II, Carol W. Greider, and Ronald A. DePinho, "Essential Role of Mouse Telomerase in Highly Proliferative Organs," *Nature* 392 (1998): 569–574.

6. Personal communication to Elizabeth H. Blackburn, cited in Prescott and Blackburn, "Telomerase: Dr. Jekyll or Mr. Hyde?" 371.

7. Richard M. Cawthon, Ken R. Smith, Elizabeth O'Brien, Anna Sivatchenko, and Richard A. Kerber, "Association between Telomere Length in Blood and Mortality in People Aged 60 Years or Older," *Lancet* 361 (2003): 1224.

8. James R. Mitchell, Emily Wood, and Kathleen Collins, "A Telomerase Component Is Defective in the Human Disease Dyskeratosis Congenita," *Nature* 402 (1999): 551–555.

9. Hinh Ly, Lifeng Xu, Melissa Rivera, Tristram G. Parslow, and Elizabeth H. Blackburn, "A Role for a Novel 'Trans-Pseudoknot' RNA-RNA Interaction in the Functional Dimerization of Human Telomerase." *Genes and Development* 17 (2003): 1078–1083.

10. Tom J. Vulliamy, Amanda Waines, Aroon Baskaradas, Philip J. Mason, Anna Marrone, and Inderjeet Dokal, "Mutations in the Reverse Transcriptase Component of Telomerase (TERT) in Patients with Bone Marrow Failure," *Blood Cells, Molecules, and Diseases* 34 (May–June 2005): 257–263.

11. Fred Goldman, Rachida Bouarich, Shashikant Kulkarni, Sara Freeman, Hong-Yan Du, Lea Harrington, Philip J. Mason, Arturo London-Vallejo, and Monica Bessler, "The Effect of TERC Haploinsufficiency on the Inheritance of Telomere Length," *Proceedings of the National Academy of Sciences* 102 (2005): 17119–17124.

12. Jerry W. Shay and Woodring E. Wright, "Telomerase and Human Cancer," in *Telomeres*, 2d ed., ed. Titia de Lange, Vicki Lundblad, and Elizabeth Blackburn, 88–89. (Cold Spring Harbor, NY: Cold Spring Harbor Laboratory Press, 2006).

13. McEachern, telephone interview.

14. All quotations attributed to Elizabeth H. Blackburn are taken from interviews with the author on March 8, 11, 18, 22, and 25, 2005.

15. Gilley, telephone interview.

16. Moses M. Kim, Melissa A. Rivera, Inna L. Botchkina, Refaat Shalaby, Ann D. Thor, and Elizabeth H. Blackburn, "A Low Threshold Level of Expression of Mutant-Template Telomerase RNA Inhibits Human Tumor Cell Proliferation," *Proceedings of the National Academy of Sciences* 98 (2001): 7982–7987.

17. Maria A. Blasco, Han-Woong Lee, M. Prakash Hande, Enrique Samper, Peter M. Landsdorp, Ronald A. DePinho, and Carol W. Greider, "Telomere Shortening and Tumor Formation by Mouse Cells Lacking Telomerase RNA," *Cell* 91 (1997): 25–34.

18. Lynda Chin, Steven E. Artandi, Qiong Shen, Alice Tam, Shwu-Luan Lee, Geoffrey J. Gottlieb, Carol W. Greider, and Ronald A. DePinho, "P53 Deficiency Rescues the Adverse Effects of Telomere Loss and Cooperates with Telomere Dysfunction to Accelerate Carcinogenesis," *Cell* 97 (1999): 527–538.

19. Inderjeet Dokal, "Dyskeratosis Congenita in All Its Forms," *British Journal of Haematology* 110 (2000): 768–779.

20. William C. Hahn, Christopher M. Counter, Ante S. Lundberg, Roderick L. Beijersbergen, Mary W. Brooks, and Robert A. Weinberg, "Creation of Human Tumour Cells with Defined Genetic Elements," *Nature* 400 (1999): 464–468.

21. William C. Hahn, Sheila A. Stewart, Mary W. Brooks, Shoshana G. York, Elinor Eaton, Akiko Kurachi, Roderick L. Beijersbergen, Joan H. M. Knoll, Matthew Meyerson, and Robert A. Weinberg, "Inhibition of Telomerase Limits the Growth of Human Cancer Cells," *Nature Medicine* 5 (1999): 1129–1130.

22. Shang Li, Jonathan E. Rosenberg, Annemarie A. Donjacour, Inna L. Botchkina, Yun Kit Hom, Gerald R. Cunha, and Elizabeth H. Blackburn, "Rapid Inhibition of Cancer Cell Growth Induced by Lentiviral Expression of Mutant-Template Telomerase RNA Constructs and Anti-Telomerase Short-interfering RNA," *Cancer Research* 64 (2004): 4833–4840.

23. Shang Li, Julia Crothers, Christopher M. Haqq, and Elizabeth H. Blackburn, "Cellular and Gene Expression Responses Involved in the Rapid Growth Inhibition of Human Cancer Cells by RNA Interference-Mediated Depletion of Telomerase," *Journal of Biological Chemistry* 280 (2005): 23709–23717.

24. Mehdi Nosrati, Shang Li, Sepideh Bagheri, David Ginzinger, Elizabeth H. Blackburn, Robert J. Debs, and Mohammed Kashani-Sabet, "Antitumor Activity of Systemically Delivered Ribozymes Targeting Murine Telomerase RNA," *Clinical Cancer Research* 10 (2004): 4983–4990.

25. Michelle L. Brandt, "Stem Cell Primer," *Stanford Medicine* 21 (Fall 2004): 15.

26. Amy Adams, "The True Seeds of Cancer: Are Treatments Targeting the Wrong Cell?" *Stanford Medicine* 21 (Fall 2004): 24.

27. Joseph Gall, telephone interview with the author, September 3, 2004.

28. Gilley, telephone interview.

Chapter 10

1. All quotations attributed to Dana Smith are taken from an interview with the author, July 29, 2005.

2. Carol Gross, interview with the author, September 26, 2005.

3. All quotations attributed to Elizabeth H. Blackburn, unless otherwise noted, are taken from interviews with the author on September 2 and 16, 2005.

4. All quotations attributed to Vicki Lundblad are taken from telephone interviews with the author on October 11 and 18, 2005.

5. Stephen S. Hall, *Merchants of Immortality: Chasing the Dream of Human Life Extension* (New York: Houghton Mifflin, 2003), 130.

6. All quotations attributed to Titia de Lange are taken from a telephone interview with the author, October 21, 2005.

7. All quotations attributed to John C. Prescott are taken from a telephone interview with the author, January 10, 2006.

8. All quotations attributed to David Gilley are taken from a telephone interview with the author, January 11, 2006.

9. All quotations attributed to Carol Greider are taken from a telephone interview with the author, January 26, 2006.

10. All quotations attributed to Janis Shampay are taken from a telephone interview with the author, November 1, 2005.

11. Tet Matsuguchi, interview with the author, July 22, 2005.

12. All quotations attributed to Dorothy Shippen are taken from a telephone interview with the author, December 15, 2004.

13. All quotations attributed to J. Michael Cherry are taken from a telephone interview with the author, October 20, 2005.

14. Keay Davidson, "Harvard President under Microscope," *San Francisco Chronicle*, January 31, 2005, 14 (table).

15. Ibid.

16. Jo Handelsman, Nancy Cantor, Molly Carnes, et al., "More Women in Science," *Science* 309 (2005): 1190. Figures are provided in a table, "Women Ph.D.'s and Faculty, Top 50 Departments in Selected Disciplines."

17. Committee on Science, Engineering, and Public Policy, National Research Council, *Beyond Bias and Barriers: Fulfilling the Potential of Women in Academic Science and Engineering* (Washington, DC: National Academies Press, 2006), S–2.

18. Virginia Valian, *Why So Slow?* (Cambridge, MA: MIT Press, 1998), 233.

19. Tanya Schevitz, "Where UC Struggles, CSU Succeeds," *San Francisco Chronicle*, May 19, 2005, A1.

20. Robin Wilson, "Rigid Tenure System Hurts Young Professors, Especially Women, Officials from Top Universities Say," *Chronicle of Higher Education*, September 26, 2005, n.p.

21. Phyllis L. Carr, et al., "Relation of Family Responsibilities and Gender to the Productivity and Career Satisfaction of Medical Faculty," *Annals of Internal Medicine* 129 (1998): 532–538.

22. Wilson, "Rigid Tenure System Hurts Young Professors."

23. Schevitz, "Where UC Struggles, CSU Succeeds."

24. Charlotte Schubert and Gunjan Sinha, "A Lab of Her Own," *Nature Medicine* 10 (February 2004): 115.

25. Quoted in Robin Wilson, "Mary-Claire King: 'The Biggest Obstacle Is How to Have Enough Hours in the Day,'" *Chronicle of Higher Education*, June 10, 2005, n.p.

26. Committee on Science, Engineering, and Public Policy, *Beyond Bias and Barriers*, S-2.

27. Schubert and Sinha, "A Lab of Her Own"; the original MIT study can be found at <http://web.mit.edu/fnl/women/women.html>.

28. Jennifer Jacobson, "NIH Grants Go Much More Often to Men Than Women, a New Study for Congress Finds," *Chronicle of Higher Education*, September 14, 2005, n. p.

29. Handelsman, Cantor, Carnes, et al., "More Women in Science," 1191.

30. Belden, Russonello, and Stewart, *The Climate for Women on the Faculty at UCSF: Report of Findings from a Survey of Faculty Members*, Office of the Chancellor, University of California at San Francisco, January 2002, 3.

31. Ibid., 28.

32. Jennifer Sheridan, Jo Handelsman, and Molly Carnes, *Current Perspectives of Women in Science and Engineering at UW-Madison: WISELI Town Hall Meeting Report*, 4.

33. Valian, *Why So Slow?* 273–274.

34. Ibid., 268.

35. Belden, Russonello, and Stewart, *The Climate for Women on the Faculty at UCSF*, 2–3.

36. Shirley Tilghman, "Changing the Demographics: Recruiting, Retaining, and Advancing Women Scientists in Academia" (lecture, Earth Institute ADVANCE Program, Columbia University, March 24, 2005, New York).

37. Valian, *Why So Slow?* 274. In chapter 7, Valian discusses in detail the nature of gender schema.

38. Piper Fogg, "Princeton Gives Automatic Tenure Extension to New Parents," *Chronicle of Higher Education*, August 19, 2005, n.p.

39. Schevitz, "Where UC Struggles, CSU Succeeds."

40. Natalie Angier, "A Clue to Longevity Found at Chromosome Tip," *New York Times* Science Times, June 9, 1992, n.p.

41. Jean Marx, "Telomeres: Chromosome End Game Draws a Crowd," *Science* 29 (2002): 2348–2351.

42. Harrison Echols, *Operators and Promoters: The Story of Molecular Biology and Its Creators*, ed. Carol A. Gross (Berkeley: University of California Press, 2001), 66.

43. Quoted in Peggy Orenstein, "Why Science Must Adapt to Women: An Elite Survivor Assesses the Hidden Costs of Exclusion," *Discover* 23 (November 2002), 61.

44. Tilghman, "Changing the Demographics."

45. Committee on Women in Science and Engineering, Panel for the Study of Gender Differences in the Career Outcomes of Science and Engineering PhDs, National Research Council, *From Scarcity to Visibility: Gender Differences in the Careers of Doctoral Scientists and Engineers*, ed. J. Scott Long (Washington, DC: National Academies Press, 2001), 37.

46. McEachern, telephone interview with the author, January 24, 2006.

47. Maxine Singer, "Shaping the Future for Women in Science," in *Career Advice for Life Scientists*, ed. Elizabeth Marincola (Washington, DC: American Society for Cell Biology, 2002), 102–103.

48. Jeffrey Brainard, "Recipients of Prestigious NIH Awards Include Women This Time," *Chronicle of Higher Education*, September 30, 2005, n.p.

49. Cathy A. Trower and Richard P. Chait, "Faculty Diversity: Too Little for Too Long," *Harvard Magazine* (March–April 2002), n.p.

50. Quoted in Jeff Miller, "The Female Factor, Part One: Women and Science at Mission Bay," *UCSF Magazine* 23 (May 2003) 1:17.

Chapter 11

1. All quotations attributed to Elizabeth H. Blackburn, unless otherwise noted, are taken from interviews with the author on July 11, 28, and 29, 2005.

2. Carol Greider, telephone interview with the author, January 26, 2006.

3. Henry Kelly, Ivan Oelrich, Steven Aftergood, and Benn H. Tanennbaum, *Flying Blind: The Rise, Fall, and Possible Resurrection of Science Policy Advice in the United States*, occasional paper no. 2 (Washington, DC: Federation of American Scientists, December 2004), 20.

4. Ibid., 22.

5. Mary B. Mahowald, "The President's Council on Bioethics, 2002–2004: An Overview," *Perspectives in Biology and Medicine* 48 (Spring 2005): 161–162.

6. Scott Shepard, "Environmental Groups Vow to Use Bush Record as Club," *Atlanta Journal-Constitution*, April 18, 2001, n.p.

7. Craig Pittman, "Global Warming Report Warns: Seas Will Rise," *St. Petersburg Times*, October 24, 2001, B3.

8. Kelly, Oelrich, Aftergood, and Tanennbaum, *Flying Blind*, 21.

9. Ibid., 22.

10. Ibid., 19.

11. Kenneth Chang, "Scientists Say They Were Questioned on Politics," *New York Times*, July 9, 2004, A13.

12. Rick Weiss, "Bush Unveils Bioethics Council: Human Cloning, Tests on Cloned Embryos Will Top Agenda of Panel's First Meeting," *Washington Post*, January 17, 2002, A21.

13. William Safire, "The Crimson Birthmark," editorial, *New York Times*, January 21, 2002, A15.

14. Quoted in Stephen S. Hall, "Human Cloning: President's Bioethics Council Delivers," *Science* 297 (2002): 322–323.

15. Nigel M. De S. Cameron, "Defender of Dignity: Leon Kass, Head of the President's Council on Bioethics, Hopes to Thwart the Business-Biomedical Agenda," *Christianity Today* 46 (June 10, 2002).

16. Weiss, "Bush Unveils Bioethics Council."

17. "Charles Krauthammer: Biography." *Time*, n.p.

18. Charles Krauthammer, "Cloning and Stem-Cell Research," *Townhall.com*, July 30, 2001, n.p.

19. Weiss, "Bush Unveils Bioethics Council."

20. Elizabeth H. Blackburn, "A 'Full Range' of Bioethical Views Just Got Narrower," *Washington Post*, March 7, 2004, B2.

21. William F. May, "The President's Council on Bioethics: My Take on Some of Its Deliberations," *Perspectives in Biology and Medicine* 48 (Spring 2005): 238.

22. Paul McHugh, "Statement," in *Human Cloning and Human Dignity: An Ethical Inquiry*, President's Council on Bioethics (Washington, DC: Government Printing Office, July 2002), 286.

23. Senate Appropriations Subcommittee on Labor, Health and Human Services, and Education, "Capitol Hill Hearing Testimony," January 24, 2002, *Federal Document Clearing House, Congressional Testimony* (Washington, DC: Federal Document Clearing House, Inc., 2002).

24. Hall, "Human Cloning: President's Bioethics Council Delivers," 323.

25. Amy Goldstein, "President Presses Senate to Ban All Human Cloning," *Washington Post*, April 11, 2002, A1.

26. Hall, "Human Cloning: President's Bioethics Council Delivers," 323.

27. President's Council on Bioethics, Transcripts, April 25, 2002, Session 4: "Human Cloning 10: Ethics of Cloning for Biomedical Research."

28. Quoted in Hall, "Human Cloning: President's Bioethics Council Delivers," 323.

29. President's Council on Bioethics, *Human Cloning and Human Dignity: An Ethical Inquiry*.

30. Mahowald, "The President's Council on Bioethics, 2002–2004: An Overview," 166.

31. May, "The President's Council on Bioethics: My Take on Some of Its Deliberations," 238.

32. Ibid.

33. Hall, "Human Cloning: President's Bioethics Council Delivers," 324.

34. Ibid.

35. President's Council on Bioethics, *Human Cloning and Human Dignity: An Ethical Inquiry*.

36. Michael Gazzaniga, "Statement," in President's Council on Bioethics, *Human Cloning and Human Dignity: An Ethical Inquiry*, 255.

37. President's Council on Bioethics, Transcripts, July 11, 2002, Session 1: "Human Cloning (Final): Council's Report to the President."

38. Janet D. Rowley, Elizabeth Blackburn, Michael S. Gazzaniga, and Daniel W. Foster, "Harmful Moratorium on Stem Cell Research," *Science* 297 (2002): 1957.

39. Quoted in Rick Weiss, "Bush Panel Has Two Views on Embryonic Cloning," *Washington Post*, July 11, 2002, A5.

40. Quoted in Hall, "Human Cloning: President's Bioethics Council Delivers," 322.

41. President's Council on Bioethics, Transcripts, July 25, 2003, Session 6: "Beyond Therapy: A Progress Report."

42. Ibid.

43. Elizabeth Marincola, telephone interview with the author, January 6, 2006.

44. Later, Blackburn would publicly list these instances in an editorial coauthored with Janet Rowley, "Reason as Our Guide," *PloS Biology*, March 5, 2004, n.p.

45. President's Council on Bioethics, "I. The Meaning of 'Ageless Bodies,'" in *Beyond Therapy: Biotechnology and the Pursuit of Happiness* (Washington, DC: President's Council on Bioethics, October 2003), n.p.

46. Quoted in Jeffrey Brainard, "A New Kind of Bioethics: Eschewing the Academic Mainstream, Bush Panel Focuses on Technology's Dangers," *Chronicle of Higher Education*, May 21, 2004, n.p.

47. Marincola, telephone interview.

48. Ibid.

Chapter 12

1. Henry Kelly, Ivan Oelrich, Steven Aftergood, and Benn H. Tanennbaum, *Flying Blind: The Rise, Fall, and Possible Resurrection of Science Policy Advice in the United States,* occasional paper no. 2 (Washington, DC: Federation of American Scientists, December 2004), 23.

2. All quotations attributed to Elizabeth H. Blackburn, unless otherwise noted, are taken from interviews with the author on June 22 and July 12, 2004.

3. Genetics and Public Policy Center, *Cloning: A Policy Analysis* (Washington, DC: Genetics and Public Policy Center, 2004), 49.

4. Ibid.

5. Kelly, Oelrich, Aftergood, and Tanennbaum, *Flying Blind,* 23.

6. Michelle L. Brandt, "The Great Stem Cell Divide: The Science and Politics of Stem Cell Research," *Stanford Medicine* 21 (Fall 2004): 16.

7. Karen Kaplan, "Approved Stem Cell Lines Contaminated, New Report Says," *San Francisco Chronicle,* January 25, 2005, A7.

8. "Stem Cells Scientists Can Use Are Tainted," *Washington Post* news service, in *San Francisco Chronicle,* October 29, 2004, A5.

9. President's Council on Bioethics, Transcripts, July 24, 2003, Session 1: "The 'Research Imperative': Is Research a Moral Obligation?"

10. Ibid.

11. President's Council on Bioethics, Transcripts, June 20, 2002, Session 3: "Human Cloning 12: Public Policy Options."

12. Nicholas Wade, "Group of Scientists Drafts Rules on Ethics for Stem Cell Research," *The Ledger,* April 27, 2005, n.p.

13. President's Council on Bioethics, Transcripts, July 24, 2003, Session 3: "Stem Cell Research: Recent Scientific and Clinical Developments."

14. Ibid.

15. Amy Adams, "Teaching Old Cells New Tricks," *Stanford Medicine* 21 (Fall 2004): 32–35.

16. President's Council on Bioethics, Transcripts, July 24, 2003, Session 3.

17. President's Council on Bioethics, Transcripts, September 4, 2003, Session 4: "Stem Cells: Moving Research from the Bench toward the Bedside: The Role of Nongovernmental Activity."

18. President's Council on Bioethics, Transcripts, September 4, 2003, Session 1: "The Meaning of Federal Funding."

19. President's Council on Bioethics, Transcripts, September 4, 2003, Session 2: "Stem Cells: The Administration's Funding Policy: Legal and Moral Obligations."

20. Ibid.

21. Brandt, "The Great Stem Cell Divide," 18.

22. Sabin Russell, " 'Adult' Stem Cells Could Skirt Embryos' Ethical Dilemmas," *San Francisco Chronicle*, June 25, 2005, A1, A10.

23. President's Council on Bioethics, Transcripts, June 13, 2003, Session 5: "Biotechnology and Public Policy: Discussion Document on the U.S. Regulatory Landscape—Part I (ART, Preimplantation Genetic Diagnosis)."

24. President's Council on Bioethics, Transcripts, October 16, 2003, Session 2: "Toward a 'Richer Bioethics': Chimeras and the Boundaries of the Human." (Gazzaniga made similar remarks at the September 5, 2003, meeting, session 6.)

25. Rick Weiss, "Bioethics Panel Calls for Ban on Radical Reproductive Procedures," *Washington Post*, January 16, 2004, A2.

26. Elizabeth H. Blackburn, "Thoughts of a Former Council Member," *Perspectives in Biology and Medicine* 48 (Spring 2005): 177–178.

27. Blackburn and Rowley's critique was published online March 5, 2004, in *PloS Biology*.

28. Gareth Cook, "President's Panel Skewed Facts, Two Scientists Say," *Boston Globe*, March 6, 2004, n.p.

29. Ibid.

30. Jeffrey Brainard, "A New Kind of Bioethics: Eschewing the Academic Mainstream, Bush Panel Focuses on Technology's Dangers," *Chronicle of Higher Education*, May 21, 2004, n.p.

31. Quoted in Cook, "President's Panel Skewed Facts."

32. Quoted in Paul Elias, "Scientist Lauded after Government Fires Her," Associated Press Online, March 20, 2004.

33. Elizabeth Marincola, telephone interview with the author, January 6, 2006.

34. Joseph Gall, telephone interview with the author, September 3, 2004.

35. Rick Weiss, "Bush Ejects Two from Bioethics Council; Changes Renew Criticism That the President Puts Politics ahead of Science," *Washington Post*, February 28, 2004, A6.

36. Yvette Pearson, "Playing Politics with Bioethics: Now That's Repugnant," *Journal of Philosophy, Science, and Law* 4 (May 2004), n. p.

37. Leon R. Kass, "We Don't Play Politics with Science," *Washington Post*, March 3, 2004, A27.

38. Elizabeth H. Blackburn, "A 'Full Range' of Bioethical Views Just Got Narrower," *Washington Post*, March 7, 2004, B2.

39. Weiss, "Bush Ejects Two from Bioethics Council."

40. Arthur C. Caplan, "Bush's Bioethics Council Lacks Balance," *Skin and Allergy News* 35.6 (June 2004): 26.

41. Weiss, "Bush Ejects Two from Bioethics Council."

42. Marincola, telephone interview.

43. Elizabeth H. Blackburn, "Bioethics and the Political Distortion of Biomedical Science," *New England Journal of Medicine* 350 (2004): 1380.

44. This conversation is reported in ibid.

45. William F. May, "The President's Council on Bioethics: My Take on Some of Its Deliberations," *Perspectives in Biology and Medicine* 48 (Spring 2005): 237.

46. Quoted in Farhad Manjoo, "Thou Shalt Not Make Scientific Progress," *Salon.com*, March 25, 2004.

47. Stephanie Ebbert, "Scientists' Union Says Bush Administration Is Misusing Data," *Boston Globe,* July 9, 2004, A8.

48. Guy Gugliotta and Rick Weiss, "President's Science Policy Questioned; Scientists Worry That Any Politics Will Compromise Their Credibility," *Washington Post*, February 19, 2004, A2.

49. Andrew C. Revkin, "Bush vs. the Laureates: How Science Became a Partisan Issue," *New York Times,* October 19, 2004, F1.

50. Duncan Campbell, "White House Cuts Global Warming from Report: Environmental Study Censored, Say Critics," *Guardian* (London), June 20, 2003, foreign pages, 15.

51. Union of Concerned Scientists, "U.S. Fish and Wildlife Service Survey Summary."

52. Ricardo Alonso-Zaldivar, "Women's Health Chief Quits over Pill Flap," *San Francisco Chronicle*, September 1, 2005, A3.

53. Michael Specter, "Political Science," *New Yorker*, March 13, 2006, 63.

54. Quoted in Ebbert, "Scientists' Union Says Bush Administration Is Misusing Data."

55. Quoted in Revkin, "Bush vs. the Laureates."

56. Quoted in John McCain and Peter Likins, "Politics vs. the Integrity of Research," *Chronicle of Higher Education*, September 2, 2005, B20.

57. Union of Concerned Scientists, *Restoring Scientific Integrity in Policymaking.*

58. Revkin, "Bush vs. the Laureates"; Rick Weiss, "9/11 Response Hurting Science, ACLU Says," *Washington Post*, June 22, 2005, A3.

59. Kelly, Oelrich, Aftergood, and Tanennbaum, *Flying Blind*, 21.

60. President's Council on Bioethics, Transcripts, July 24, 2003, Session 3.

61. Kelly, Oelrich, Aftergood, and Tanennbaum, *Flying Blind*, 2.

62. Blackburn, "A 'Full Range' of Bioethical Views Just Got Narrower."

Chapter 13

1. David Glenn, "Ethics Questions Prompt South Korean Stem-Cell Pioneer to Resign, Clouding Consortium's Future," *Chronicle of Higher Education*, November 28, 2005, n.p.

2. Richard Monastersky, "Cell Divisions," *Chronicle of Higher Education*, January 6, 2006, n.p.

3. Quoted in Rick Weiss, "Stem Cell Field Rocked by Scam of Star Scientist," *Washington Post* news service, reprinted in *San Francisco Chronicle*, December 24, 2005, A5.

4. Rick Weiss, "None of Stem Cell Lines Scientists Said He Created Exists," *Washington Post* news service, reprinted in *San Francisco Chronicle*, December 30, 2005, A16.

5. Monastersky, "Cell Divisions."

6. Rick Weiss, "Stem Cell Field Rocked by Scam of Star Scientist."

7. University of Pittsburgh, *Summary Investigative Report on Allegations of Possible Scientific Misconduct on the Part of Gerald P. Schatten, Ph.D.* (Pittsburgh, PA: University of Pittsburgh Medical Center Research Integrity Panel, February 8, 2006), 7–8.

8. Rick Weiss, "Scientist Involved in Stem-Cell Cloning Denies Faking Data," *Washington Post* news service, reprinted in *San Francisco Chronicle*, December 17, 2005, A12.

9. American Society for Cell Biology, "Awards."

10. Royal Netherlands Academy of Arts and Sciences, "Dr. A. H. Heineken Prize for Medicine: Elizabeth H. Blackburn," Heineken Lectures 2004 (Amsterdam: Royal Netherlands Academy of Arts and Sciences, 2005), 13.

11. American Association for Cancer Research, "Awards."

12. Franklin Institute, "Announcing the Franklin Institute Awards" (advertisement), *Natural History* 114 (2005): 36–37.

13. Lawrence K. Altman, "Psychiatrist Is among Five Chosen for Medical Award, *New York Times*, September 17, 2006, n.p.; Lasker Foundation, "Lasker Awards."

14. Lasker Foundation, "Lasker Awards."

15. Jason M. Breslow, "Five Professors Named Winners of Lasker Awards for Medical Research," *Chronicle of Higher Education*, September 18, 2006, n. p.

16. Lasker Foundation, "This Year's Winners, Basic Medical Research" (Blackburn).

17. Lasker Foundation, "This Year's Winners, Basic Medical Research" (Szostak).

18. Lasker Foundation, "This Year's Winners, Basic Medical Research" (Greider).

19. All quotations attributed to Elizabeth H. Blackburn are taken from interviews with the author on September 4 and 16, 2005, and December 28 and 29, 2005.

20. Jeffrey P. Gardner, Sheng Xu Li, Sathanur Srinivasan, Wei Chen, Masayuki Kimura, Xiaobin Lu, Gerald S. Berenson, and Abraham Aviv, "Rise in Insulin Resistance Is Associated with Escalated Telomere Attrition," *Circulation* 111 (2005): 2171–2177.

21. Elissa S. Epel, Elizabeth H. Blackburn, Jue Lin, Firdaus Dhabarr, Nancy E. Adler, Jason D. Morrow, and Richard M. Cawthon, "Accelerated Telomere Shortening in Response to Life Stress," *Proceedings of the National Academy of Sciences* 101 (2004): 17312–17315.

22. David Perlman, "Early Aging Tied to Chronic Stress," *San Francisco Chronicle*, November 30, 2004, A1, A2.

23. Maureen Dowd, "Jingle Bell Schlock," *New York Times*, December 5, 2004, op-ed page.

24. Quoted in Rob Stein, "Study Is First to Confirm That Stress Speeds Aging," *Washington Post*, November 30, 2004, A1.

25. Elissa S. Epel, Jue Lin, F. H. Wilhelm, O. M. Wolkowitz, Richard M. Cawthon, Nancy E. Adler, C. Dolbier, W. B. Mendes, and Elizabeth H. Blackburn, "Cell Aging in Relation to Stress Arousal and Cardiovascular Disease Risk Factors," *Psychoneuroendocrinology* 31 (2006): 277–287.

26. Kavita Y. Sarin, Peggie Cheung, Daniel Gilison, Eunice Lee, Ruth I. Tennen, Estee Wang, Maja K. Artandi, Anthony E. Oro, and Steven E. Artandi, "Conditional Telomerase Induction Causes Proliferation of Hair Follicle Stem Cells," *Nature* 436 (2005): 1048–1052.

27. Lou Bergeron, "Behind Method for Activating Adult Stem Cells, a Shaggy-Haired Mouse Story," *Stanford Report*, August 24, 2005, n.p.

28. Elizabeth H. Blackburn, "Cell Biology: Shaggy Mouse Tales," *Nature* 436 (2005): 923.

29. Daniel L. Levy and Elizabeth H. Blackburn, "Counting of Rif1p and Rif2p on *Saccharomyces cerevisiae* Telomeres Regulates Telomere Length," *Molecular and Cellular Biology* 24 (2004): 10857–10867.

30. Lifeng Xu and Elizabeth H. Blackburn, "Human RIF1 Protein Binds Aberrant Telomeres and Aligns along Anaphase Midzone Microtubules," *Journal of Cell Biology* 167 (2004): 819–830.

31. John Sedat, interview with the author, March 5, 2006.

32. Lawrence Summers, "Remarks at NBER Conference on Diversifying the Science and Engineering Workforce," Office of the President, Harvard University.

33. Quoted in Michael Dobbs, "Harvard Chief's Comments on Women Assailed," *Washington Post*, January 19, 2005, A2.

34. Lawrence Summers, "Letter to the Faculty regarding NBER Remarks," Office of the President, Harvard University.

35. Keay Davidson, "Harvard President under Microscope," *San Francisco Chronicle*, January 31, 2005, A4.

36. Summers, "Remarks at NBER Conference."

37. Quoted in Davidson, "Harvard President under Microscope."

Bibliography

Adams, Amy. "The True Seeds of Cancer: Are Treatments Targeting the Wrong Cell?" *Stanford Medicine* 21 (Fall 2004): 24–26.

———. "Teaching Old Cells New Tricks." *Stanford Medicine* 21 (Fall 2004): 32–35.

Ainsworth, Diane. "Women in Science: A Scarcity of Females Suggests That Science, Engineering Is Still Difficult Terrain for Women." *Berkeleyan*, June 6, 2001, n. p. Available at <http://www.berkeley.edu/news/berkeleyan/2001/06/06_women.html> (accessed September 29, 2004).

Alonso-Zaldivar, Ricardo. "Women's Health Chief Quits over Pill Flap." *San Francisco Chronicle*, September 1, 2005, A3.

Altman, Lawrence K. "Psychiatrist Is among Five Chosen for Medical Award." *New York Times*, September 17, 2006, n.p. Available at <http: www.nytimes.com/> (accessed September 18, 2004).

American Association for Cancer Research. "Awards." Available at <http://www.aacr.org/> (accessed January 17, 2006).

American Society for Cell Biology. "About Us." Available at <http://www.ascb.org/> (accessed January 19, 2006).

———. "ASCB Profile: Carol W. Greider." Available at <http://www.ascb.org/news/vol22no5/profile.htm> (accessed January 19, 2006).

———. "Awards." Available at <http://www.ascb.org/> (accessed January 19, 2006).

———. "Past Officers." Available at <http://www.ascb.org/information/past officers.html> <accessed January 19, 2006>.

Angier, Natalie. "A Clue to Longevity Found at Chromosome Tip." *New York Times* Science Times, June 9, 1992, n. p. Available at <http://www.nytimes.com/> (accessed March 14, 2006).

Artavanis-Tsakonas, Spyros. Telephone interview with the author, September 9, 2004.

Association of American Medical Colleges. *Enhancing the Environment for Women in Academic Medicine.* Washington, DC: Association of American Medical Colleges, 1996.

Barinaga, Marcia. "Academic Employment: UCSF Researchers Leave, Charging Bias." *Science* 288 (2000): 26–27.

Belden, Russonello, and Stewart. *The Climate for Women on the Faculty at UCSF: Report of Findings from a Survey of Faculty Members.* Office of the Chancellor, University of California at San Francisco, January 2002. Available at <http://chancellor.ucsf.edu/CWF/r-final.pdf> (accessed August 9, 2005).

Bergeron, Lou. "Behind Method for Activating Adult Stem Cells, a Shaggy-Haired Mouse Story." *Stanford Report*, August 24, 2005, n.p. Available at <http://news-service.stanford.edu/news/2005/august/mice-082405.html> (accessed January 17, 2006).

Bernards, Andre, Paul A. Michels, Carsten R. Lincke, and Piet Borst. "Growth of Chromosome Ends in Multiplying Trypanosomes." *Nature* 303 (1983): 592–597.

Birmingham, Karen. "Elizabeth Blackburn." *Nature Medicine* 7 (2001): 520. Available at <http://www.nature.com/cgi-taf/Dynapage.taf?file=nm/journal/v7/n5/full/nm0501_520.html> (accessed March 14, 2006).

Blackburn, Elizabeth H. "A 'Full Range' of Bioethical Views Just Got Narrower." *Washington Post*, March 7, 2004, B2. Available at <http://www.washingtonpost.com/> (accessed March 7, 2004).

———. "Appendix: A Personal Account of the Discovery of Telomerase." In *Telomeres*, 2d ed., edited by Titia de Lange, Vicki Lundblad, and Elizabeth Blackburn, 551–553. Cold Spring Harbor, NY: Cold Spring Harbor Laboratory Press, 2006.

———. "Bioethics and the Political Distortion of Biomedical Science." *New England Journal of Medicine* 350 (2004): 1379–1380.

———. "Cell Biology: Shaggy Mouse Tales." *Nature* 436 (2005): 923.

———. "A History of Telomere Biology." In *Telomeres*, 2d ed., edited by Titia de Lange, Vicki Lundblad, and Elizabeth Blackburn, 1–20. Cold Spring Harbor, NY: Cold Spring Harbor Laboratory Press, 2006.

———. Interviews with the author, June 12, 22, and 26, 2004; July 2, 12, and 19, 2004; August 6, 2004; September 3, 10, and 17, 2004; November 12, 2004; December 15, 17, 21, 22, and 24, 2004; January 4, 5, 7, 11, 12, 14, and 17, 2005; February 15 and 18, 2005; March 8, 11, 18, 22, 25, and 29, 2005; June 8 and 21, 2005; July 11, 28, and 29, 2005; September 2, 4, and 16, 2005; December 28 and 29, 2005, San Francisco.

———. Private communication to the author, July 7, 2004.

———. "Structure and Function of Telomeres." *Nature* 350 (1991): 569–573.

———. "Telomerase." In *The RNA World II*, edited by Raymond F. Gesteland and John F. Atkins, 609–635. Cold Spring Harbor, NY: Cold Spring Harbor Laboratory Press, 1999.

———. "Telomerase RNA Structure and Function." In *RNA Structure and Function*, edited by R. W. Simons and Marian Grunberg-Manago, 669–693. Cold Spring Harbor, NY: Cold Spring Harbor Laboratory Press, 1998.

———. "Telomeres: Do the Ends Justify the Means?" *Cell* 37 (1984): 7–8.

———. "Telomere States and Cell Fates." *Nature* 408 (2000): 53–56.

———. "Thoughts of a Former Council Member." *Perspectives in Biology and Medicine* 48 (Spring 2005): 172–180.

Blackburn, Elizabeth H., Marsha L. Budarf, Peter B. Challoner, J. Michael Cherry, Elizabeth A. Howard, A. L. Katzen, Wei-Chun Pan, and T. Ryan. "DNA Termini in Ciliate Macronuclei." In *Cold Spring Harbor Symposia on Quantitative Biology*, vol. 47 (part 2), 1195–1207. Cold Spring Harbor, NY: Cold Spring Harbor Laboratory Press, 1983.

Blackburn, Elizabeth H., and San-San Chiou. "Non-nucleosomal Packaging of a Tandemly Repeated DNA Sequence at Termini of Extrachromosomal DNA Coding for rRNA in *Tetrahymena*." *Proceedings of the National Academy of Sciences* 78 (1981): 2263–2267.

Blackburn, Elizabeth H., and Joseph Gall. "A Tandemly Repeated Sequence at the Termini of the Extrachromosomal Ribosomal RNA Genes in *Tetrahymena*." *Journal of Molecular Biology* 120 (1978): 33–53.

Blackburn, Elizabeth H., Carol W. Greider, and Jack W. Szostak. "Telomeres and Telomerase: The Path from Maize, *Tetrahymena*, and Yeast to Human Cancer and Aging." *Nature Medicine* 12, no. 10 (2006): viii–xii.

Blackburn, Elizabeth H., and Janet Rowley. "Reason as Our Guide." *PloS Biology*, March 5, 2004, n. p. Available at <http://www.plosbiology.org/> (accessed June 24, 2005).

Blasco, Maria A., Han-Woong Lee, M. Prakash Hande, Enrique Samper, Peter M. Landsdorp, Ronald A. DePinho, and Carol W. Greider. "Telomere Shortening and Tumor Formation by Mouse Cells Lacking Telomerase RNA." *Cell* 91 (1997): 25–34.

Brainard, Jeffrey. "A New Kind of Bioethics: Eschewing the Academic Mainstream, Bush Panel Focuses on Technology's Dangers." *Chronicle of Higher Education*, May 21, 2004, n. p. Available at <http://chronicle.com/weekly/v50/i37/37a02201.htm> (accessed September 13, 2005).

———. "NIH Program Seeks to Speed Grant Process for New Applicants." *Chronicle of Higher Education*, December 5, 2005, n.p. Available at <http://chronicle.com/daily/2005/12/20051202504n.htm> (accessed December 6, 2005).

———. "NIH Would Receive No Increase for 2007." *Chronicle of Higher Education*, February 7, 2006, n.p. Available at <http://chronicle.com/daily/2006/02/2006020704n.htm> (accessed February 7, 2006).

———. "Recipients of Prestigious NIH Awards Include Women This Time." *Chronicle of Higher Education*, September 30, 2005, n.p. Available at <http://chronicle.com/daily/2005/09/2005093003n.htm> (accessed September 30, 2005).

Brandt, Michelle L. "The Great Stem Cell Divide: The Science and Politics of Stem Cell Research." *Stanford Medicine* 21 (Fall 2004): 13–20.

———. "Stem Cell Primer." *Stanford Medicine* 21 (Fall 2004): 15.

Breslow, Jason M. "Five Professors Named Winners of Lasker Awards for Medical Research." *Chronicle of Higher Education*, September 18, 2006, n.p. Available at <http://chronicle.com/daily/2006/09/2006091806n.htm> (accessed September 18, 2006).

Cameron, Nigel M. De S. "Defender of Dignity: Leon Kass, Head of the President's Council on Bioethics, Hopes to Thwart the Business-Biomedical Agenda." *Christianity Today* 46 (June 10, 2002), n.p. Available at <http://www.infotrac.galegroup.com/> (accessed June 24, 2005).

Campbell, Duncan. "White House Cuts Global Warming from Report: Environmental Study Censored, Say Critics." *Guardian* (London), June 20, 2003, foreign pages, 15. Available at <http://www.lexis-nexis.com> (accessed June 24, 2005).

Caplan, Arthur C. "Bush's Bioethics Council Lacks Balance." *Skin and Allergy News* 35.6 (June 2004): 26. Available at <http://www.infotrac.galegroup.com/> (accessed October 22, 2006).

Carr, Phyllis L., et al. "Relation of Family Responsibilities and Gender to the Productivity and Career Satisfaction of Medical Faculty." *Annals of Internal Medicine* 129 (1998): 532–538.

Cawthon, Richard M., Ken R. Smith, Elizabeth O'Brien, Anna Sivatchenko, and Richard A. Kerber. "Association between Telomere Length in Blood and Mortality in People Aged 60 Years or Older." *Lancet* 361 (2003): 1224.

Cech, Thomas R. "Beginning to Understand the End of the Chromosome." *Cell* 116 (2004): 273–279.

Cech, Thomas R., and Enriqueta Bond. "Managing Your Own Lab." *Science* 304 (2004): 1717.

Chan, Simon, and Elizabeth H. Blackburn. "Telomerase and ATM/Tel1p Protect Telomeres from Nonhomologous End Joining." *Molecular Biology of the Cell* 11 (2003): 1379–1387.

Chang, Kenneth. "Scientists Say They Were Questioned on Politics." *New York Times*, July 9, 2004, A13. Available at <http://www.nytimes.com/> (accessed June 24, 2005).

"Charles Krauthammer: Biography." *Time*, n. p. Available at <http://www.time.com/time/columnist/krauthammer/article/0,9565,559573,00.html> (accessed September 3, 2004).

Chen, Jiunn-Liang, and Carol W. Greider. "Telomerase RNA Structure and Function: Implications for Dyskeratosis Congenita." *Trends in Biochemial Sciences* 29 (April 2004): 183–192.

Cherry, J. Michael. Telephone interview with the author, October 20, 2005.

Chin, Lynda, Steven E. Artandi, Qiong Shen, Alice Tam, Shwu-Luan Lee, Geoffrey J. Gottlieb, Carol W. Greider, and Ronald A. DePinho. "P53 Deficiency Rescues the Adverse Effects of Telomere Loss and Cooperates with Telomere Dysfunction to Accelerate Carcinogenesis." *Cell* 97 (1999): 527–538.

Committee on Science, Engineering, and Public Policy, National Research Council. *Beyond Bias and Barriers: Fulfilling the Potential of Women in Academic Science and Engineering*. Washington, DC: National Academies Press, 2006. Available at <http://www.nap.edu> (accessed September 29, 2006).

Committee on Women in Science and Engineering, Panel for the Study of Gender Differences in the Career Outcomes of Science and Engineering PhDs, National Research Council. *From Scarcity to Visibility: Gender Differences in the Careers of Doctoral Scientists and Engineers*. Edited by J. Scott Long. Washington, DC: National Academies Press, 2001. Available at <http://www.nap.edu> (accessed October 14, 2005).

Cook, Gareth. "President's Panel Skewed Facts, Two Scientists Say." *Boston Globe*, March 6, 2004, n.p. Available at <http://www.boston.com/news/globe/> (accessed June 24, 2005).

Cooke, Howard J., and B. A. Smith. "Variability at the Telomeres of the Human X/Y Pseudoautosomal Region." In *Cold Spring Harbor Symposium on Quantitative Biology*, vol. 51 (part 1), 213–219. Cold Spring Harbor, NY: Cold Spring Harbor Laboratory Press, 1986.

Counter, Christopher M., Ariel A. Avilion, Catherine E. LeFeuvre, Nancy G. Stewart, Carol W. Greider, Calvin B. Harley, and Sylvia Bacchetti. "Telomere Shortening Associated with Chromosome Instability Is Arrested in Immortal Cells Which Express Telomerase Activity." *EMBO Journal* 11 (1992): 1921–1929.

Daniell, Ellen. "Facing Disaster: Ellen's Story." In *Every Other Thursday: Stories and Strategies from Successful Women Scientists*. New Haven, CT: Yale University Press, 2006.

Davidson, Barrie, Elizabeth H. Blackburn, and Theo Dopheide. "Chorismate Mutase-Prephenate Dehydratase from *Escherichia coli* K–12." *Journal of Biological Chemistry* 247, no. 14 (July 25, 1972): 4441–4446.

Davidson, Keay. "Harvard President under Microscope." *San Francisco Chronicle*, January 31, 2005, A4.

de Chadarevian, Soraya. *Designs for Life: Molecular Biology after World War II*. Cambridge: Cambridge University Press, 2002.

de Lange, Titia. Telephone interview with the author, October 21, 2005.

de Lange, Titia, Vicki Lundblad, and Elizabeth Blackburn, eds. *Telomeres,* 2d ed. Cold Spring Harbor, NY: Cold Spring Harbor Laboratory Press, 2006.

Dobbs, Michael. "Harvard Chief's Comments on Women Assailed." *Washington Post,* January 19, 2005, A2. Available at <http://www.washingtonpost.com/> (accessed January 19, 2005).

Dokal, Inderjeet. "Dyskeratosis Congenita in All Its Forms." *British Journal of Haematology* 110 (2000): 768–779.

Dowd, Maureen. "Jingle Bell Schlock." *New York Times,* December 5, 2004, op-ed page. Available at <http://www.nytimes.com/> (accessed December 17, 2004).

Dunn, Kyla, "A Look at . . . Patents and Biotech," *Washington Post,* October 1, 2000, B3. Available at <http://www.washingtonpost.com> (accessed March 19, 2007).

Ebbert, Stephanie. "Scientists' Union Says Bush Administration Is Misusing Data." *Boston Globe,* July 9, 2004, A8. Available at <http://boston.com/news/globe/> (accessed June 24, 2005).

Echols, Harrison. *Operators and Promoters: The Story of Molecular Biology and Its Creators.* Edited by Carol A. Gross. Berkeley: University of California Press, 2001.

Elias, Paul. "Scientist Lauded after Government Fires Her." Associated Press Online, March 20, 2004. Available at <http://www.lexis-nexis.com/> (accessed July 20, 2005).

Epel, Elissa S., Elizabeth H. Blackburn, Jue Lin, Firdaus Dhabarr, Nancy E. Adler, Jason D. Morrow, and Richard M. Cawthon. "Accelerated Telomere Shortening in Response to Life Stress." *Proceedings of the National Academy of Sciences* 101 (2004): 17312–17315.

Epel, Elissa S., Jue Lin, F. H. Wilhelm, O. M. Wolkowitz, Richard M. Cawthon, Nancy E. Adler, C. Dolbier, W. B. Mendes, and Elizabeth H. Blackburn. "Cell Aging in Relation to Stress Arousal and Cardiovascular Disease Risk Factors." *Psychoneuroendocrinology* 31 (2006): 277–287.

First and Second Committees on Women Faculty in the School of Science. "A Study on the Status of Women Faculty in Science at MIT." Available at <http://web.mit.edu/fnl/women/women.html> (accessed July 10, 2007).

Fogg, Piper. "Princeton Gives Automatic Tenure Extension to New Parents." *Chronicle of Higher Education,* August 19, 2005, n.p. Available at <http://chronicle.com/daily/2005/09/2005081901n.htm> (accessed August 22, 2005).

Franklin Institute. "Announcing the Franklin Institute Awards (advertisement)." *Natural History* 114 (2005): 36–37.

Frenck, Robert W., Jr., Elizabeth H. Blackburn, and Kevin M. Shannon. "The Rate of Telomere Sequence Loss in Human Leukocytes Varies with Age." *Proceedings of the National Academy of Sciences* 95 (1998): 5607–5610.

Gall, Joseph. *A Pictorial History: Views of a Cell.* Bethesda, MD: American Society for Cell Biology, 1996.

————. Telephone interview with the author, September 3, 2004.

Gardner, Jeffrey P., Sheng Xu Li, Sathanur Srinivasan, Wei Chen, Masayuki Kimura, Xiaobin Lu, Gerald S. Berenson, and Abraham Aviv. "Rise in Insulin Resistance Is Associated with Escalated Telomere Attrition." *Circulation* 111 (2005): 2171–2177.

Gazzaniga, Michael. "Statement." In *Human Cloning and Human Dignity: An Ethical Inquiry*, President's Council on Bioethics. Washington, DC: Government Printing Office, July 2002, Available at <http://www.bioethics. gov/reports/humancloningandhumandignity/index.html> (accessed August 3, 2005).

Gehring, Gillian. "Mixing Motherhood and Science." *Physics World*, March 2002. Available at <http://physicsweb.org/article/world/15/3/3> (accessed July 22, 2004).

Genetics and Public Policy Center. *Cloning: A Policy Analysis.* Washington, DC: Genetics and Public Policy Center, 2004.

Geron Corporation, "'Immortalizing Enzyme' Detected in Many Types of Cancer, Reported in *Science*," December 23, 1994 (press release), n. p. Available at <http://www.geron.com/pressiview.asp?id=585> (accessed March 17, 2007).

Gilley, David. Telephone interview with the author, January 11, 2006.

Gilley, David, M. S. Lee, and Elizabeth H. Blackburn. "Altering Specific Telomerase RNA Template Residues Affects Active Site Function." *Genes and Development* 9 (1995): 2214–2226.

Glenn, David. "Ethics Questions Prompt South Korean Stem-Cell Pioneer to Resign, Clouding Consortium's Future." *Chronicle of Higher Education*, November 28, 2005, n.p. Available at <http://chronicle.com/daily/2005/11/ 2005112801n.htm> (accessed November 28, 2005).

Goldman, Fred, Rachida Bouarich, Shashikant Kulkarni, Sara Freeman, Hong-Yan Du, Lea Harrington, Philip J. Mason, Arturo London-Vallejo, and Monica Bessler. "The Effect of TERC Haploinsufficiency on the Inheritance of Telomere Length." *Proceedings of the National Academy of Sciences* 102 (2005): 17119–17124.

Goldstein, Amy. "President Presses Senate to Ban All Human Cloning." *Washington Post*, April 11, 2002, A1. Available at <http://www.washingtonpost.com> (accessed July 20, 2005).

Gottschling, Dan. E., and Thomas R. Cech. "Chromatin Structure of the Molecular Ends of *Oxytricha* Macronuclear DNA: Phased Nucleosomes and a Telomeric Complex." *Cell* 38 (1984): 501–510.

Gottschling, Dan. E., and Virginia A. Zakian. "Telomere Proteins: Specific Recognition and Protection of the Natural Termini of *Oxytricha* Macronuclear DNA." *Cell* 47 (1986): 195–205.

Greider, Carol W. Telephone interview with the author, January 26, 2006.

Greider, Carol W., and Elizabeth H. Blackburn. "Identification of a Specific Telomere Terminal Transferase Activity in *Tetrahymena* Extracts." *Cell* 43 (1985): 405–413.

———. "Telomeres, Telomerase, and Cancer." *Scientific American* 274 (February 1996): 92–97.

———. "The Telomere Terminal Transferase of *Tetrahymena* Is a Ribonucleoprotein Enzyme with Two Kinds of Primer Specificity." *Cell* 51 (1987): 887–898.

———. "A Telomeric Sequence in the RNA of *Tetrahymena* Telomerase Required for Telomere Repeat Synthesis." *Nature* 337 (1989): 331–337.

———. "Tracking Telomerase." *Cell* S116 (2004): S83–S86.

Griffith, Jack D., Laurey Corneau, Soraya Rosenfield, Rachel M. Stansel, Alessandra Bianchi, Heidi Moss, and Titia de Lange. "Mammalian Telomeres End in Large Duplex Loop." *Cell* 97 (1999): 503–514.

Gross, Carol. Interview with the author, September 26, 2005, San Francisco.

Gugliotta, Guy, and Rick Weiss. "President's Science Policy Questioned; Scientists Worry That Any Politics Will Compromise Their Credibility." *Washington Post*, February 19, 2004, A2. Available at <http://www.washingtonpost.com/> (accessed June 24, 2005).

Haglund, Keith, and Jennifer Steinberg. "The Landmarks Interviews: Means That Justified the Ends," *Journal of NIH Research* 8 (September 1996): 58.

Hahn, William C., Christopher M. Counter, Ante S. Lundberg, Roderick L. Beijersbergen, Mary W. Brooks, and Robert A. Weinberg. "Creation of Human Tumour Cells with Defined Genetic Elements." *Nature* 400 (1999): 464–468.

Hahn, William C., Sheila A. Stewart, Mary W. Brooks, Shoshana G. York, Elinor Eaton, Akiko Kurachi, Roderick L. Beijersbergen, Joan H. M. Knoll, Matthew Meyerson, and Robert A. Weinberg. "Inhibition of Telomerase Limits the Growth of Human Cancer Cells." *Nature Medicine* 5 (1999): 1129–1130.

Hall, Stephen S. "Human Cloning: President's Bioethics Council Delivers." *Science* 297 (2002): 322–324.

———. *Merchants of Immortality: Chasing the Dream of Human Life Extension.* New York: Houghton Mifflin Co., 2003.

Handelsman, Jo, Nancy Cantor, Molly Carnes, et al. "More Women in Science." *Science* 309 (2005): 1190.

Hansen, Stephen, Amanda Brewster, Jana Asher, and Michael Kisielewski. *The Effects of Patenting in the AAAS Community*, 2d ed. Washington, DC: American Association for the Advancement of Science, 2006. Available at <http://sippi.aas.org/Pubs> (accessed November 1, 2006).

Harley, Calvin B., A. Bruce Futcher, and Carol W. Greider. "Telomeres Shorten during Aging of Human Fibroblasts." *Nature* 345 (1990): 458–460.

Heineken International. "Elizabeth H. Blackburn." *Laureates*. Available at <http://www.heinekeninternational.com/prizes/laureates/2004> (accessed January 26, 2005).

Hird, Frank. Letter to Mrs. M. Blackburn, March 10, 1983. Private papers of Elizabeth H. Blackburn.

Jacobson, Jennifer. "NIH Grants Go Much More Often to Men Than Women, a New Study for Congress Finds." *Chronicle of Higher Education*, September 14, 2005, n.p. Available at <http:// chronicle.com/daily/2005/09/2005091404n .htm> (accessed September 14, 2005).

Joint Steering Committee for Public Policy. "About JSC." Available at <http://www.jscpp.org/about.cfm> (accessed January 9, 2006).

Kaplan, Karen. "Approved Stem Cell Lines Contaminated, New Report Says." *San Francisco Chronicle*, January 25, 2005. Reprinted from the *Los Angeles Times*.

Kass, Leon R. "We Don't Play Politics with Science." *Washington Post*, March 3, 2004, A27. Available at <http://www.washingtonpost.com/> (accessed July 20, 2005).

Kean, Sam. "Scientists Spend Nearly Half Their Time on Administrative Tasks, Survey Finds." *Chronicle of Higher Education*, July 7, 2006, n.p. Available at <http://chronicle.com/daily/2006/07/2006070702n.htm> (accessed February 7, 2006).

Kelly, Henry, Ivan Oelrich, Steven Aftergood, and Benn H. Tanennbaum. *Flying Blind: The Rise, Fall, and Possible Resurrection of Science Policy Advice in the United States,* occasional paper no. 2. Washington, DC: Federation of American Scientists, December 2004. Available at <http://www.fas.org/> (accessed December 7, 2004).

Kim, Moses M., Melissa A. Rivera, Inna L. Botchkina, Refaat Shalaby, Ann D. Thor, and Elizabeth H. Blackburn. "A Low Threshold Level of Expression of Mutant-Template Telomerase RNA Inhibits Human Tumor Cell Proliferation." *Proceedings of the National Academy of Sciences* 98 (2001): 7982–7987.

Krauskopf, Anat, and Elizabeth H. Blackburn. "Rap1 Protein Regulates Telomere Turnover in Yeast." *Proceedings of the National Academy of Sciences* 95 (1998): 12486–12491.

Krauthammer, Charles. "Cloning and Stem-Cell Research." Townhall.com, July 30, 2001, n.p. Available at <http://www.Townhall.com/columnists/charleskrauthammer/ck20010730.shtml> (accessed July 22, 2004).

Kruger, Kelly, Paula J. Grabowski, Arthur J. Zaug, Julie Sands, Daniel E. Gottschling, and Thomas R. Cech. "Self-splicing RNA: Autoexcision and Autocyclization of the Ribosomal RNA Intervening Sequence of *Tetrahymena*." *Cell* 31 (1982): 147–157.

Lasker Foundation. "Lasker Awards." Available at <http://www .laskerfoundation.org/awards/awards.html> (accessed October 22, 2006).

———. "This Year's Winners, Basic Medical Research" (Blackburn). Available at <http://www.laskerfoundation.org/awards/library/2006_basic_blackburn.shtml> (accessed October 22, 2006).

———. "This Year's Winners, Basic Medical Research" (Greider). Available at <http://www.laskerfoundation.org/awards/library/2006_basic_greider.shtml> (accessed October 22, 2006).

———. "This Year's Winners, Basic Medical Research" (Szostak). Available at <http://www.laskerfoundation.org/awards/library/2006_basic_szostak.shtml> (accessed October 22, 2006).

Lavett, Diane K. Telephone interview with the author, January 30, 2005.

Lee, Han-Woong, Maria A. Blasco, Geoffrey J. Gottlieb, James W. Horner II, Carol W. Greider, and Ronald A. DePinho. "Essential Role of Mouse Telomerase in Highly Proliferative Organs." *Nature* 392 (1998): 569–574.

Levy, Daniel L., and Elizabeth H. Blackburn. "Counting of Rif1p and Rif2p on *Saccharomyces cerevisiae* Telomeres Regulates Telomere Length." *Molecular and Cellular Biology* 24 (2004): 10857–10867.

Li, Shang, Julia Crothers, Christopher M. Haqq, and Elizabeth H. Blackburn. "Cellular and Gene Expression Responses Involved in the Rapid Growth Inhibition of Human Cancer Cells by RNA Interference-Mediated Depletion of Telomerase." *Journal of Biological Chemistry* 280 (2005): 23709–23717.

Li, Shang, Jonathan E. Rosenberg, Annemarie A. Donjacour, Inna L. Botchkina, Yun Kit Hom, Gerald R. Cunha, and Elizabeth H. Blackburn. "Rapid Inhibition of Cancer Cell Growth Induced by Lentiviral Expression of Mutant-Template Telomerase RNA Constructs and Anti-Telomerase Short-interfering RNA." *Cancer Research* 64 (2004): 4833–4840.

Lin, Jue, Hinh Ly, Arif Hussain, Mira Abraham, Siran Pearl, Tristram G. Parslow, Yehuda Tzfati, and Elizabeth H. Blackburn. "A Universal Telomerase RNA Core Structure Includes Structure Motifs Required for Binding the Telomerase Reverse Transcriptase Protein." *Proceedings of the National Academy of Sciences* 101 (2004): 14713–14718.

Lingner, Joachim, Timothy R. Hughes, Andrej Shevchenko, Matthias Mann, Victoria Lundblad, and Thomas R. Cech. "Reverse Transcriptase Motifs in the Catalytic Subunit of Telomerase." *Science* 276 (1997): 561–567.

Lundblad, Vicki. Telephone interviews with the author, October 11 and 18, 2005.

Lundblad, Vicki, and Elizabeth H. Blackburn. "An Alternate Pathway for Yeast Telomere Maintenance Rescues est1-Senescence." *Cell* 73 (1993): 347–360.

Lundblad, Vicki, and Jack W. Szostak. "A Mutant with a Defect in Telomere Elongation Leads to Senescence in Yeast." *Cell* 57 (1989): 633–643.

Ly, Hinh, Lifeng Xu, Melissa Rivera, Tristram G. Parslow, and Elizabeth H. Blackburn. "A Role for a Novel 'Trans-Pseudoknot' RNA-RNA Interaction in the Functional Dimerization of Human Telomerase." *Genes and Development* 17 (2003): 1078–1083.

Mahowald, Mary B. "The President's Council on Bioethics, 2002–2004: An Overview." *Perspectives in Biology and Medicine* 48 (Spring 2005): 159–171.

Manjoo, Farhad. "Thou Shalt Not Make Scientific Progress." *Salon.com*, March 25, 2004. Available at <http://www.lexis-nexis.com/> (accessed July 20, 2005).

Marcand, Stephane, Eric Gilson, and David Shore. "A Protein-counting Mechanism for Telomere Length Regulation in Yeast." *Science* 275 (1997): 986–990.

Marincola, Elizabeth. Telephone interview with the author, January 6, 2006.

Marsden, Katherine. Interview with the author, October 3, 2004, San Francisco.

Marx, Jean. "Telomeres: Chromosome End Game Draws a Crowd." *Science* 29 (2002): 2348–2351.

Matsuguchi, Tet. Interview with the author, July 22, 2005, San Francisco.

Mattern, Karin A., Susan J. J. Swiggers, Alex L. Nigg, Bob Lowenberg, Adriaan B. Houtsmuller, and J. Mark J. M. Zijlmans. "Dynamics of Protein Binding to Telomeres in Living Cells: Implications for Telomere Structure and Function." *Molecular and Cellular Biology* 24 (2004): 5587–5594.

May, William F. "The President's Council on Bioethics: My Take on Some of Its Deliberations." *Perspectives in Biology and Medicine* 48 (Spring 2005): 229–239.

McCain, John, and Peter Likins. "Politics vs. the Integrity of Research." *Chronicle of Higher Education*, September 2, 2005, B20.

McClintock, Barbara. "Cytological Observations of Deficiencies Involving Known Genes, Translocations and an Inversion in *Zea mays*." *Missouri Agricultural Experiment Station Research Bulletin* 163 (1931): 1–30.

———. "The Fusion of Broken Ends of Sister Half-Chromatids Following Chromatid Breakage at Meiotic Anaphases," *Missouri Agricultural Experiment Station Research Bulletin* 290 (1938): 1–48.

———. Letter to Elizabeth H. Blackburn, March 11, 1983. Private papers of Elizabeth H. Blackburn. Also available online in the Barbara McClintock Papers, "Profiles in Science," National Library of Medicine, <http://profiles.nlm.nih.gov/LL/B/B/D/W> (accessed March 18, 2006).

McEachern, Michael J. Telephone interview with the author, January 24, 2006.

McEachern, Michael J., and Elizabeth H. Blackburn. "Runaway Telomere Elongation Caused by Telomerase RNA Gene Mutations." *Nature* 376 (1995): 403–409.

McHugh, Paul. "Statement." In *Human Cloning and Human Dignity: An Ethical Inquiry*, President's Council on Bioethics. Washington, DC: Government Printing Office, July 2002. Available at <http://bioethics.gov/reports/humancloningandhumandignity/index.html> (accessed August 3, 2005).

Medical Research Council. "Ratio of Women to Men, 1970s and 1980s" (unnumbered table) in *Medical Research Council Handbook*. London: Medical Research Council, 1990.

Miller, Jeff. "The Female Factor, Part One: Women and Science at Mission Bay." *UCSF Magazine* 23 (May 2003) 1:17.

Mitchell, James R., Emily Wood, and Kathleen Collins. "A Telomerase Component Is Defective in the Human Disease Dyskeratosis Congenita." *Nature* 402 (1999): 551–555.

Monastersky, Richard. "Cell Divisions." *Chronicle of Higher Education*, January 6, 2006, n.p. Available at <http://chronicle.com/weekly/v52/i18/18a02601.htm> (accessed January 4, 2006).

———. "The Number That's Devouring Science." *Chronicle of Higher Education*, October 14, 2005, n.p. Available at <http://chronicle.com/free/v52/i08/08a01201.htm> (accessed October 16, 2005).

Nobel Foundation. "Nobel Chemistry Laureates 1980: Fred Sanger, Autobiography." *Nobel e Museum*. Available at <http://www.nobel.se/chemistry/laureates/1980/sanger-autobio.html> (accessed September 10, 2004).

Nosrati, Mehdi, Shang Li, Sepideh Bagheri, David Ginzinger, Elizabeth H. Blackburn, Robert J. Debs, and Mohammed Kashani-Sabet. "Antitumor Activity of Systemically Delivered Ribozymes Targeting Murine Telomerase RNA." *Clinical Cancer Research* 10 (2004): 4983–4990.

Office of the President, "University of California, Universitywide New Appointments of Ladder Rank Faculty: 1984–85 through 2002–03," University of California at Berkeley, Academic Advancement (unnumbered table), n.p. Available at <http://www.ucop.edu/acadadv/datamgmt/welcome.html> (accessed September 10, 2004).

Orenstein, Peggy. "Why Science Must Adapt to Women: An Elite Survivor Assesses the Hidden Costs of Exclusion." *Discover* 23 (November 2002): 58–86.

Pearson, Yvette. "Playing Politics with Bioethics: Now That's Repugnant." *Journal of Philosophy, Science, and Law* 4 (May 2004), n.p. Available at <http://www.psljournal.com/archives/all/playingPolitics.cfm> (accessed November 16, 2004).

Perlman, David. "Early Aging Tied to Chronic Stress." *San Francisco Chronicle*, November 30, 2004, A1, A2.

Pierson, George W., ed. *A Yale Book of Numbers: Historical Statistics of the College and University, 1701–1976.* New Haven, CT: Yale University Office of Institutional Research, 1977. Available at <http://www.yale.edu/oir/pierson_original.htm> (accessed September 10, 2004).

Pittman, Craig. "Global Warming Report Warns: Seas Will Rise." *St. Petersburg Times*, October 24, 2001, B3. Available at <http://lexis-nexis.com/> (accessed June 24, 2005).

Prescott, John C. Telephone interview with the author, January 10, 2006.

Prescott, John C., and Elizabeth H. Blackburn. "Telomerase: Dr. Jekyll or Mr. Hyde?" *Current Opinion in Genetics and Development* 9 (1999): 372.

———. "Telomerase RNA Mutations in *Saccharomyces cerevisiae* Alter Telomerase Action and Reveal Nonprocessivity *in vivo* and *in vitro*." *Genes and Development* 11 (1997): 528–540.

President's Council on Bioethics. *Beyond Therapy: Biotechnology and the Pursuit of Happiness.* Washington, DC: Government Printing Office, October 2003. Available at <http://www.bioethics.gov/reports/beyondtherapy/index.html> (accessed August 18, 2005).

———. *Human Cloning and Human Dignity: An Ethical Inquiry.* Washington, DC: Government Printing Office, July 2002. Available at <http://www.bioethics.gov/reports/humancloningandhumandignity/index.html> (accessed August 3, 2005).

———. Transcripts, April 25, 2002, Session 4: "Human Cloning 10: Ethics of Cloning for Biomedical Research." Available at <http://www.bioethics.gov/transcripts/apri02/apri25session4.html> (accessed July 29, 2005).

———. Transcripts, June 13, 2003, Session 5: "Biotechnology and Public Policy: Discussion Document on the U.S. Regulatory Landscape—Part I (ART, Preimplantation Genetic Diagnosis)." Available at <http://www.bioethics.gov/transcripts/jun03/session5/html> (accessed August 9, 2005).

———. Transcripts, June 20, 2002, Session 3: "Human Cloning 12: Public Policy Options." Available at <http://www.bioethics.gov/transcripts/jun02/session3.html> (accessed July 29, 2005).

———. Transcripts, July 11, 2002, Session 1: "Human Cloning (Final): Council's Report to the President." Available at <http://www.bioethics.gov/transcripts/jul02/session 1.html> (accessed July 29, 2005).

———. Transcripts, July 24, 2003, Session 1: "The 'Research Imperative': Is Research a Moral Obligation?" Available at <http://www.bioethics.gov/transcripts/july03/session1.html> (accessed July 21, 2005).

———. Transcripts, July 24, 2003, Session 3: "Stem Cell Research: Recent Scientific and Clinical Developments." Available at <http://www.bioethics.gov/transcripts/july03/session3.html> (accessed August 3, 2005).

———. Transcripts, July 25, 2003, Session 6: "Beyond Therapy: A Progress Report." Available at <http://www.bioethics.gov/transcripts/july03/session6.html> (accessed August 3, 2005).

———. Transcripts, September 4, 2003, Session 1: "The Meaning of Federal Funding." Available at <http://www.bioethics.gov/transcripts/sep03/session1.html> (accessed August 3, 2005).

———. Transcripts, September 4, 2003, Session 2: "Stem Cells: The Administration's Funding Policy: Legal and Moral Obligations." Available at <http://www.bioethics.gov/transcripts/sep03/session2.html> (accessed August 3, 2005).

———. Transcripts, September 4, 2003, Session 4: "Stem Cells: Moving Research from the Bench toward the Bedside: The Role of Nongovernmental Activity." Available at <http://www.bioethics.gov/transcripts/sep03/session4.html> (accessed August 3, 2005).

————. Transcripts, October 16, 2003, Session 2: "Toward a 'Richer Bioethics': Chimeras and the Boundaries of the Human." Available at <http://www.bioethics.gov/transcripts/oct03/session6.html> (accessed December 15, 2005).

Public Library of Science. "Board of Directors." Available at <http://www.plos.org/>.

Revkin, Andrew C. "Bush vs. the Laureates: How Science Became a Partisan Issue." *New York Times*, October 19, 2004, F1. Available at <http://www.nytimes.com/> (accessed June 24, 2005).

Romero, Daniel P., and Elizabeth H. Blackburn. "A Conserved Secondary Structure for Telomerase RNA." *Cell* 67 (1991): 343–353.

Rowley, Janet D., Elizabeth Blackburn, Michael S. Gazzaniga, and Daniel W. Foster. "Harmful Moratorium on Stem Cell Research." *Science* 297 (2002): 1957.

Royal Netherlands Academy of Arts and Sciences. "Dr. A. H. Heineken Prize for Medicine: Elizabeth H. Blackburn," Heineken Lectures 2004. Amsterdam: Royal Netherlands Academy of Arts and Sciences, 2005.

Royal Society of London. "List of New Fellows and Foreign Members for 2005." Available at <http://www.royalsoc.ac.uk/page.asp?id=2217> (accessed December 21, 2005).

Russell, Sabin. " 'Adult' Stem Cells Could Skirt Embryos' Ethical Dilemmas." *San Francisco Chronicle*, June 25, 2005, A1, A10.

Safire, William. "The Crimson Birthmark." Editorial, *New York Times*, January 21, 2002, A15. Available at <http://www.nytimes.com/> (accessed June 24, 2005).

Sarin, Kavita Y., Peggie Cheung, Daniel Gilison, Eunice Lee, Ruth I. Tennen, Estee Wang, Maja K. Artandi, Anthony E. Oro, and Steven E. Artandi. "Conditional Telomerase Induction Causes Proliferation of Hair Follicle Stem Cells." *Nature* 436 (2005): 1048–1052.

Schevitz, Tanya. "Where UC Struggles, CSU Succeeds." *San Francisco Chronicle*, May 19, 2005, A1, A10.

Schubert, Charlotte, and Gunjan Sinha. "A Lab of Her Own." *Nature Medicine* 10 (February 2004): 115.

Sedat, John. Interview with the author, March 5, 2006, San Francisco.

Seto, Anita G., April J. Livengood, Yehuda Tzfati, Elizabeth H. Blackburn, and Thomas R. Cech. "A Bulged Stem Tethers Est1p to Telomerase RNA in Budding Yeast." *Genes and Development* 16 (2002): 2800–2812.

Shampay, Janis. Telephone interview with the author, November 1, 2005.

Shampay, Janis, and Elizabeth H. Blackburn. "Generation of Telomere-Length Heterogeneity in *Saccharomyces cerevisiae*." *Proceedings of the National Academy of Sciences* 85 (1988): 534–538.

Shampay, Janis, Jack W. Szostak, and Elizabeth H. Blackburn. "DNA Sequences of Telomeres Maintained in Yeast." *Nature* 310 (1984): 154–157.

Shay, Jerry W., and Woodring E. Wright. "Telomerase and Human Cancer." In *Telomeres*, 2d ed., edited by Titia de Lange, Vicki Lundblad, and Elizabeth Blackburn, 88–89. Cold Spring Harbor, NY: Cold Spring Harbor Laboratory Press, 2006.

Shepard, Scott. "Environmental Groups Vow to Use Bush Record as Club." *Atlanta Journal-Constitution*, April 18, 2001, n.p. Available at <http://lexis-nexis.com/> (accessed June 24, 2005).

Sheridan, Jennifer, Jo Handelsman, and Molly Carnes. *Current Perspectives of Women in Science and Engineering at UW-Madison: WISELI Town Hall Meeting Report.* Available at <http://www.wiseli.engr.wisc.edu/Products/MoreWomen.htm> (accessed August 9, 2005).

Shippen, Dorothy (formerly Shippen-Lentz). Telephone interview with the author, December 15, 2004.

Shippen-Lentz, Dorothy, and Elizabeth H. Blackburn. "Functional Evidence for an RNA Template in Telomerase." *Science* 247 (1990): 546–552.

———. "Telomere Terminal Transferase Activity from *Euplotes crassus* Adds Large Numbers of TTTTGGGG Repeats onto Telomeric Primers." *Molecular and Cellular Biology* 9 (1989): 2761–2764.

Singer, Maxine. "Shaping the Future for Women in Science." In *Career Advice for Life Scientists*, edited by Elizabeth Marincola, 101–105. Washington, DC: American Society for Cell Biology, 2002.

Smith, Christopher D., Dana L. Smith, Joe L. DeRisi, and Elizabeth H. Blackburn. "Telomeric Protein Distributions and Remodeling through the Cell Cycle in *S. cerevisiae.*" *Molecular Biology of the Cell* 14 (2003): 556–570.

Smith, Dana. Interview with the author, July 29, 2005, San Francisco.

Specter, Michael. "Political Science." *New Yorker*, March 13, 2006, 58–69.

Stein, Rob. "Study Is First to Confirm That Stress Speeds Aging." *Washington Post*, November 30, 2004, A1. Available at <http://washingtonpost.com/> (accessed December 3, 2004).

"Stem Cells Scientists Can Use Are Tainted." *Washington Post* news service, in *San Francisco Chronicle*, October 29, 2004, A5.

Stipp, David. "The Hunt for the Youth Pill: From Cell-immortalizing Drugs to Cloned Organs, Biotech Finds New Ways to Fight against Time's Toll." *Fortune*, October 11, 1999, 199.

Summers, Lawrence. "Letter to the Faculty regarding NBER Remarks." Office of the President, Harvard University. Available at <http://www.president.harvard.edu/speeches/2005/facletter.html> (accessed February 18, 2005).

———. "Remarks at NBER Conference on Diversifying the Science and Engineering Workforce." Office of the President, Harvard University. Available at <http://www.president.harvard.edu/speeches/2005/nber.html> (accessed February 18, 2005).

Szostak, Jack W. "The Beginning of the Ends." *Nature* 337 (1989): 303–304.

Szostak, Jack W., and Elizabeth H. Blackburn. "Cloning Yeast Telomeres on Linear Plasmid Vectors." *Cell* 29 (1982): 245–255.

Tilghman, Shirley. "Changing the Demographics: Recruiting, Retaining, and Advancing Women Scientists in Academia." Lecture, Earth Institute ADVANCE Program, Columbia University, March 24, 2005, New York. Available at <http://www.earthinstitute.columbia.edu/advance/documents/Tilghman _032405_ADV_CU.pdf> (accessed March 14, 2006).

Torassa, Ulysses. "Gender Bias Dispute Flares at UCSF." *San Francisco Examiner*, March 26, 2000, n.p. Available at <http://www.sfgate.com/> (accessed on July 14, 2006).

Trower, Cathy A., and Richard P. Chait. "Faculty Diversity: Too Little for Too Long." *Harvard Magazine* (March–April 2002), n.p. Available at <http://www.harvardmagazine.com/on-line/030218.html> (accessed July 20, 2004).

Truett, Martha. Telephone interview with the author, September 8, 2004.

Union of Concerned Scientists. *Restoring Scientific Integrity in Policymaking*. Available at <http://www.ucsusa.org/global_environment/rsi/page.cfm?pageID= 1320> (accessed July 22, 2005).

———. "U.S. Fish and Wildlife Service Survey Summary. "Available at <http://www.ucsusa.org/global_environment/rsi/page.cfm?pageID=1601> (accessed July 22, 2005).

University of California at San Francisco. "Making History: Facts." Available at <http://pub.ucsf.edu/missionbay/faq> (accessed December 28, 2005).

———. "Making History: Mission Bay Milestones." Available at <http://pub .ucsf.edu/missionbay/history/milestones.php> (accessed December 28, 2005).

University of Pittsburgh. *Summary Investigative Report on Allegations of Possible Scientific Misconduct on the Part of Gerald P. Schatten, Ph.D.* Pittsburgh, PA: University of Pittsburgh Medical Center Research Integrity Panel, February 8, 2006. Available at <http://newsbureau.upmc.com/TX/SchattenPanelRelease .htm> (accessed February 15, 2004).

U.S. Congress. Senate. Appropriations Subcommittee on Labor, Health and Human Services, Education. "Capitol Hill Hearing Testimony," January 24, 2002. In *Federal Document Clearing House, Congressional Testimony*. Washington, DC: Federal Document Clearing House, Inc., 2002.

Valian, Virginia. *Why So Slow?* Cambridge, MA: MIT Press, 1998.

Vulliamy, Tom J., Amanda Waine, Aroon Baskaradas, Philip J. Mason, Anna Marrone, and Inderjeet Dokal. "Mutations in the Reverse Transcriptase Component of Telomerase (TERT) in Patients with Bone Marrow Failure." *Blood Cells, Molecules, and Diseases* 34 (May–June 2005): 257–263.

Wade, Nicholas. "Group of Scientists Drafts Rules on Ethics for Stem Cell Research." *The Ledger*, April 27, 2005, n.p. Available at <http:theledger.com/> (accessed August 17, 2005).

Walmsley, Richard W., Clarence S. Chan, Bik-Kwoon Tye, and Thomas D. Petes. "Unusual DNA Sequences Associated with the Ends of Yeast Chromosomes." *Nature* 310 (1984): 157–160.

Walmsley, Richard W., Jack W. Szostak, and Thomas D. Petes. "Is There Left-handed DNA at the Ends of Yeast Chromosomes?" *Nature* 302 (1983): 84–86.

Waters, Beverly, ed. *A Yale Book of Numbers, 1976–2000.* New Haven, CT: Yale University Office of Institutional Research, 2001. Available at <http://www .yale.edu/oir/pierson_update.htm> (accessed September 10, 2004).

Watson, James. *The Double Helix: A Personal Account of the Discovery of the Structure of DNA.* New York: Atheneum, 1968.

Weiss, Rick. "Bioethics Panel Calls for Ban on Radical Reproductive Procedures." *Washington Post,* January 16, 2004, A2. Available at <http://washingtonpost.com/> (accessed July 20, 2005).

———. "Bush Ejects Two from Bioethics Council; Changes Renew Criticism That the President Puts Politics ahead of Science." *Washington Post,* February 28, 2004, A6. Available at <http://www.washingtonpost.com/> (accessed June 24, 2005).

———. "Bush Panel Has Two Views on Embryonic Cloning." *Washington Post,* July 11, 2002, A5. Available at <http://www.washingtonpost.com/> (accessed June 24, 2005).

———. "Bush Unveils Bioethics Council: Human Cloning, Tests on Cloned Embryos Will Top Agenda of Panel's First Meeting." *Washington Post,* January 17, 2002, A21. Available at <http://www.washingtonpost.com/> (accessed June 24, 2005).

———. "9/11 Response Hurting Science, ACLU Says." *Washington Post,* June 22, 2005, A3. Available at <http://www.washingtonpost.com/> (accessed June 24, 2005).

———. "None of Stem Cell Lines Scientists Said He Created Exists." *Washington Post* news service, reprinted in *San Francisco Chronicle,* December 30, 2005, A16.

———. "Scientist Involved in Stem-Cell Cloning Denies Faking Data." *Washington Post* news service, reprinted in *San Francisco Chronicle,* December 17, 2005, A12.

———. "Stem Cell Field Rocked by Scam of Star Scientist." *Washington Post* news service, reprinted in *San Francisco Chronicle,* December 24, 2005, A5.

Wenneras, Christine, and Agnes Wold. "Nepotism and Sexism in Peer-Review." *Nature* 387 (1997): 341–342.

Wilson, Robin. "Mary-Claire King: 'The Biggest Obstacle Is How to Have Enough Hours in the Day.' " *Chronicle of Higher Education,* June 10, 2005, n.p. Available at <http://chronicle.com/weekly/v51/i40/40a00901.htm> (accessed June 7, 2005).

————. "Rigid Tenure System Hurts Young Professors, Especially Women, Officials from Top Universities Say." *Chronicle of Higher Education*, September 26, 2005, n.p. Available at <http://chronicle.com/daily/2005/09/2005> (accessed October 2, 2005).

————. "Women in the National Academy." *Chronicle of Higher Education*, June 10, 2005, A8.

Xu, Lifeng, and Elizabeth H. Blackburn. "Human RIF1 Protein Binds Aberrant Telomeres and Aligns along Anaphase Midzone Microtubules." *Journal of Cell Biology* 167 (2004): 819–830.

Yaeger, Peter C., Eduardo Orias, Wen-Ling Shaiu, Drena D. Larson, and Elizabeth H. Blackburn. "The Replication Advantage of a Free Linear rRNA Gene Is Restored by Somatic Recombination in *Tetrahymena thermophila*." *Molecular and Cellular Biology* 9 (1989): 452–460.

Yu, Guo-Liang, John D. Bradley, Laura D. Attardi, and Elizabeth H. Blackburn. "In Vivo Alteration of Telomere Sequences and Senescence Caused by Mutated *Tetrahymena* Telomerase RNAs." *Nature* 344 (1990): 126–132.

Zaug, Arthur J., and Thomas R. Cech. "The Intervening Sequence RNA of *Tetrahymena* Is an Enzyme." *Science* 231 (1986): 470–475.

Zhu, Jiyue, He Wang, J. Michael Bishop, and Elizabeth H. Blackburn. "Telomerase Extends the Lifespan of Virus-Transformed Human Cells without Net Telomere Lengthening." *Proceedings of the National Academy of Sciences* 96 (1996): 3723–3728.

Zijlmans, J. Mark. "The Role of Telomeres and Telomerase in Cancer." Paper presented at the conference of the American Association for Cancer Research, December 2000, San Francisco.

Index

Rhodes, Daniela, 325
Ribonuclease A (RNase A), 99–100
Ribonucleoprotein (RNP), 99, 107, 114
Ribosomal RNA (rRNA), 39, 44, 99, 117, 185–186
Rif1, 199, 211, 324–325, 326–327
Rif2, 199, 211, 324–325
Ring chromosome, 45–46, 47, 53, 54
Risk factors for cardiovascular disease, 318–319
Rivera, Melissa, 219, 226–228, 229–230
RNA. *See* Messenger RNA; Mutant-template RNA of telomerase; Ribosomal RNA; RNA component of telomerase; RNA template of telomerase; Transfer RNA
RNA component of telomerase
 binding regions for TERT and ancillary proteins, 191 (fig. 8.2), 192
 conserved sequences in, 185–190
 effects of mutations in, 190–192
 effects of mutations in pseudoknot, 219–220
 existence first demonstrated in vitro in *Tetrahymena*, 100, 107
 gene for, isolated in humans, 147
 gene for, isolated in *Tetrahymena*, 112
 role in enzyme's activity, 192–195 (*see also* RNA template of telomerase)
 universal model for structure of, 189–190
 varying size in different organisms, 188–189
RNA interference, 230
RNase A. *See* Ribonuclease A
RNA sequencing, 26
RNA structure, 118, 186, 187 (fig. 8.1)
 in telomerase, 186–190, 192 (*see also* Helix; Pseudoknot)

RNA template of telomerase. *See also* Mutant-template RNA of telomerase
 alignment region of, 112, 119
 first clue to existence of, 92
 first sequenced in *Tetrahymena*, 112
 primer recognition properties of, 102, 106–107, 112–113
 role in constructing tandem repeat patterns, 97, 100
 role in elongation and translocation, 112–114, 113 (fig. 5.3)
 singular role in enzymatic activity, 195
RNP. *See* Ribonucleoprotein
Romero, Daniel, 186, 188
Rowley, Janet
 appointed to bioethics council, 270
 coauthored editorial in *PLoS Biology*, 295–296
 coauthored editorial in *Science*, 278
 coauthored letter to *Science*, 282–284
 criticism of council proceedings, 276, 281, 302
 defense of public funding for research, 288
 opposition to moratorium, 277
 recipient of Dorothy P. Landon Prize for Translational Cancer Research, 311
 response to E.B.'s dismissal, 298
Royal Society of London, 103, 153
rRNA. *See* Ribosomal RNA

Saccharomyces cerevisiae, 76, 102, 107, 210, 324
Salk Institute, 242, 309
Sandel, Michael, 273, 277, 279
Sanger, Fred
 influence on E.B., 20–21, 28, 34–35, 53
 Nobel Prizes, 23, 34
 research, 22–25, 34, 80
 social milieu of lab, 20, 24, 25, 27–29